U0174907

王中林精选论文摘要集

Abstracts Collection of Zhong Lin Wang's Selected Papers

王中林 著

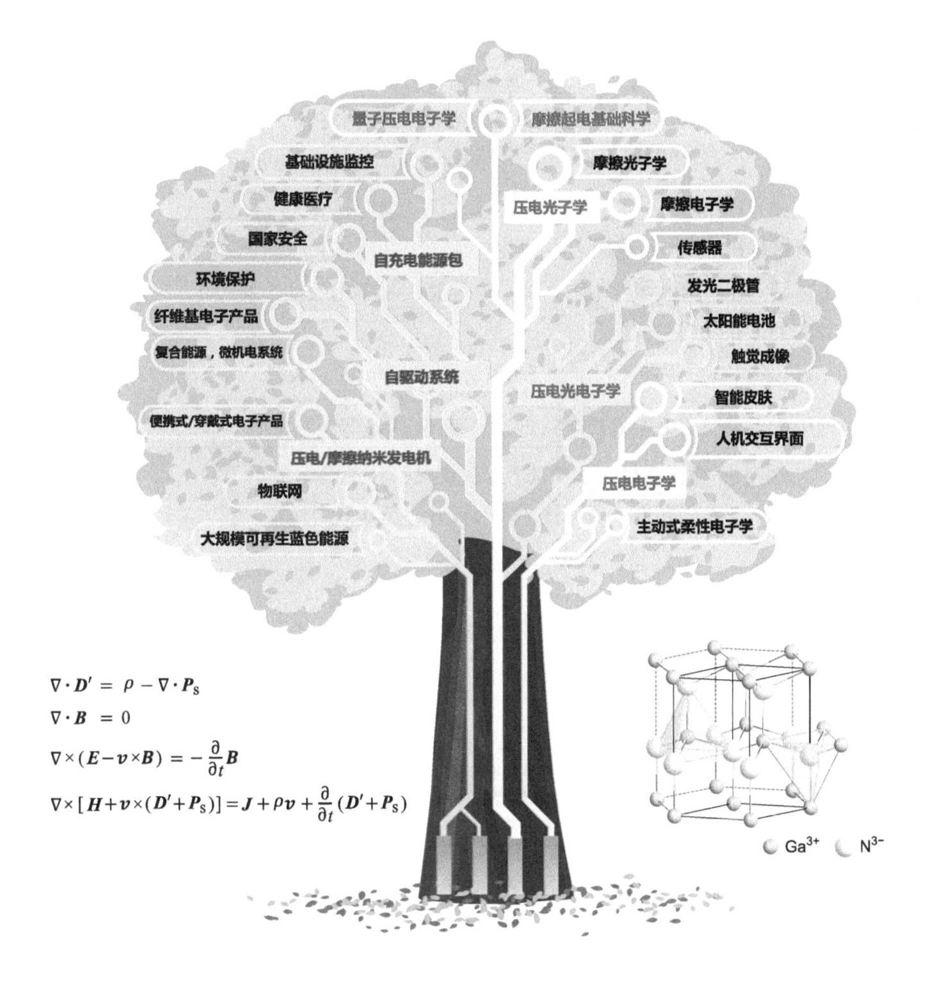

科学出版社

北京

内 容 简 介

发表科学论文是科研工作者进行学术交流和科学传承的重要途径。在过去40年的研究历程中，作者从事过从物理学、电子显微学到材料学，再到能源学等多个学科和领域的研究，历经了从理论到实验，从科学到技术的交叉研究历程。通过梳理该论文摘要集所收录的代表作，作者希望能达到以下几个目的：一是能够呈现出作者过去40年来的研究历程和思维方式的演变和突变过程；二是给出如何创立一个新领域的系统性布局和发展一个交叉学科的创新性思维；三是还原在2006年以前的二十多年间，为找到一个属于作者自己的崭新领域而苦苦寻求及知识积累的漫长过程；四是体现在研究方向上，在遵循“一棵树”的主导理念下，如何实现“三年一小变、五年一大变”的目标；最后是反映科研工作追求螺旋式上升、不断提高的理念。收录的论文基于以下五个标准：提出新的科学思想或理念，提出新的物理效应，开启一个新的研究领域，在建立一个研究领域过程中具有标志性的突出贡献，展现原创技术在一些领域的首次示范应用。在筛选这些代表作中，作者并没有考虑发表这些文章的期刊的影响因子，而是斟酌这些文章对学术领域和实际产业是否具有实质性贡献。希望这部论文摘要集能够为青年学生和年轻学者展现出作者“从0到1”原创性研究的思维方法和分享作者“十年磨一剑”与“千里走单骑”的科研历程。

本书适用于高校和科研院所从事物理、化学、材料、电子和新能源等领域或交叉领域的研究人员参考阅读。

图书在版编目（CIP）数据

王中林精选论文摘要集 / 王中林著. —北京：科学出版社，2022.1

ISBN 978-7-03-071337-7

Ⅰ. ①王… Ⅱ. ①王… Ⅲ. ①科学技术-文集 Ⅳ. ①N53

中国版本图书馆CIP数据核字（2022）第013195号

责任编辑：李明楠 / 责任校对：杜子昂

责任印制：吴兆东 / 封面设计：蓝正设计

科学出版社 出版

北京东黄城根北街16号

邮政编码：100717

http://www.sciencep.com

北京建宏印刷有限公司 印刷

科学出版社发行 各地新华书店经销

*

2022年1月第 一 版 开本：889×1194 1/16

2023年1月第二次印刷 印张：23

字数：745 000

定价：198.00元

（如有印装质量问题，我社负责调换）

前　　言

科学研究一般可以分为三个层次：建立一个领域的原创研究；解决一个科学问题的基础研究；解决关键技术瓶颈的攻关研究。基础科学是无国界的，交流基础科学发现和进展的最重要途径就是发表有意义的科学论文。在过去近四十年的学习与研究历程中，对于科学研究我总是抱有极强的好奇心和不灭的热情，锲而不舍、坚持不懈，形成异于常人的思维方式，精于创新。

日前，我从已发表的近两千篇 SCI 期刊学术论文中，精选了近三百篇代表作，将其细分研究领域和方向后，按发表的时间顺序以摘要集的形式汇于本书，且每篇辅以相关的背景、意义等概要说明。附每篇论文的英文索引、简单中文背景说明、一张提纲图以及电子文本的链接和二维码。论文精选基于以下标准：提出新的科学思想或理念；提出新物理效应；开启一个新的研究领域；在建立一个研究领域过程中具有标志性的突出贡献；首次展示原创技术在一些领域的示范应用。在筛选这些代表作中，我并没有考虑这些文章发表期刊的影响因子，而是根据上面几个准则考虑这些文章对学术领域和实际产业的实质性贡献。这些论文收集了我们几个重要研究阶段的代表性作品：

· **1983～1998 年**：开展了纳米颗粒的等离子元激发研究，开创了高能电子非弹性散射的动力学理论，从首次发展了反射电子成像与能量损失谱到首次提出了高角度环形暗场扫描透射（HAADF STEM）成像理论。

· **1996～2000 年**：发展了原位透射电子显微镜技术，并利用该技术首次研究了一维材料的力学和电学性能，进而提出了原位纳米力学的概念。

· **2000～2004 年**：率先开始引领世界的纳米材料合成研究，原创性地系统研制、表征氧化锌的多种纳米结构并提出了其生长机理与性能，并且开启了基于压电效应的能源与电子学研究。

· **2005 年至今**：发明了压电纳米发电机与摩擦纳米发电机，开创了基于纳米发电机的分布式能源与蓝色能源两大领域，统称为高熵能源。

· **2006 年至今**：首次提出了自驱动系统的概念，目前已经在全球范围内广泛地得到了重视和研究。

· **2007 年至今**：原创性地将压电效应与半导体效应耦合、创立了压电电子学与压电光电子学学科，发展出第三代半导体中两个独特的新物理效应：压电电子学效应，压电光电子学效应。

· **2011 年至今**：开辟了摩擦纳米发电机的四大应用领域，即微纳能源、自驱动传感、蓝色能源和高压电源等。

- **2014年至今**：深入挖掘固体-固体接触起电的物理机制，确定了电子转移与输运的基本物理模型，为解决这个存在了2600年的古老科学问题提出了崭新的理论构架。
- **2016年至今**：拓展了麦克斯韦方程组中位移电流的数学表达式，提出了基于位移电流的纳米发电机的第一性原理与理论构架。
- **2018年至今**：确定了液体-固体界面接触的电子转移过程，并修正了传统化学中的双电层理论模型。
- **2019年至今**：首次提出了适用于半导体型固体-固体和液体-固体界面的摩擦伏特效应。

通过这部论文摘要集所收录的代表作，我希望能够以此呈现出我个人的研究历程和思维方式，给出创立一个领域的系统性思维和发展一个交叉学科的创新思维，还原在2006年以前的二十多年间为了找到一个属于自己的崭新领域而苦苦寻求及知识积累的过程，指出我对研究所遵循的“三年一小变、五年一大变”的原则，以及追求螺旋式上升、不断提高的理念。我们的研究是从“0”开始，从小做大，坚持自己的特色，不追赶热点，从星星之火发展到烈焰燎原。衷心希望这部论文摘要集能够为研究生和年轻学者展示出“从0到1”原创性研究的思维方法和完成过程。

受时间、精力所限，同时因视角可能存在差别，此部论文摘要集所收录的文献并不一定全面准确，敬请读者包涵并提出批评意见。

最后衷心感谢和我一起奋斗拼搏过的学生们、博士后们和访问学者们！感谢我的合作同仁！同时感谢参与编辑本书的所有人员！我也衷心感谢科学出版社的大力支持！

谢谢！

2021年12月于北京

Preface

There are three approaches in doing scientific research: creating an original research field; understanding a specific scientific question; achieving technological innovation towards solving a bottleneck engineering problem. Since fundamental science belongs to humankind without country boundary, one of the key means for communicating scientific discoveries is via publishing high quality scientific research papers. In the last 40 years of my own research experience, I have always been enthusiastic, excited and curiosity toward scientific discoveries, and I have been taking distinguished research approaches, being persistent and consistent in advancing my own career. Among the ~2000 published peer reviewed scientific papers by our team, about 300 papers are selected and featured here as a collection of our work based on following principles: papers with new scientific ideas or concepts; papers reporting new physics effects; papers that open new research fields; papers that have made landmark contributions in developing a field; and papers with the first demonstration of a discovery in a specific technical area. Each paper is featured with an abstract, a TOC figure, the link and 2D barcode to the electronic file of the full article. The selected papers solely based on these principles without much consideration regarding to the impact factors of the journals on which the papers were published. These papers include my represent work in the following areas:

- **1983～1998:** Surface plasmon excitation of supported nanoparticles; dynamic theory of inelastic scattering in electron scattering and diffraction; imaging and spectroscopy of reflected electrons from crystal surfaces; and the theory for high-angle annular-dark-field STEM imaging;
- **1996～2000:** *In-situ* measurements of the electric transport or mechanical properties of a single one-dimensional nanomaterials in TEM, opening the field of *in-situ* nanomechanics;
- **2000～2004:** First synthesis, characterization and understanding of ZnO nanostructures, establishing the base of using piezoelectric effect for energy and electronics;
- **2005～Now:** Invented and pioneered the fields of nanoenergy, energy for the new era, distributed energy and large-scale blue energy;
- **2006～Now:** First proposed the idea of self-powered systems and expanded its impact worldwide;

- **2007～Now:** First proposed the piezotronic effect and piezo-phototronic effect, and established the piezotronics and piezo-phototronics fields for the third generation semiconductors;
- **2011～Now:** Pioneered the four application fields of triboelectric nanogenerators (TENG): micro-nano-energy sources; self-powered sensors; blue energy and high-voltage sources;
- **2014～Now:** Explored the contact-electrification effect between solid-solid, and first proposed the electron transfer model for a general case, providing a new approach for solving the science model for triboelectricity that has been known for over 2600 years;
- **2016～Now:** Expanded the scope of the displacement current in the Maxwell's equations, established the first principle theory for nanogenerators;
- **2018～Now:** Confirmed the electron transfer process in the contact-electrification between liquid and solid, modified the formation model of the electric double layer (EDL) in traditional chemistry;
- **2019～Now:** First proposed the tribovoltaic effect for solid-solid and liquid-solid interfaces in semiconductor materials.

The idea for selecting these articles are to summarize our research experience, ideas and approaches of doing science. The purpose is to present our own experience in establishing a new field and our systematic approach for starting an original research. I hope that these volumes can provide a guidance for young scientists and students for advancing their scientific careers.

I am very thankful to many of my students, postdoctoral fellows, visiting scientists and collaborators who worked with me. Thanks the editorial members who helped me in composing these volumes. Thanks the publisher for producing such a nice piece of work.

Zhong Lin Wang
August 2021

王中林简介和主要科学贡献

王中林，1961 年出生于陕西省蒲城县。1982 年毕业于西安电子科技大学，1983 年通过中美联合招收的物理研究生（CUSPEA）赴美留学，1987 年获 Arizona State University 物理学博士，先后在纽约州立大学、英国剑桥大学、美国橡树岭国家实验室和美国国家科学与技术标准局从事研究员/教授工作近十年，1995～2020 年任教于美国佐治亚理工学院，曾任该校的纳米科学中心主任、显微标准中心主任、终身校董事讲席教授，和 Hightower 终身讲席教授。2020 年从美国佐治亚理工学院早退，全职回中国工作。现担任中国科学院北京纳米能源与系统研究所首席科学家、创所所长和现任所长。中国科学院外籍院士、欧洲科学院院士和加拿大工程院院士。

国际公认的纳米能源研究领域奠基人，首次将纳米能源定义为“新时代的能源”，将分布式能源与蓝色能源两大领域定义为“高熵能源”，开创了压电电子学和压电光电子学研究的领域，对物联网、传感网络、人机界面、医学健康、穿戴式/柔性电子学、安全防护、第三代半导体光电器件、LED、光伏电池等技术的发展具有里程碑意义。凭借在微纳能源和自驱动系统领域的开创性成就，荣获 2011 年美国材料学会奖章（MRS Medal）、2014 年美国物理学会詹姆斯·马克顾瓦迪（James C. McGroddy）新材料奖、2015 年汤森路透引文桂冠奖，2018 年世界能源领域的最高奖项——“埃尼奖”，并斩获 2019 年“阿尔伯特·爱因斯坦世界科学奖”，成为首位获此殊荣的华人科学家。

在国际一流刊物发表期刊论文 2100 余篇（其中 13 篇发表于 *Science*，7 篇发表于 *Nature*，64 篇发表在相应子刊上），200 余项专利，7 部专著和 20 余部编辑书籍和会议文集。受邀做过 1000 余次学术讲演和大会特邀报告，是国际纳米能源领域著名期刊 *Nano Energy* 的创刊者与现任主编。学术论文他引 310 000 次以上，h 因子（h-index）270。2021 年度全球材料科学总引用数和 h 指数排名世界第一；2021 年度世界横跨所有领域前 10 万科学家终身科学影响力综合排名第 3 位，其中 2019 年度和 2020 年度连续两个年度排名第 1 位。

指导了 300 名博士研究生、博士后和访问学者，培养了一大批国际优秀人才。帮助扩大中国科技界在国际科学舞台的知名度和影响力，为国内培养、锻炼和输送了一批优秀的科研工作者，荣获中华人民共和国国际科学技术合作奖。

王中林终身最有影响力的科学贡献可以总结为“1-2-3-6-7”，即：

一个体系 One system

王中林近 20 年的研究坚持一棵树的理念，发展了基于纳米能源的高熵能源与新时代能源体系。

Through 20 years of effort, Wang developed the high entropy energy system based on nanoenergy and the energy system for the new era.

研究体系理念图

两大领域 Two fields

基于纳米发电机的自驱动系统与蓝色能源宏大领域；基于压电电子学与压电光电子学效应的第三代半导体的崭新领域。

Nanogenerator based self-powered systems and blue energy; Piezotronics and piezo-phototronics of the third generation semiconductors.

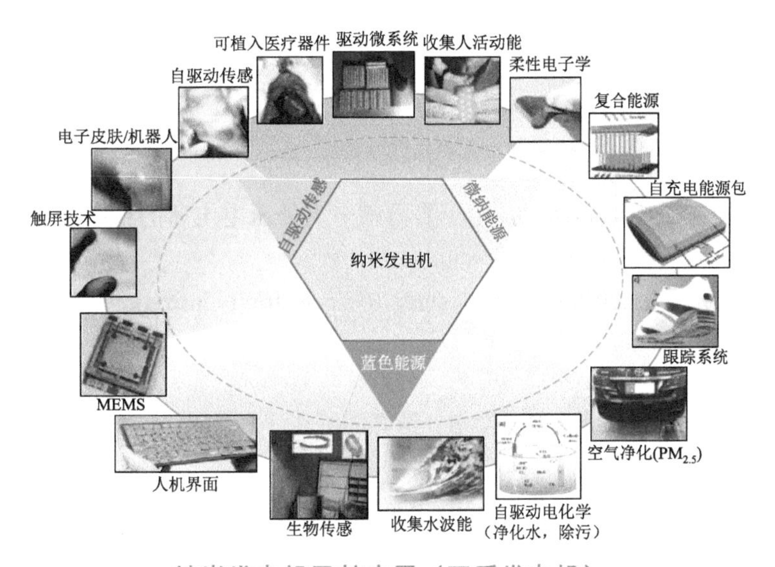

纳米发电机及其应用（王氏发电机）

Applications of nanogenerators (Wang generator)

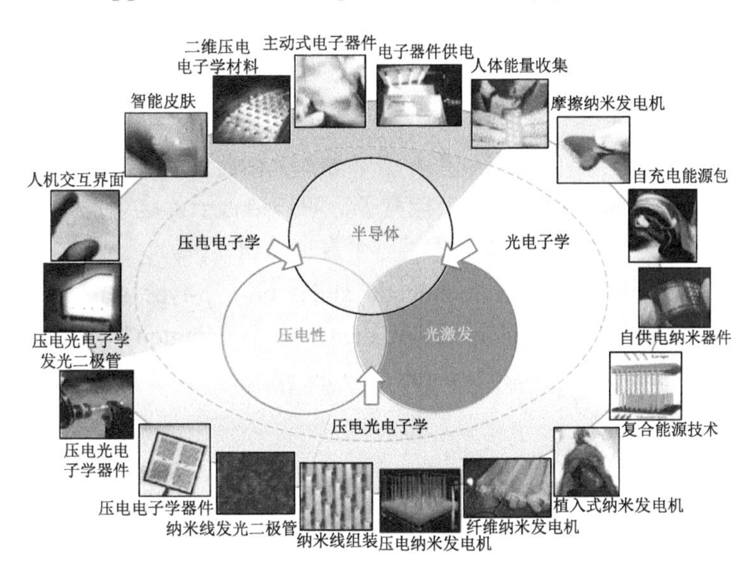

压电电子学与压电光电子学

Piezotronics and piezo-phototronics

三个学科 Three scientific majors

1. 压电电子学：利用压电电子学效应调控的半导体物理。

Piezotronics is about the devices made using piezotronic effect for tuning the carrier transport in semiconductors.

2. 压电光电子学：利用压电光电子学效应调控的半导体光电物理。

Piezo-phototronics is about the devices made using piezo-phototronic effect for tuning the carrier separation or recombination in optoelectric processes.

3. 摩擦电子学：利用接触起电所产生的电场做门电压来调制或控制半导体中载流子的输运过程而制做的电子器件。

Tribotronics is about the devices fabricated using the electrostatic potential created by triboelectrification as a "gate" voltage to tune/control charge carrier transport in semiconductors.

六个新物理效应 Six new physics effects

1. 压电电子学效应：在压电半导体中，由于界面或节点处压电电荷的存在所导致的对载流子输运的调制或控制效应。

Piezotronic effect: in piezoelectric semiconductors, piezoelectric polarization charges tuned/controlled charge transport process at interface/junction.

2. 压电光电子学效应：在压电半导体中，pn 结界面处由于压电电荷的存在所导致的对电子-空穴对的结合或分离所产生的光电调制效应。

Piezo-phototronic effect: in piezoelectric semiconductors, piezoelectric polarization charges tuned electron-hole separation or recombination in optoelectronic processes.

3. 压电光子学效应：在压电半导体中，压电半导体中压电电场所导致的光子出射现象。

Piezophotonic effect: in piezoelectric semiconductors, piezoelectric polarization potential induced photon emission process.

4. 摩擦伏特效应：当一个 n 型半导体在一个 p 型半导体上滑行时，在新形成的界面处的两个属于两边的原子相结合而形成一个键合，就会释放出一个能量子，叫作结合子，这个能量子激发 pn 结处的电子-空穴对；内建电场把结处的电子-空穴对分开就形成了一个直流输出。

Tribovoltaic effect: Once an n-type semiconductor slides on a p-type semiconductor, the newly formed atomic bond at the interface releases an energy quantum, named as "bindington", which excites electron-hole pairs at the pn junction. The electrons and holes are seperated by the internal built-in electric field, generating a direct current.

5. 热势光电子效应：对于脉冲型入射光，由于局部产生的热势效应而导致的电压、电流输出被称为热势光电子效应。该效应具有极高的光探测灵敏度。

Pyro-phototronic effect: For the pulsed incident light, local generated heating results in a strong pyro-electric effect, which gives a high output voltage and current. This effect can be used for high sensitive photon sensing.

6. 交流光伏效应：利用一个周期照射的脉冲型光，半导体 pn 节或界面给出一个交流电输出信号，其原理可能是动态变化的光照导致的半导体接触面处在非平衡状态下费米面的相对位移。该效应可能在超灵敏光探测传感器方面有独特的应用。

AC photovoltaic effect: a photovoltaic effect that generates alternating current (AC) in the nonequilibrium states when the illumination light periodically shines at the junction/interface of materials. It is suggested to be a result of the relative shift and realignment between the quasi-Fermi levels of the semiconductors adjacent to the junction/interface under the nonequilibrium conditions, which results in electron flow in the external circuit back and forth to balance the potential difference between the two electrodes.

七大科学贡献 Seven contributions in science

1. 将介质表面本征静电荷分布由于介质形体或分布变化所产生的极化密度引入电位移矢量，推导出了位移电流的另外一项（称为王氏项），拓展了麦克斯韦方程组，奠定了纳米发电机的理论基础，开启了麦克斯韦方程在能源与传感领域的新篇章。

Due to the existence of electrostatic charges on surfaces, the change in medium shape and the distribution of electrostatic charges, an additional term is first introduced in the displacement vector. As a result, an additional term in displacement current is introduced (Wang term), and the Maxwell's equations is expanded, and the theory for nanogenerators is established. This work opens the applications of Maxwell's equations in energy and sensors.

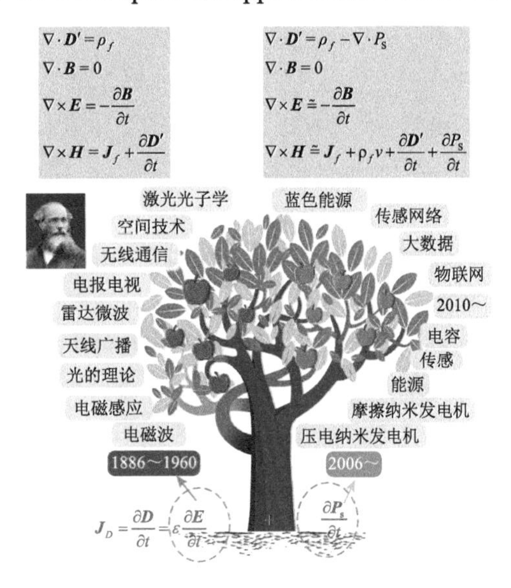

2. 统一了摩擦起电（接触起电）的物理模型，明确了原子间电子跃迁是接触起电的根本机理（被称为王氏跃迁），解决了2600年来最古老的科学问题：摩擦起电的机理。

A unified model is proposed for clearly elaborating the dominant role played by interatomic electron transition in contact electrification (called Wang transition). This model solves the debate about the origin of contact electrification, which has been known for over 2600 years.

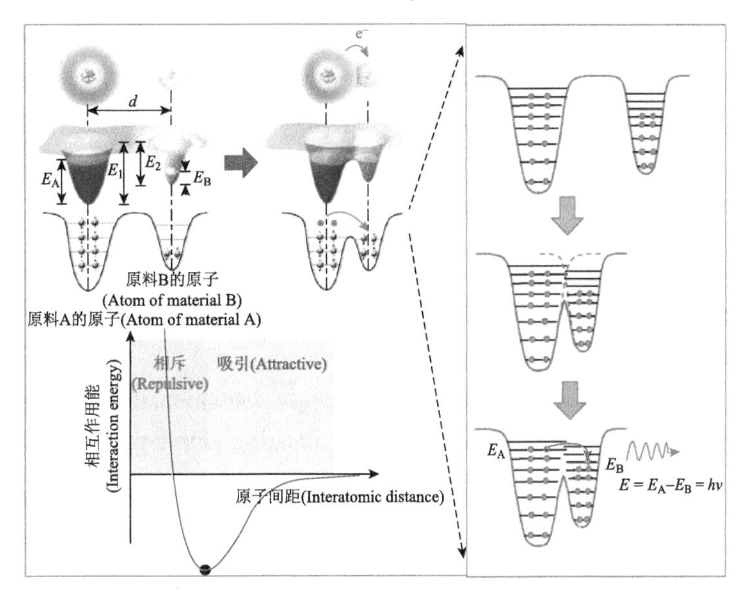

3. 提出并验证了跨原子电子跃迁是气体-液体-固体多相间接触起电的普适性机理，并首次提出了界面光谱学与接触起电催化学。

Established that interatomic/molecular electron transition is a popular and unified charge transfer mechanism among phases of gas-liquid-solid. First demonstrated the contact electrification induced interface light emission spectroscopy (CEIILES), and contact-electro-catalysis. These two areas will be further expanded in the near future.

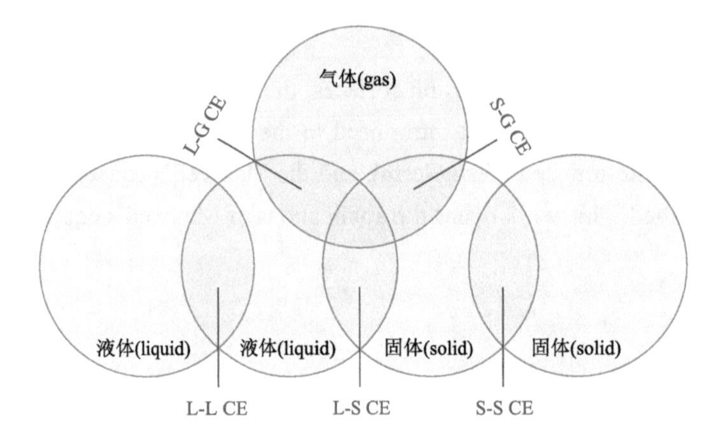

4. 确定了液体-固体接触中界面电子转移的过程，并提出了形成双电层结构的两步走机理模型（被称为王氏模型）：先是电子转移，然后是离子吸附。

Established the electron transition process between liquid-solid, and proposed a two-step model regarding to the formation of electric double layer (EDL) (called Wang model): electron transfer occurs first, then followed by ion adsorption.

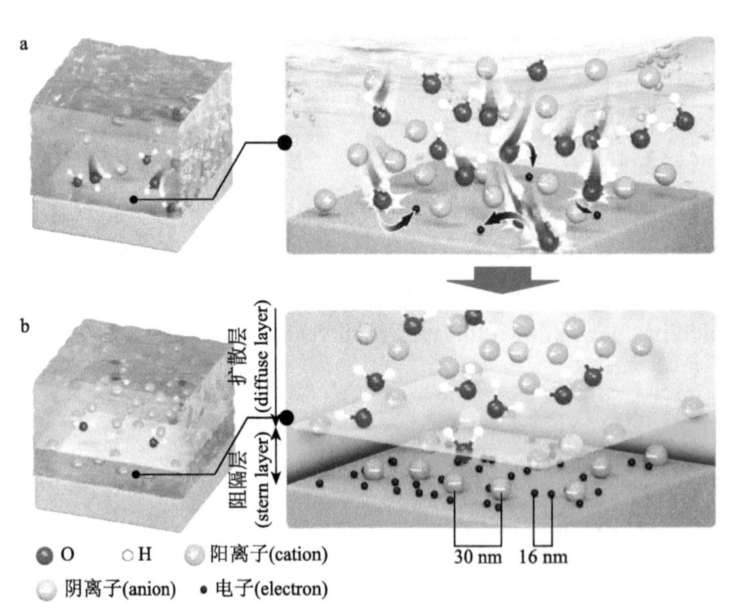

5. 系统地发展了声子散射在高能电子衍射与成像中的动力学理论，首先提出了 HAADF STEM 成像模拟理论。

Systematically and comprehensively developed the dynamic theory regarding to phonon scattering (thermal diffuse scattering) in high energy electron diffraction and imaging. First proposed the theory for simulating HAADF STEM images.

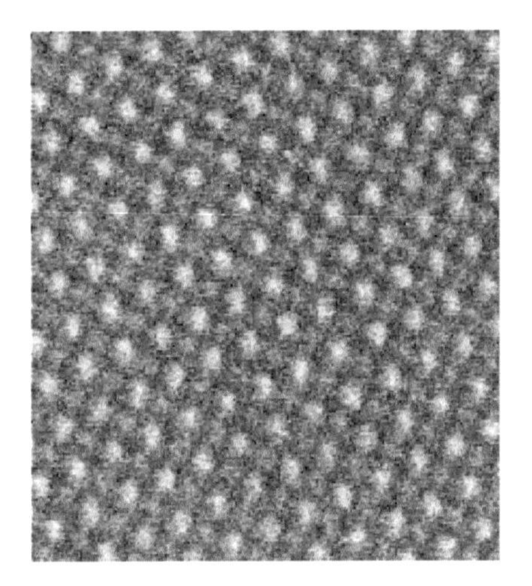

1995 年的第一本专著 *Elastic and Inelastic Scattering in Electron Diffraction and Imaging* 被 *American Scientist* 评论为“具有卓越成就和极大价值的经典之作”，被欧洲学者据其封面颜色称为“yellow bible（黄皮圣经）”。

The first book entitled of *Elastic and Inelastic Scattering in Electron Diffraction and Imaging* was regarded as an outstanding and extremely valuable classical text book by *American Scientist*. This book is regarded by European scientists as the “yellow bible” in the field.

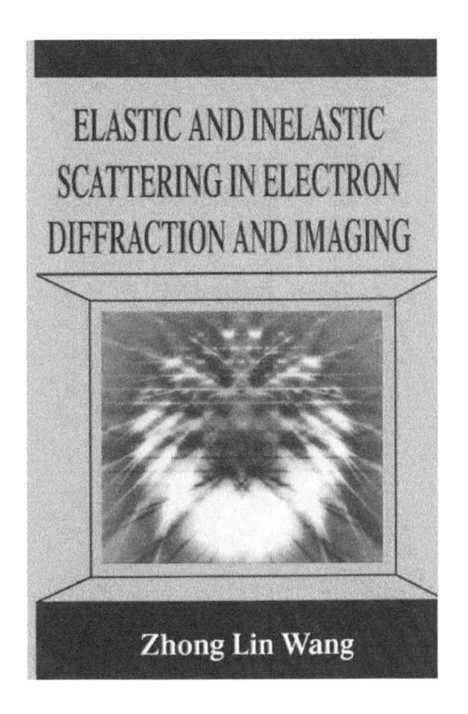

6. 发现了称单个纳米颗粒的方法—开创了电子显微镜中原位纳米测试技术。

Discovered a method for weighting a single nanoparticle, initiated *in-situ* nanomeasurements in TEM.

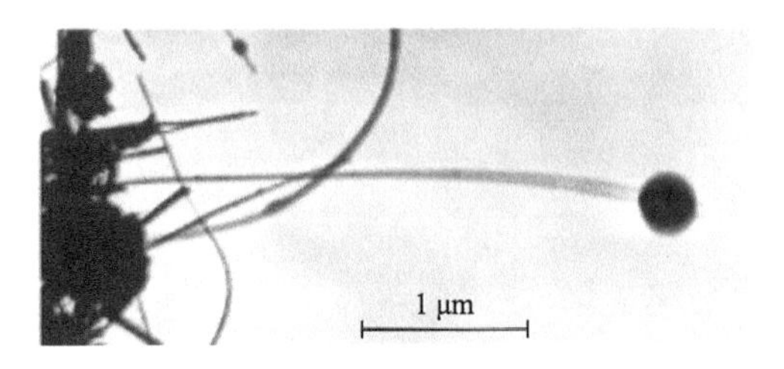

7. 首次发现氧化物纳米带，开启了研究氧化锌纳米结构的历程。

Discovered oxide nanobelts, pioneered the research on nanostructures of zinc oxide.

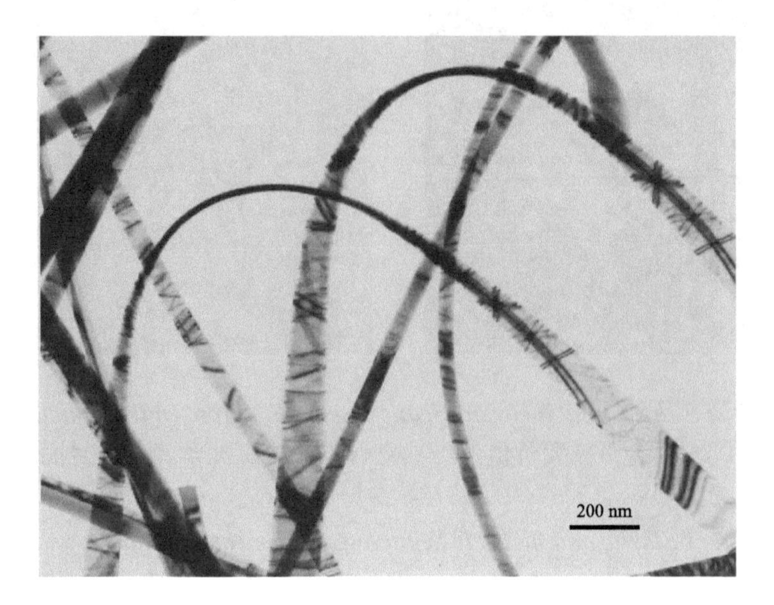

Biography and Summary of Wang's Achievements

Experience: Dr. Zhong Lin (Z.L.) Wang received his Ph.D in Physics from Arizona State University in 1987. He is the Hightower Chair in Materials Science and Engineering, Regents' Professor, and College of Engineering Distinguished Professor at the Georgia Institute of Technology. He served as a Visiting Lecturer in SUNY (1987～1988), Stony Brook, as a research fellow at the Cavendish Laboratory in Cambridge (England) (1988～1989), Oak Ridge National Laboratory (1989～1993) and at National Institute of Standards and Technology (1993～1995) before joining Georgia Tech in 1995. He is the Hightower Chair in Materials Science and Engineering and Regents' Professor at Georgia Tech. He is the director of the Beijing Institute of Nanoenergy and Nanosystems. He was also a guest professor at Uppsala University, Sweden (2017～2020).

Wang's discovery and breakthroughs in developing nanogenerators establish the principle and technological road map for harvesting mechanical energy from environment and biological systems for powering mobile sensors. He first showed that the nanogenerator is originated from the Maxwell's displacement current, revived the applications of Maxwell's equations in energy and sensors, which is 155 years later after the invention of electromagnetic wave based on displacement current. His research on self-powered nanosystems has inspired the worldwide effort in academia and industry for harvesting ambient energy for micro-nano-systems, which is now a distinct disciplinary in energy science for future sensor networks and internet of things. He coined and pioneered the fields of piezotronics and piezo-phototronics by introducing piezoelectric potential gated charge transport process in fabricating strain-gated transistors for new electronics, optoelectronics, sensors and energy sciences. The piezotronic transistors have important applications in smart MEMS/NEMS, nanorobotics, human-electronics interface and sensors. Wang also invented and pioneered the *in-situ* technique for measuring the mechanical and electrical properties of a single nanotube/nanowire inside a transmission electron microscope (TEM) (Fig.1).

Dr. Wang is a pioneer and world leader in nanoscience and nanotechnology for his outstanding creativity and productivity. He has authored and co-authored 7 scientific reference and textbooks and over 2100 peer reviewed journal articles (84 in *Nature*, *Science* and their family journals), 45 review papers and book chapters, edited and co-edited 14 volumes of books on nanotechnology, and held over 200 US and foreign patents. Wang's Google scholar gives a citation is over 310,000 with an *h*-index of 270. Dr. Wang is ranked #1 among 100,000 scientists worldwide across all fields in the years of 2019 and 2020, as #3 in the entire career scientific impacts in the year of 2021! The ranking was made based on six citation metrics (total citations; Hirsch *h*-index; coauthorship-adjusted Schreiber *hm*-index; number of citations to papers as single author; number of citations to papers as single or first author; and number of citations to papers as single, first, or last author). Microsoft Academic: https://academic. microsoft.com/authors/192562407 published over 1million scientists in the Materials Science field in the world; of those scientists,

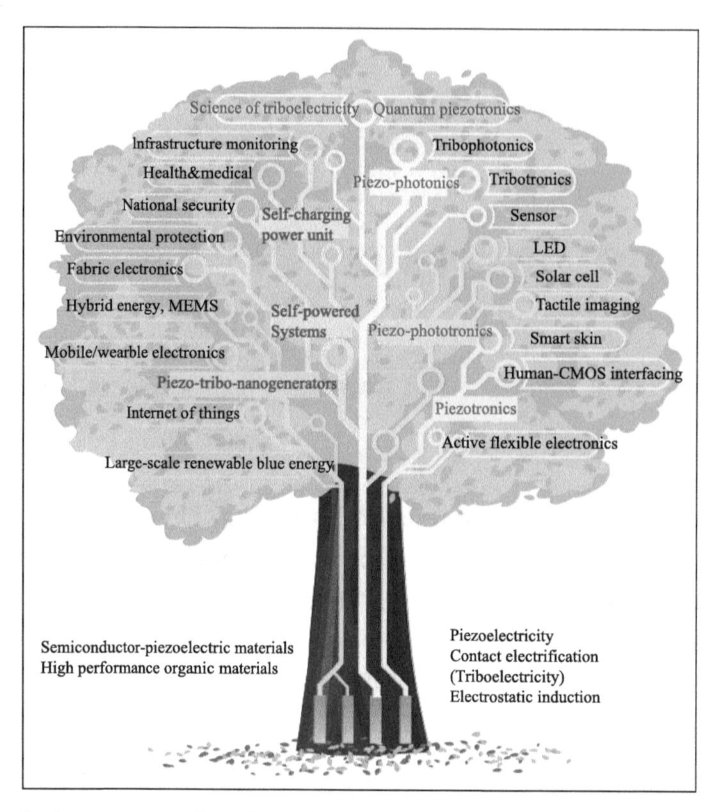

Fig.1 A "tree" approach that summarizes Wang's major original and pioneer contributions in science and technology as well as broad impacts

Wang is #1 in Salience, #1 in Citations and #1 in *h*-index. Wang is ranked #14 in Google scholar public profiles across all fields.

Honors and awards: Dr. Wang has received numerous international honors and awards. They include: 2020 Celsius Lecture Laureate, Uppsala University, Sweden; 2019 Albert Einstein World Award of Science; 2019 Diels-Planck lecture award; 2018 ENI award in Energy Frontiers (the "Nobel prize" for energy); American Chemical Soc. Publication most prolific author (2017); Global Nanoenergy Prize (2017)**,** The NANOSMAT Society, UK (2017); Distinguished Research Award, Pan Wen Yuan foundation (2017); Outstanding Achievement in Research Innovation award, Georgia Tech (2016); Distinguished Scientist Award from (US) Southeastern Universities Research Association (2016); Thomas Router Citation Laureate in Physics (2015); World Technology Award (Materials) (2014); Distinguished Professor Award (Highest faculty honor at Georgia Tech) (2014); NANOSMAT Prize (United Kingdom) (2014); China International Science and Technology Collaboration Award, China (2014); The James C. McGroddy Prize in New Materials from American Physical Society (2014); ACS Nano Lectureship (2013); Edward Orton Memorial Lecture Award, American Ceramic Society (2012); MRS Medal from Materials Research Society. (2011); Dow Lecture, Northwestern University (2011); Hubei Province Bianzhong Award, China (2009); Purdy Award, American Ceramic Society (2009); John M. Cowley Distinguished Lecture, Arizona State University (2012); Lee Hsun Lecture Award, Institute of Metal Research (CAS), China (2006); NanoTech Briefs, Top50 Award (2005); Sigma Xi Sustain Research Awards, Georgia Tech (2005); Georgia Tech Faculty Outstanding Research Author Award (2004); S.T. Li Prize for Distinguished Achievement in Science and Technology (2001); Outstanding Research Author Award, Georgia Tech (2000); Burton Medal, Microscopy Society. of America (1999); Outstanding Oversea Young Scientists Award from NSFC China (1998); NSF CAREER (1998).

Dr. Wang was elected as a foreign member of Chinese Academy of Sciences in 2009, member of European Academy of Sciences in 2002, international fellow of Canadian Academy of Engineering 2019, fellow of American Physical Society in 2005, fellow of AAAS (American Association for the Advancement of Science) in 2006, fellow of Materials Research Society of America in 2008, fellow of Microscopy Society of America in 2010, fellow of the World Innovation Foundation in 2002, fellow of Royal Society of Chemistry, and fellow of World Technology Network 2014. He is an honorable professor of over 10 universities in China and many countries of Europe.

Dr. Wang is the founding editor and chief editor of an international journal *Nano Energy*, which now has an impact factor of 16.6.

Dr. Wang's breakthrough researches in the last 15 years have been featured by over 50 media world wide including *CNN, BBC, FOX News, New York Times, Washington Post, Reuters, NPR radio, Time, National Geographic, Discovery, New Scientists, and Scientific America*. Dr. Wang is the #25 in the list of the world's greatest scientists.

Scientific contributions

Pioneered the field of nanogenerators and self-powered sensors

1. Wang invented the piezoelectric nanogenerators and initiated the field of nanoenergy

Developing novel technologies for wireless nanodevices and nanosystems are of critical importance for *in-situ*, real-time and implantable biosensors, environmental science, personal electronics and national security. It is highly desired for wireless devices to be self-powered without using battery; otherwise 90% of internet of things would be impossible. A groundbreaking research by Wang in 2006 is the invention of the piezoelectric nanogenerators for self-powered nanodevices [*Science*, 312 (2006), 242]. He demonstrated an innovative approach for effectively converting mechanical energy into electric energy by piezoelectric zinc oxide nanowire arrays. This research opens up the area of nanoenergy, a field that uses nanomaterials and nanodevices for high efficient harvesting of energy from ambient environment, which is now a focal area of research for applications in sensor networks, mobile electronics and internet of things. This research was chosen as the world top 10 most outstanding discoveries in science by the Chinese Academy of Sciences. Wang was featured by *Science Watch* in Dec. 2008 issue for his pioneer work in nanogenerator.

2. Wang pioneered the original idea of self-powered systems

Wang developed the first microfiber-nanowire hybrid nanogenerator [*Science*, 316 (2007), 102; *Nature*, 451 (2008), 809-813; *Nature Nanotechnology*, 4 (2009), 34], establishing the basis of using textile fibers for harvesting mechanical energy. The principle and technology demonstrated converts mechanical movement energy (such as body movement, muscle contractions, blood pressure), vibration energy (such as acoustic/ultrasonic wave), and hydraulic energy (such as flow of body fluid, blood flow, contraction of blood vessel) into electric energy that may be sufficient for self-powering nanodevices and nanosystems. Wang has made ground-breaking progress in scale up the output of the nanogenerators through three-dimensional integration [*Nano Letters*, 10 (2010), 5025; *Nano Letters*, 10 (2010), 3151], clearly demonstrating its outstanding potential for powering sensors and personal electronics. The prototype technology established by the nanogenerator sets a platform for developing self-powering nanosystems with important applications in implantable *in-vivo* biosensors, wireless and remote sensors, nanorobotics, MEMS and sonic wave detection [*Scientific American*, January issue (2008), 82]. The nanogenerator is selected by *New Scientist* as the top 10 most potential technologies in the coming 30 years, which will be as important as the invention of cell phone, is among the top 20 featured nanotechnologies by *Discovery Magazine* in 2010, and is the top 10 scientific discoveries by *Physics World* in 2012. The fiber based nanogenerator was selected as the top 10 most important emerging technologies in 2008 by the British *Physics World*, *MIT Technology Review*, and *Beijing Daily* newspaper(Fig.2).

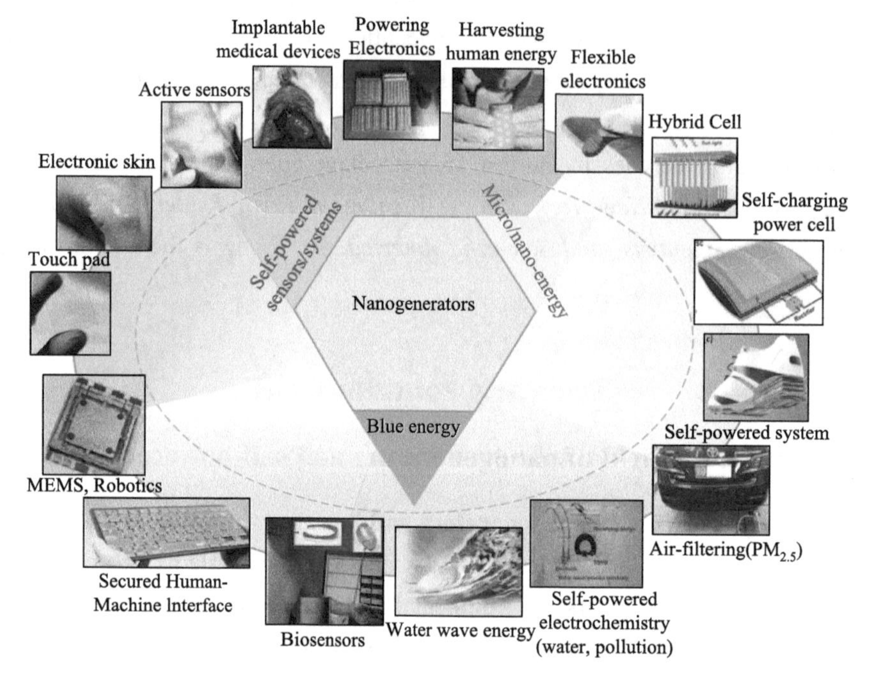

Fig.2 Nanogenerator invented by Wang's group and its applications

3. Wang invented triboelectric nanogenerator for internet of things

The principle of electromagnetic generator (EMG) that coverts mechanical energy into electricity was first introduced by Faraday in 1831, which is most efficiently at high frequency (>10～60 Hz). As for low frequency mechanical triggering (<1～5 Hz) in our living environment with random amplitude, EMG is inapplicable for harvesting such energy owing to extreme low output. Therefore, our energy system was established only by using high-quality, regulated energy that has a high operation frequency, because EMG is the only technology available for converting mechanical energy in the last 180 years. However, as the world is advancing into the era of internet of things and artificial intelligence, distributed power sources are desperately needed. In 2011, Wang has made a discovery of utilizing the conjunction of triboelectrification effect and electrostatic induction for electricity generation using organic thin film materials. The triboelectric nanogenerator (TENG) is a simple, low cost and effective approach for power generation using human motion, which is fabricated by stacking two polymer sheets made of materials having distinctly different triboelectric characteristics, with metal films deposited on the top and bottom of the assembled structure. Later, Wang has systematically invented the modes and theories for the TENG to meet a variety of needs. TENG has been demonstrated to exhibit an unprecedented conversion efficiency of 50%～85%, an area power density of 500 W/m^2 [*Nature Communication*, 5 (2014), 3456; *Nano Letters*, 12 (2012), 3109; *Nano Letters*, 12 (2012), 4960; *Nano Letters*, 12 (2012), 6339; *Nano Letters*, 13 (2013), 847; *Adv. Mater.* 26, (2014), 3788]. TENGs have the revolutionary applications for harvesting energy from human activities, rotating tires, mechanical vibration and more, with great applications in self-powered systems for personal electronics, environmental monitoring, medical science and even large-scale power. Therefore, he is referred to as the "father of nanogenerator". US based IDTechEx predited that TENG will have multi-billion $ of marked in a broad range of fields.

4. Wang developed triboelectric nanogenerator as a new energy technology for potentially large-scale blue energy

Wang invented the four working modes of the TENG using which it becomes technologically feasible to harvest energy from the tide and waves in ocean, which is unrealistic using the traditional generators. By constructing a unit of TENG in the size of a baseball that gives an output power of 1～10 mW as driven by water wave energy, connecting and integrating many such units into a "fishing net" structure [*Faraday Discussions*, 176 (2014), 447]. His prediction shows that using the size of state of Georgia (152, 488 km^2) and 10 m of depth of water, the energy harvested by TENG grid could be enough to power the world (16 TW). This type of energy is stable and has less dependence on day or night, good weather *vs.* poor weather in comparison to solar energy, so that it has the potential to be integrated with major power grid. This research opens a new chapter in exploring the blue energy to meet the needs of large-scale power [*Nature*, 542 (2017), 159].

5. Wang expanded and reformulated the Maxwell's equations as the first principle theory of nanogenerators (*Materials Today*, 2017)

In the Maxwell's displacement current, the main term $\varepsilon\partial \boldsymbol{E}/\partial t$ given by Maxwell in 1861 gave the birth of electromagnetic wave in 1886, which is the foundation of wireless communication, radar and later the information technology. Wang's study indicates that the second term $\partial \boldsymbol{Ps}/\partial t$ added by him in the Maxwell's displacement current is directly related to the output electric current of the nanogenerator if the surface polarization is introduced, meaning that the nanogenerators are the applications of Maxwell's displacement current in energy and sensors (Fig.3), lighting the applications of Maxwell's equation in energy and sensors (2006～). He showed that the classical electromagnetic generator is the result of time variation of magnetic field $\boldsymbol{B}$; while the nanogenerator is the result of time variation of the surface polarization density $\boldsymbol{P}_s$ [*Nano Energy*, 68 (2020), 104272]. More importantly, he has found that the nanogenerator has a killer application at low triggering frequency (<5 Hz), so that it is the choice for harvesting irregular energy from our living environment, which, however, cannot be accomplished using electromagnetic generator.

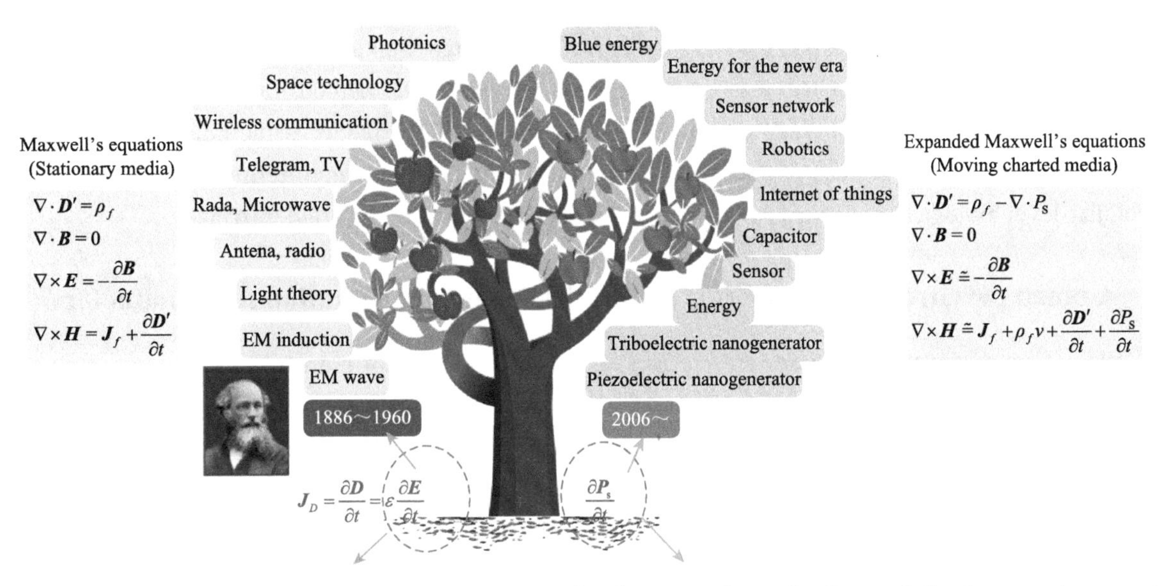

Fig.3 Contribution of Wang's nanogenerators to the theory and applications of displacement current in energy and sensors. The two first term was introduced in 1861 by Maxwell, which is responsible for the electromagnetic wave; the second term due to non-electric field induced polarization was introduced by Wang which is responsible for the first principle theory of nanogenerators

6. Wang proposed the mechanism of contact-electrification for nanogenerators

Triboelectrification is the scientific base for nanogenerators. Although the contact electrification (triboelectrification) (CE) has been documented since 2600 years ago, its scientific understanding remains inconclusive, unclear and un-unified. As for solid-solid cases, Wang found that electron transfer is the dominant mechanism for CE between solid-solid pairs [*Adv. Mater.*, 30 (2018), 1706790; *Adv. Mater.*, 30 (2018), 1803968]. Electron transfer occurs only when the interatomic distance between the two materials is shorter than the normal bonding length (typically ~0.2 nm) in the region of repulsive forces. A strong electron cloud overlap (or wave function overlap) between the two atoms/molecules in the repulsive region leads to electron transition between the atoms/molecules, owing to the reduced interatomic potential barrier. The role played by contact/friction force is to induce strong overlap between the electron clouds (or wave function in physics, bonding in chemistry). The electrostatic charges on the surfaces can be released from the surface by electron thermionic emission and/or photon excitation, so these electrostatic charges may not remain on the surface if sample temperature is higher than ～300～400℃. Wang's study solves a science problem that has been around for 2600 years [*Materials Today*, 30 (2019), 34-51], and it will have huge impact on the materials design for TENG.

7. Wang revised the mechanism about the formation of the electric double layer at liquid-solid interface based on contact-electrification effect

The electron transfer model for contact-electrification could be extended to liquid-solid, liquid-gas and even liquid-liquid cases [*Materials Today*, 30 (2019), 34-51]. As for the liquid-solid case, molecules in the liquid would have electron cloud overlap with the atoms on the solid surface at the very first contact with a virginal solid surface, and electron transfer is required in order to create the first layer of electrostatic charges on the solid surface. This step only occurs for the very first contact of the liquid with the solid. Then, ion adsorption is the second step on surface, followed by a redistribution of the ions in solution considering electrostatic interactions with the charged solid surface. The ration of electron-transfer to ion-transfer at liquid-solid interface depends on materials (*Nature Commun*, 2020; *Adv. Mater.*, 2019, 1905696). In general, electron transfer due to the overlapping electron cloud under mechanical force/pressure is proposed as the dominant mechanism for initiating CE between solids, liquids and gases. Wang provides not only the first systematic understanding about the physics of CE, but also demonstrates that the triboelectric nanogenerator (TENG) is an effective method for studying the nature of CE between any materials. This newly proposed electric double layer (EDL) model at the liquid-solid interface can largely impact the interface science in chemical, electrochemistry, chemical engineering, particle separation and even cellular-level biological processes due to electron-ion interaction at liquid-solid interfaces.

Coined piezotronics and piezo-phototronics for third generation semiconductors

1. Wang first discovered the piezotronic effect and coined the field of piezotronics

Owing to the polarization of ions in a crystal that has non-central symmetry, a piezoelectric potential *(piezopotential)* is created in the material by applying a stress. This internal field created inside of a ZnO nanowire can effectively tune the Schottky barrier height between the nanowire and its metal contact, which can effectively tune and gate the charge carrier transport process across the interface. This is the *piezotronic effect* first proposed by Wang in 2007 [*Advanced Materials*, 19 (2007), 889], based on which piezoelectric field effect transistor, piezoelectric diode and strain gated logic operations have been developed by Wang. The electronics fabricated by using the piezopotential as a gate voltage is coined *piezotronics* [*Science*, 340 (2013), 952]. The design of piezotronics fundamentally changes the design of traditional CMOS transistor in three ways: the gate electrode is eliminated so that the piezotronic transistor only has two leads; the externally applied gate voltage is replaced by an internally created piezopotential so that the device is controlled by the strain applied to the semiconductor nanowire rather than gate voltage; the transport of the charges is controlled by the contact at the drain (source)-nanowire interface rather than the channel width. Wang is also the first who demonstrated the piezotronic effect in 2D materials [*Nature*, 514 (2014), 470]. Piezotronics has applications in human-computer interfacing, smart MEMS, nanorobotics and sensors. Piezotronics was chosen as the top 10 emerging technology in 2009 by the *MIT Technology Review.*

2. Wang first discovered the piezo-photonic effect

Wang first discovered the piezo-photonic effect, which is about the piezoelectric field induced photon emission from a semiconductor material under mechanical straining [*Adv. Funct. Mater*, (18) 2008, 3553; *Nano Today*, (5) 2010, 540]. Wang demonstrated the first mechanoluminescent ZnS:Mn nanoparticles based, self-powered pressure sensor matrix for securer signature collections by recording both the handwritten signatures and the force/pressure applied by the signers at each pixel during signing, leading to an invention of high security electronic signature system (*Adv. Mater.*, DOI: 10.1002/adma.201405826). This large-area, flexible sensor matrix can *in-situ* map two-dimensional pressure distributions ranging from 0.6～50 MPa either statically or dynamically within 10 ms, with a spatial resolution of 100m (254 dpi). Utilizing strain-induced piezoelectric polarization charges to tune the band structure and facilitate the detrapping of electrons within ZnS:Mn nanoparticles. This device is applicable to real-time pressure mapping, smart sensor networks, high-level security systems, human-machine interfaces and artificial skins (Fig.4).

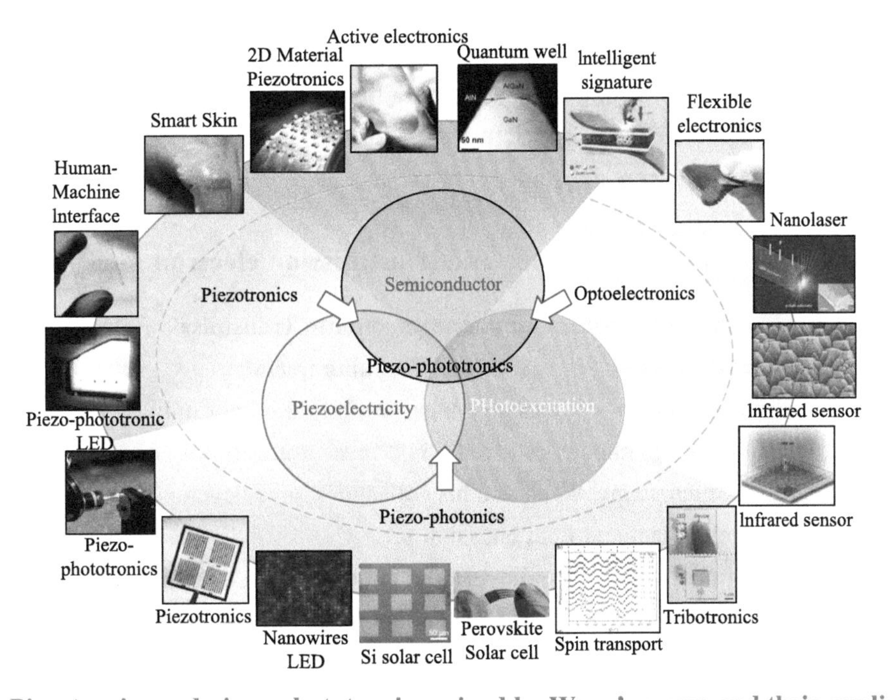

Fig.4 Piezotronics and piezo-phototronics coined by Wang's group and their applications

3. Wang first discovered the piezo-phototronic effect

Due to the polarization of ions in a crystal that has non-central symmetry, a piezoelectric potential *(piezopotential)* is created in the crystal under stress. The presence of polarization charges at an interface can largely tune the local band structure as well as shift the charge depletion zone at a pn junction, which can be effectively used to enhance the separation or recombination of charge carriers at the junction as excited by photon.

This is the *piezo-phototronic effect* first introduced by Wang in 2009 for tuning and controlling optoelectronic processes by strain induced piezopotential. Using this effect, his team has demonstrated individual-nanowire light-emitting-diode (NW-LED) based pressure/force sensor arrays for mapping strain at an unprecedented resolution of 2.7 μm and density of 6350 dpi [*Nature Photonics*, 7 (2013), 752-758], high sensitive UV sensors [*Adv. Mater.*, 24 (2012), 1410], largely enhancing LED efficiency [*Nano Letters*, 11 (2011), 4012; *Nano Letters*, 13 (2013), 607], and high performance solar cells [*Nano Letters*, 12 (2012), 3302]. Piezo-phototronic effect is a newly found physics effect, which has a broad range of applications in optimizing the performance of optoelectronic devices.

4. Wang discovered oxide nanobelts [*Science*, 209 (2001), 1947]

The nanobelts are a new class of one-dimensional nanostructures denoting a wide range of semiconducting oxides with cations of different valence states and materials of distinct crystallographic structures. This landmark paper is among the list of the top 30 most influential papers published in *Science* in the last 10 years, the top10 most cited paper in materials science in last decade. The rational approach outlined in this work has subsequently served to nucleate a large body of studies by other researchers worldwide. As a result, ZnO is the most exciting type of one-dimensional nanostructures for oxides that holds equal importance to Si nanowires and carbon nanotubes. Wang has been the world leader in studying of ZnO nanostructures.

5. Wang first proposed the growth processes of novel oxide nanostructures

Owing to the positive and negative ionic charges on the zinc- and oxygen-terminated ZnO basal planes, respectively, a spontaneous polarization normal to the nanobelt surface is induced. As a result, helical nanosprings/nanocoils are formed by rolling up single crystalline nanobelts and nanorings [*Science*, 303 (2004), 1348; *Science*, 309 (2005), 1700]. These are the first papers that described the spontaneous polarization-induced novel nanostructures and they open a new direction of research for studying piezoelectric properties at nano-scale.

Contributions to nanoscience and transmission electron microscopy

1. Wang pioneered the field of *in-situ* nanomeasurements in transmission electron microscopy on the mechanical, electrical and field emission properties of 1D nanomaterials

Characterizing the physical properties of carbon nanotubes is limited not only by the purity of the specimen but also by the size distribution of the nanotubes. Traditional measurements rely on scanning probe microscopy. Based on transmission electron microscopy, Wang and his colleagues have developed a series of unique techniques for measuring the mechanical, electrical and field emission properties of individual nanotubes in 1999. His *in-situ* TEM technique is not only an imaging tool that allows a direct observation of the crystal and surface structures of the material at atomic-resolution, but also an *in-situ* apparatus that can be effectively used to carry out nano-scale property measurements [*Science*, 283 (1999), 1513]. A nanobalance technique and a novel approach toward nanomechanics have been demonstrated [*Phys. Rev. Letts.*, 85 (2000), 622], which was selected by APS as the breakthrough in nanotechnology in 1999. This study creates a new field of *in-situ* nanomeasurements in materials science and mechanics.

2. Wang has made fundamental contribution to materials science and electron microscopy

His textbook entitled of *Functional and Smart Materials—Structural Evolution and Structure Analysis* (Plenum Press, 1998) is "a unique, cutting-edge text on smart materials ... it is recommended as an adjunct to device design books used for engineers as well as scientists during the development of smart devices and structures" [*Physics Today* , Nov. (1998), 70]. His textbook on *Elastic and Inelastic Scattering in Electron Diffraction and Imaging* (Plenum Press, 1995) is "a noteworthy achievement and a valuable contribution to the literature" (*American Scientist*, 1996). His textbook on *Reflected Electron Microscopy And Spectroscopy For Surface Analysis* (Cambridge University Press, 1996) is "a book that any materials science or physics library should be holding" (*MRS Bulletin*, Oct., 1998).

目　　录 Contents

I.3 电子非弹性散射的基本理论 Fundamental Theory of Electron Inelastic Scattering

I.4 透射显微镜中的原位测量 *In-situ* Nanomeasurements in TEM

I.5 纳米材料与 ZnO 纳米结构 Nanomaterials and ZnO Nanostructures

I.6 纳米器件 Nanodevices

I.7 压电纳米发电机 Piezoelectric Nanogenerators

II.6 摩擦纳米发电机的综述介绍 Review Papers on Triboelectric Nanogenerator

II.7 摩擦纳米发电机：等效电路理论 Triboelectric Nanogenerators—Equivalent Circuit Theory

Ⅱ.9 摩擦纳米发电机：技术标准 Triboelectric Nanogenerators—Technical Standards

Ⅱ.10 摩擦纳米发电机：功率优化与管理 Triboelectric Nanogenerators—Power Maximization and Management

Ⅱ.11 固体-固体接触起电的机理 The Origins of Solid-Solid Contact-Electrification

Ⅱ.12 液体-固体接触起电的机理 The Origins of Liquid-Solid Contact-Electrification

Ⅲ.3　摩擦纳米发电机：高压电压源 Triboelectric Nanogenerators—As a High Voltage Source

Ⅲ.4　摩擦纳米发电机：蓝色能源 Triboelectric Nanogenerators—For Blue Energy

Ⅲ.5 复合式能量回收系统 Hybridized Energy Systems

Ⅲ.6 热势纳米发电机与传感 Pyroelectric Nanogenerators and Sensors

Ⅲ.7 自驱动传感与系统 Self-Powered Sensors and Systems

第四部分 Part Ⅳ

Ⅳ.1 压电电子学：原理与器件 Piezotronics—Principle and Devices

Ⅳ.2 压电电子学：应变传感器 Piezotronics—Strain Sensors

Ⅳ.3 压电电子学：理论 Piezotronics—Theory

Ⅳ.5 压电光电子学：理论 Piezo-Phototronics—Theory

Ⅳ.6 压电光子学 Piezophotonics

Ⅳ.7 摩擦电子学 Tribotronics

第一部分

Part Ⅰ

I.1 电子显微学基础技术

Fundamental Techniques of Electron Microscopy

1. Surface Plasmon Excitation for Supported Metal Particles

Z.L. Wang *and* **J.M. Cowley**

Ultramicroscopy, 21 (1987) 77-94.

简介：等离子体激元的研究在最近这十年来是热点之一，其主要原因是现在可以合成结构可控而量大的纳米材料，因此可以用束斑大的激光来激发。但最早的等离子体激元的研究是在电子显微镜中利用电子能量损失谱（EELS）进行的，因为电子束可以聚焦到 0.3 nm 直径，可以选择性地激发单个小颗粒。考虑到纳米颗粒不可能悬空，该文章是首次考虑衬底对单个小颗粒的等离子体激元的影响，文中的理论预言得到了很好的实验验证。在 1987 年纳米颗粒还不叫 nanoparticle，而是叫小颗粒（small particle）。因此，在纳米还没有叫“纳米”之前，王中林就在做纳米的研究了。

ABSTRACT Classical dielectric theory has been used to study the effects of a support on the surface plasmon excitation frequencies for small metal particles. Two experimentally accessible cases have been considered and the theoretical results have been compared with observations. For small spherical Al particles, coated with a thin oxide layer and half-buried in an infinite support of Al, four surface plasmon frequencies (ω_1 to ω_4) are predicted. Comparison with results for isolated particles and planar interfaces suggests the identification of ω_3 as the normal small particle surface plasmon; ω_1 and ω_2 are generated by the presence of the support and ω_4 is related to the oxide shell. The ω_2 and ω_3 frequencies are observed in accord with the predictions. For the second case considered, the Al support is replaced by a dielectric, AlF_3. The support then contributes a new peak at about 5 eV which has been observed. The observations were made with the EELS attachment to a conventional analytical electron microscope.

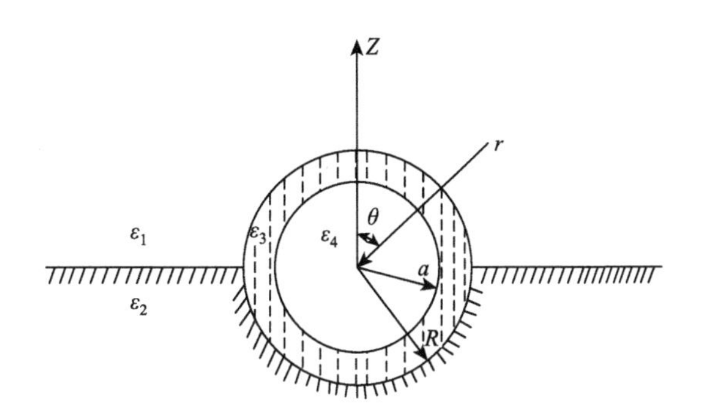

扫码阅全文

http://www.binn.cas.cn/book/202106/W020210602507072888778.pdf

2. Size and Shape Dependence of the Surface Plasmon Frequencies for Supported Metal Particle Systems

Z.L. Wang *and* **J.M. Cowley**

Ultramicroscopy, 23 (1987) 97-108.

简介：该文章首次考虑小颗粒的形貌对等离子体激元的影响。文中的理论预言和实验观察符合得很好，不但等离子体激元的能量可以被测到，它的激发概率也可以通过 EELS 被定量测到。

ABSTRACT The effect of the size and shape of small particles on plasmon excitations has been investigated, as possible indicators for the physical properties of supported metal catalyst particles. The plasmon frequencies of a small metal sphere, which is covered with an oxide shell and is half embedded in some kind of support, arc calculated using hydrodynamic theory as functions of the radius of the particles. The presence of the support introduces new plasmon frequencies. In addition to frequencies close to those for isolated particles, there are three more frequencies if the support is of the same metal as the particle, and one frequency is added if the support is an insulator. All the surface plasmon frequencies show the same type of increase for particle sizes smaller than about 4nm, but the magnitude of the increase varies considerably. Classical dielectric theory has been used to calculate the surface plasmon excitation frequencies for the supported hemispheres. A general formula has been derived and applied to some cases of special geometry, It has been shown theoretically that only, the two neighboring even and odd modes are coupled in the supported hemisphere system. A shift of 3 eV for the surface plasmon for the supported hemisphere metal particle (11 eV) with respect to that in the supported metal sphere particle (8 eV～has been predicted and observed in the specimen AL/AlF_3.

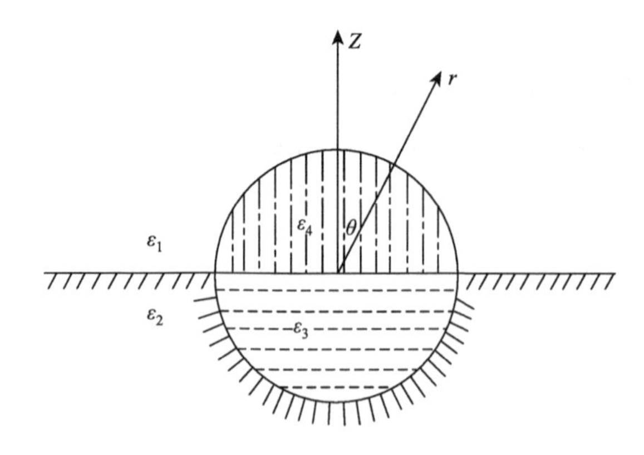

扫码阅全文

http://www.binn.cas.cn/book/202106/W020210602507073325166.pdf

3. Valence Electron Excitations and Plasmon Oscillations in Thin Films, Surfaces, Interfaces and Small Particles

Z. L. Wang

Micron, 27 (1996) 265-299; 28 (1997) 505-506.

简介：本文首次从麦克斯韦方程组出发推导出了包括相对论效应来计算等离子体激元的激发概率方程式。把薄膜、表面、界面和小颗粒的离子体激元激发的研究做了系统性和完整性的总结。本文是首篇对电子激发等离子体激元理论和实验做全面描述的文献。

ABSTRACT Plasmon oscillation is a collective excitation of electrons in a valence band of a solid material. The motion and polarization of valence electrons under the impact of a fast moving charged particle directly reflect the solid state properties of the material. Oscillations of surface charges depend sensitively on dielectric properties of the material and, more importantly, on the geometrical configuration of the media. The advances of electron microscopy techniques have made it possible to study local excitations from each individual particle smaller than a few manometers in diameter. Dielectric response theory has shown remarkable success in describing the observed valence-loss spectra and resonance modes. This review gives a systematic description on the classical electron energy-loss theory and its applications in characterizing interband transition and plasmon excitations in thin films, surfaces, interfaces, isolated particles and supported particles of different geometrical configurations. These fundamental studies are important for characterizing many advanced nanophase and nanostructured materials of technological importance. This article is focused on quantitative calculation of valence-loss spectra acquired from different geometrical configurations of dielectric objects. The classical energy-loss theory is equivalent to the quantum mechanical theory, provided all the scattered electrons are collected by the spectrometer. The hydrodynamic model is also described to include fluctuation of electron density in metallic particles smaller than 10 nm in diameter. Applications of valence electron excitation spectroscopy are demonstrated using numerous experimental results.

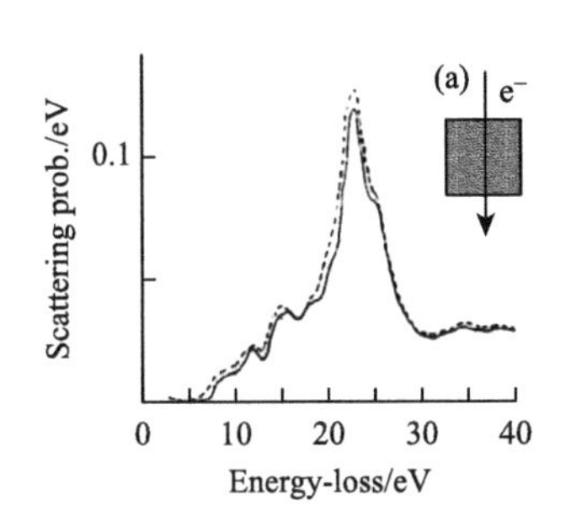

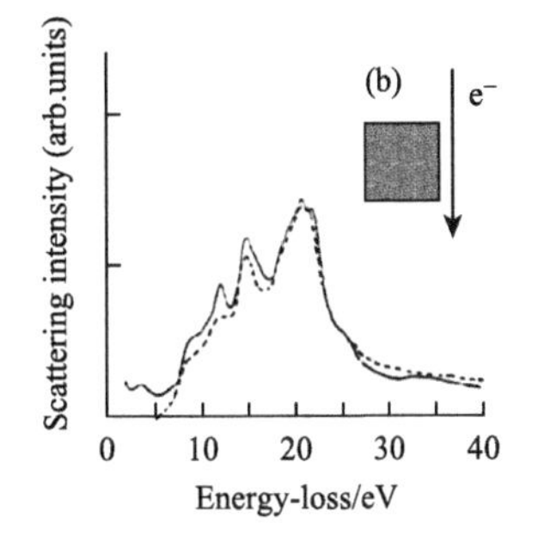

扫码阅全文

http://www.binn.cas.cn/book/202106/W020210602507073634602.pdf

4. Reflection Electron Energy Loss Spectroscopy (REELS): A Technique for the Study of Surfaces

Z.L. Wang* *and* **J.M. Cowley**

Surface Science, 193 (1988) 501-512.

简介：本文首次发明了高能电子的反射电子能量损失谱，可以探索表面顶层几个原子层的电子和化学结构。

ABSTRACT Electron energy loss spectroscopy is used in the glancing-angle reflection mode in conjunction with reflection electron microscopy as a technique for studying the composition and atomic coordination in surface layers of crystals. The resonance propagation of the electron beam along the top few atomic layers of the surface in the RHEED geometry results in the excitation of the surface states in MgO(100) and ensures sensitivity to the surface composition and structure. Results of ELNES and EXELFS analyses of the oxygen K edge for MgO(100) cleavage faces, and also the indications of an increase in O/Mg ratio at the surface relative to the bulk MgO, are consistent with a model of O atoms absorption over Mg ions.

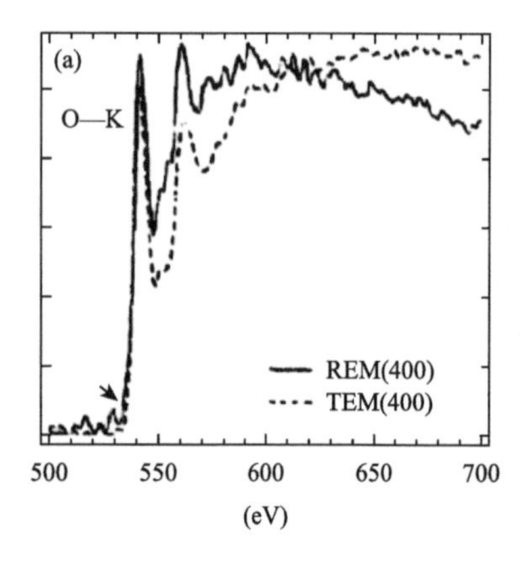

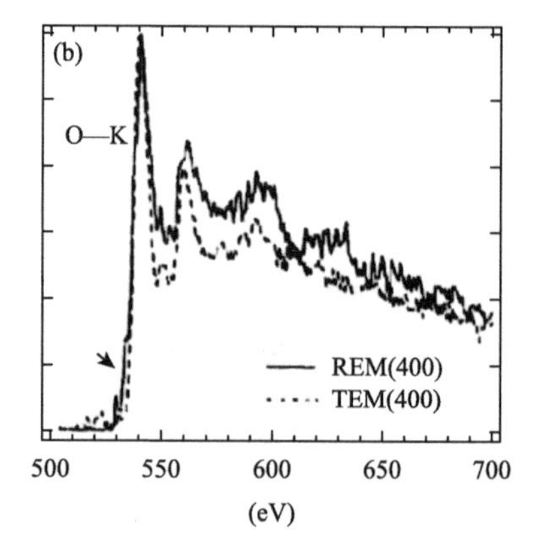

扫码阅全文

http://www.binn.cas.cn/book/202106/W020210602507074968273.pdf

注：作者名字右上角 * 表示合著论文中该作者为通讯作者。余同。

5. Electron Inelastic Plasmon Scattering and its Resonance Propagation at Crystal Surfaces in RHEED

Z.L. Wang*, **J. Liu** *and* **J.M. Cowley**

Acta Crystallographica Section A, 45 (1989) 325-333.

简介：本文首次从理论计算和实验观察得到了高能电子在表面顶层几个原子层的共振传播过程。

ABSTRACT The modified multislice theory [Wang (1989). *Acta Cryst.* A45, 193-199] has been employed to calculate the electron reflection intensity with and without considering the plasmon diffuse scattering in the geometry of reflection high-energy electron diffraction (RHEED). It has been shown that the inelastic scattering can greatly enhance the reflectance of a surface, depending critically on the incident conditions of the electrons. At some incidences, the inelastic resonance reflection is enhanced, which is considered as the "true" surface resonance state. This happens within a very narrow angular range (<1mrad). For "true" resonance states, the inelastic intensity is much stronger than for other conditions as shown both theoretically and experimentally. The enhancement of the reflection intensity may not be the proper criterion for identifying the "true" surface resonance. Besides the surface plasmon peaks, an "extra" peak, located at 4-5eV, is observed in the reflection electron energy-loss spectroscopy (REELS) study of the "true" resonance of GaAs (110) surface. This is considered as a characteristic of the resonance propagations of the electrons along the surface and may result from the generation of resonance radiation.

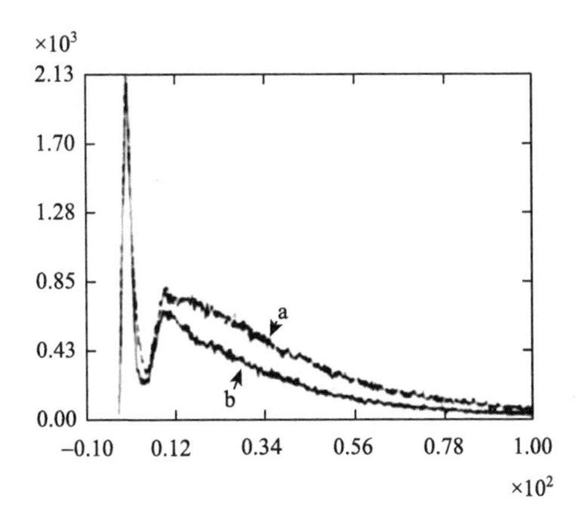

扫码阅全文

http://www.binn.cas.cn/book/202106/W020210602507075285964.pdf

6. Mapping the Valence States of Transition-Metal Elements Using Energy-Filtered Transmission Electron Microscopy

Z.L. Wang*, J. Bentley *and* N.D. Evans

Journal of Physical Chemistry B, 103 (1999) 751-753 plus cover.

简介：利用能量损失谱和透射显微镜的结合，根据钴和锰等元素的 L-edge white line 的强度比率，本文首次利用能量过滤的透射显微镜得到了过渡金属价态分布和变化的高分辨图像，奠定了在实空间得到纳米级价态分布的图像。

ABSTRACT The properties of transition-metal oxides are related to the presence of elements with mixed valences, such as Mn and Co. Spatial mapping of the valence-state distribution of transition-metal elements is a challenge to existing microscopy techniques. In this letter, using the valence-state information provided by the white lines observed in electron energy-loss spectroscopy in a transmission electron microscope (TEM), an experimental approach is demonstrated to map the valence-state distributions of Mn and Co using the energy-filtered TEM. An optimum spatial resolution of～2nm has been achieved for two-phase Co oxides with sharp boundaries. This provides a new technique for quantifying the valence states of cations in magnetic oxides.

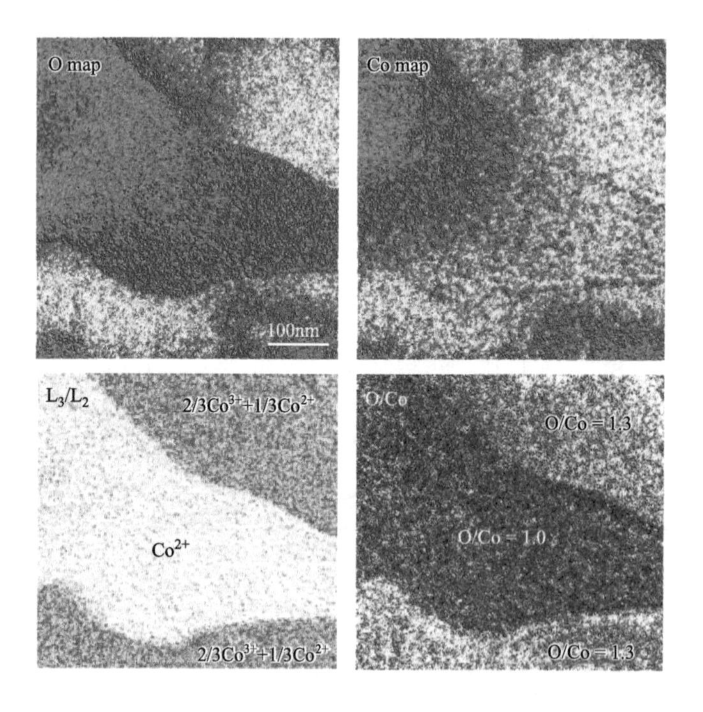

扫码阅全文

http://www.binn.cas.cn/book/202106/W020210602507075669383.pdf

7. EELS Analysis of Cation Valence States and Oxygen Vacancies in Magnetic Oxides

Z.L. Wang*, J.S. Yin, Y.D. Jiang

Micron, 31(2000) 571-580.

简介：利用能量损失谱中元素的单电子激发峰，本文首次提出了利用钴和锰元素的 L-edge white line 的强度比率来确定所具有的价态，为分析这些元素的电子态提出了一个新方法。

ABSTRACT Transition metal oxides are a class of materials that are vitally important for developing new materials with functionality and smartness. The unique properties of these materials are related to the presence of elements with mixed valences of transition elements. Electron energy-loss spectroscopy (EELS) in the transmission electron microscope is a powerful technique for measuring the valences of some transition metal elements of practical importance. This paper reports our current progress in applying EELS for quantitative determination of Mn and Co valences in magnetic oxides, including valence state transition, quantification of oxygen vacancies, refinement of crystal structures, and identification of the structure of nanoparticles.

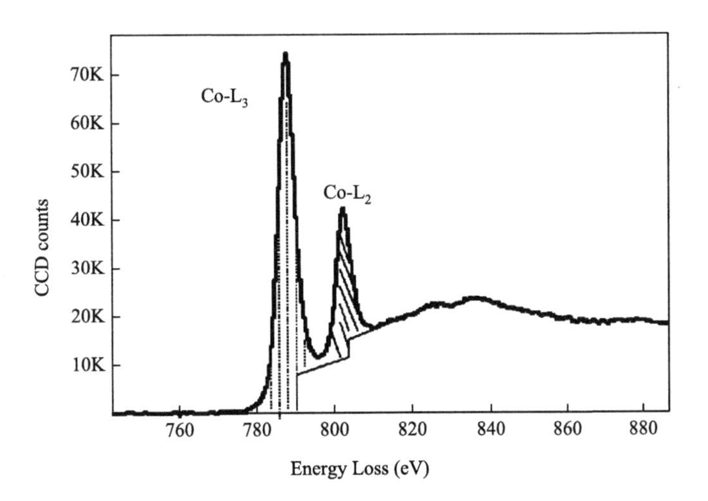

扫码阅全文

http://www.binn.cas.cn/book/202106/W020210602507075755759.pdf

1.2 HAADF STEM 成像理论

Theory of High-Angle Annular Dark-Field STEM Imaging

1. Simulating High-Angle Annular Dark-Field STEM Images Including Inelastic Thermal Diffuse Scattering

Z.L. Wang * *and* **J.M. Cowley**

Ultramicroscopy, 31 (1989) 437-454.

简介：本文首次提出晶体中声子的漫散射（热漫散射）是 HAADF STEM 成像中的主要贡献者，并提出了相应的动力学计算方法。本文也纠正了被 Pennycook 等认为 HAADF STEM 成像是由于卢瑟福（Rutherford）散射的说法，开创了定量解释 HAADF STEM 成像的先河。今天 HAADF STEM 技术是研究纳米材料特别是二维材料、催化材料的有力工具。

ABSTRACT A generalized multislice theory is employed to simulate the annular dark-field (ADF) images, in scanning transmission electron microscopy (STEM), taking into account the inelastic thermal diffuse scattering (TDS). It is shown that the ADF image has the following characteristics if the first-order Laue zone (FOLZ) (or high-order Laue zones (HOLZ)) is excluded: (1) there is no contrast reversal with change of focus; (2) there is no contrast reversal with change of specimen thickness and (3) the ADF image shows highly localized scattering from each atomic column and can provide some chemical information for a specimen having two or more atoms with very different atomic numbers. It is shown that the TDS may take a dominant role in determining the ADF imaging contrast, depending strongly on the atomic number, the sizes (or shapes) of the atoms and their relative vibration amplitudes. TDS may increase the visibility of light elements. A full dynamical calculation including TDS is required to interpret the ADF imaging contrast. It is therefore not proper to consider the ADF imaging contrast as purely z dependent. The HOLZ can introduce strong diffraction contrast in the ADF image.

$$(\nabla^2 + k_0^2)\boldsymbol{\Psi}_0 = \sum_m \frac{2m}{\hbar^2} H'_{0m}(\boldsymbol{r})\boldsymbol{\Psi}_m$$

$$(\nabla^2 + k_0^2)\boldsymbol{\Psi}_n = \sum_m \frac{2m}{\hbar^2} H'_{nm}(\boldsymbol{r})\boldsymbol{\Psi}_m$$

扫码阅全文

http://www.binn.cas.cn/book/202106/W020210609399277463447.pdf

2. Dynamic Theory of High-Angle Annular Dark Field STEM Lattice Images for a Ge/Si Interface

Z.L. Wang* *and* **J.M. Cowley**

Ultramicroscopy, 32 (1990) 275-289.

简介：本文完善了晶体中声子的漫散射（热漫散射）是 HAADF STEM 成像中的主要贡献者的理论构架。通过解析推导和数值计算，提出了计算 HAADF STEM 成像的普适性声子散射模型。这个理论后来被称为 Wang-Cowley 理论。

ABSTRACT A dynamical multislice approach for simulating the annular-dark-field (ADF) scanning transmission electron microscopy (STEM) images is presented. The ADF imaging contrast is dominated by thermal diffuse scattering (TDS) if the diffraction contrast from the high-order Laue zones (HOLZ) is excluded. The simulation of ADF images can be greatly simplified if the assumption is made of incoherent imaging theory such as is used in optics, that is, the image intensity is a thickness integration of the convolution of the electron probe intensity distribution at depth z with the localized inelastic generation function $|G|^2$. This result may also be derived on the basis of coherent elastic scattering theory. The calculation according to the simplified theory can give almost exactly the same contrast as obtained from full dynamical inelastic calculations for thin specimens. The TDS can give atomic resolution lattice images if a small probe, less than the size of the unit cell, is used. A programming flow-chart for simulating ADF images is given. Calculations for Ge/Si(100) interfaces have shown that the ADF imaging contrast of the atoms is not affected by correlations of vibrations of neighboring atoms. It is pointed out that the contribution from elastic large-angle scattering will become important only if the atomic vibration amplitudes are less than about 0.03Å. The ADF image production may not be considered as incoherent imaging if this is the case.

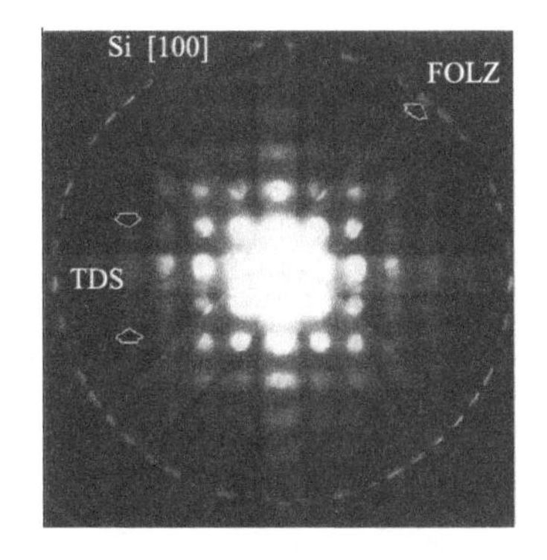

扫码阅全文

http://www.binn.cas.cn/book/202106/W020210609399278504234.pdf

I.3 电子非弹性散射的基本理论

Fundamental Theory of Electron Inelastic Scattering

1. Dynamical Inelastic Scattering in High-Energy Electron Diffraction and Imaging: A New Theoretical Approach

Z.L. Wang*

Physical Review B, 41 (1990) 12818-12837.

简介：王中林的导师 J.M. Cowley 教授 1957 年提出计算高分辨显微像的多片层弹性动力学散射理论，该理论是电子显微学的重要科学基础。本文王中林从 n x m 耦合的薛定谔方程矩阵组出发，首次推导出了多片层学非弹性散射的动力学理论，奠定了包括非弹性散射过程的电子衍射与成像理论，大大延伸和开拓了 Cowley 教授的理论。

ABSTRACT A generalized multislice theory is proposed from quantum mechanics to approach the multiple elastic and multiple inelastic scattering of high-energy electrons in a solid. The nonperiodic structure of crystals can be introduced in the calculations for the scattering geometries of transmission electron microscopy and reflection electron microscopy. Detailed applications of this generalized theory will be given for calculating (a) the energy-filtered-plasmon energy-loss diffraction patterns and images, (b) the energy-filtered diffraction patterns from atomic inner-shell losses, and (c) the contribution of thermal diffuse scattering to the high-angle annular-dark-field (ADF) scanning-transmission-electron-microscopy (STEM) lattice images. An "incoherent" imaging theory is presented for simulating the ADF STEM images and the detailed calculations are addressed for Ge/Si interfaces.

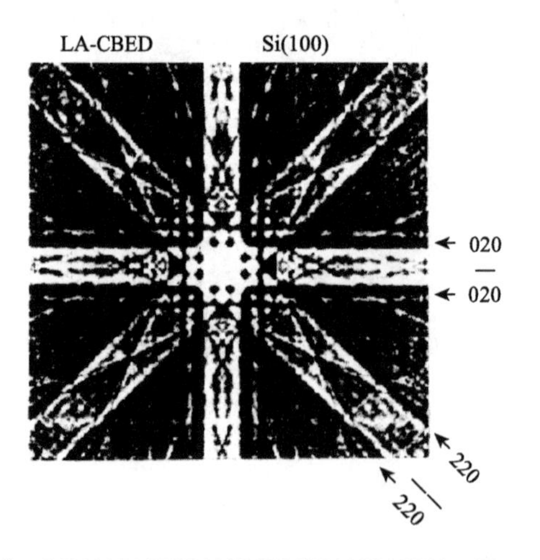

扫码阅全文

http://www.binn.cas.cn/book/202106/W020210609403002810092.pdf

2. Theory of Phase Correlations in Localized Inelastic Electron Diffraction and Imaging

Z.L. Wang* *and* **J. Bentley**

Ultramicroscopy, 38 (1991) 181-213.

简介：本文从 n x m 耦合的薛定谔方程矩阵组出发，考虑了声子散射中不同原子之间振动的相位关联，首次推导出理解电子衍射中热散射所引入的衍射斑点沿着某些方向拉长的规律以及实验验证，为理解电子衍射中非弹性散射提供了新的方法。

ABSTRACT A dynamical theory is proposed starting from the coupled wave equations for describing inelastic electron scattering in perfect crystals and crystals with defects. The phase correlations of the localized inelastic scattering occurring at different atomic sites can be statistically evaluated before any numerical calculations, essentially providing an easy way for treating coherence, incoherence or partial coherence in dynamical electron diffraction. The time average of electron diffraction from the lattices of different thermal vibration configurations can be performed in just a single calculation. It is shown that the phase correlations of atomic vibrations result in streaking of electron diffraction spots. The possible streaks in 2D reciprocal space $\boldsymbol{\tau}$ can be predicted by examining $\sum_{l}\sum_{l_1}\cos\left[\boldsymbol{\tau}\cdot(\boldsymbol{R}_l-\boldsymbol{R}_{l_1})\right]$ for-a monatomic system, where $\boldsymbol{R}_l$ and $\boldsymbol{R}_{l_1}$ are the 3D positions of the lth atom within the unit cell and its first nearest neighbors located in the same atomic plane perpendicular to the incidence beam direction, respectively. The TDS streaks are located on lines satisfying $\boldsymbol{\tau}\cdot(\boldsymbol{R}_l-\boldsymbol{R}_{l_1})=0$. This rule can also be applied to multiatomic cases if the differences made by atomic structure factors are considered. Also this rule should allow the determination of strong atomic interaction relationships from the orientation of TDS in diffraction patterns. A dynamical diffraction theory is described to calculate the intensities of thermal diffuse scattering (TDS) streaks. Theoretical schemes are described for calculating electron energy-filtered images and diffraction patterns after exciting an atomic inner-shell, and *Z*-contrast images in scanning transmission electron microscopy (STEM). Some experimental evidence is shown to demonstrate spot streaking in electron diffraction and the corresponding theoretical interpretations.

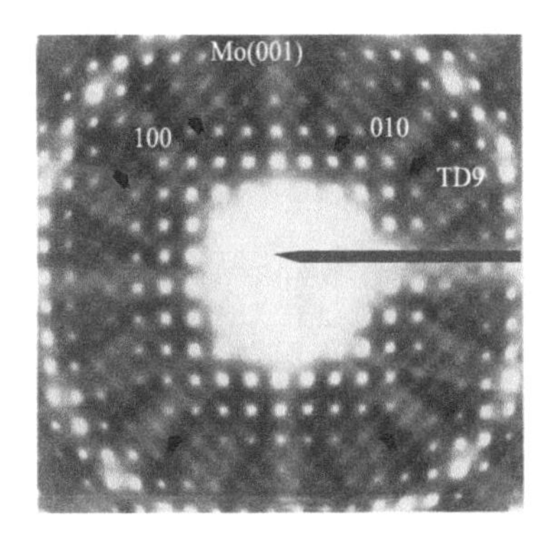

扫码阅全文

http://www.binn.cas.cn/book/202106/W020210609403003276794.pdf

3. Dynamics of Thermal Diffuse Scattering in High-Energy Electron Diffraction and Imaging: Theory and Experiments

Z.L. Wang

Philosophical Magazine Part B, 65 (1992) 559-587.

简介：本文从电子散射的动力学理论出发，首次计算得到了在有原子晶格热振动情况下薄膜的电子衍射图，并和实验观测做了定量的比较。该文是发展电子与声子相互作用的动力学理论的一个里程碑性工作。

ABSTRACT A dynamical theory is proposed to calculate the diffraction patterns of electrons which have undergone thermal diffuse scattering (TDS). The lattice dynamics and electron diffraction dynamics are comprehensively combined in this approach. Full dynamical simulations are presented for Mo(001). Under the single-inelastic-scattering approximation and in the near-zone-axis case, the sharpness of the TDS streaks is determined by the phonon dispersion relationships of the acoustic branches, the optical branches contribute only a diffuse background, and dynamical scattering effects can change the intensity distribution of TDS electrons but have almost no effect on the sharpness of TDS streaks. The TDS streaks are defined by the q_x-q_y curves which satisfy $\omega_j(\boldsymbol{q}) = 0$, where $\omega_j(\boldsymbol{q})$ is the phonon dispersion relationship determined by the two-dimensional atomic vibrations in the (hkl) plane perpendicular to the incident beam direction $\mathbf{B} = [hkl]$. The directions of TDS streaks predicted according to these procedures are consistent with those predicted according to the $\boldsymbol{q}\cdot[\boldsymbol{r}(l)-\boldsymbol{r}(l_1)]=0$ rule given by Wang and Bentley, where the summation of l_1 is restricted to the first-nearest neighbours of the lth atom that are located in the same atomic plane as the lth atom perpendicular to the incidence beam direction. Atomic resolution images can be formed by TDS electrons in transmission electron microscopy. The image theory is equivalent to the incoherent imaging theory if the spherical aberration coefficient C_s for the objective lens is small. The image resolution may be higher than that formed by pure elastically scattered electrons, but the "inclined-incidence effect" (i.e. the correction of momentum transfer $\boldsymbol{q}$ to the electron wave-vector $\boldsymbol{K}$) in phonon scattering may distort the image. The phase coupling of vibrating atoms does not affect calculations for images but does affect diffraction patterns. Thus the image simulation can be performed based on the Einstein model as long as the correct vibration amplitude for each atom is used. Equivalent results are obtained for the TDS electrons based on either the quantum inelastic scattering theory or the "frozen"-lattice model of the semiclassical approach if the temperature is not much higher than room temperature.

扫码阅全文

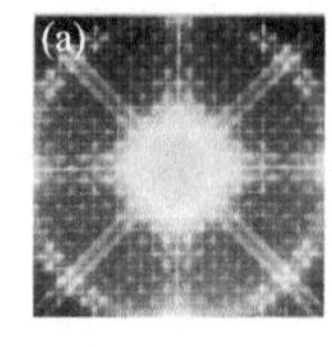

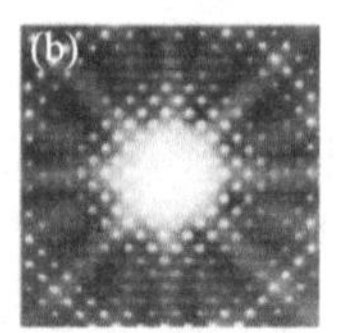

http://www.binn.cas.cn/book/202106/W020210609403004069468.pdf

4. Thermal Diffuse Scattering in High-Resolution Electron Holography

Z.L. Wang

Ultramicroscopy, 52 (1993) 504-511.

简介：本文首次从理论上证明了电子全息成像是一个完美的弹性散射的过滤器，即所有非弹性散射的电子包括声子散射的电子都被过滤掉。这篇文章发表后，王中林的导师 J.M. Cowley 教授还发文表示不同意这个观点，但后来德国科学家 Lichte 实验证明王中林的结论是正确的。

ABSTRACT A theory for wave function simulations with inclusion of thermal diffuse scattering (TDS) in electron holography is outlined. The inelastic thermal diffuse scattered electrons do not contribute to the sidebands but introduce an absorption factor in the simulated elastic wave.The TDS absorption effects are significant in the simulated amplitude but not in the phase and become important only when the image resolution approaches atomic level (<1.6Å). Valence, single electron and even phonon inelastically scattered electrons do not affect the sidebands of the off-axis holograms, thus the reconstructed wave function from a hologram is an essentially zero-loss, energy-filtered image of the specimen. In contrast, inelastically scattered electrons including the TDS electrons do contribute to conventional high-resolution transmission electron microscopy (HRTEM) images. This simply means that quantitative analysis of electron holograms may be easier than the analysis of HRTEM images.

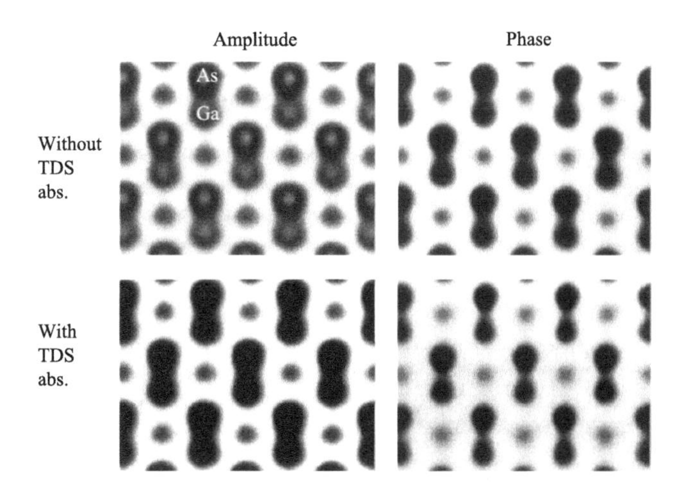

扫码阅全文

http://www.binn.cas.cn/book/202106/W020210609403004576273.pdf

5. Statistical Multiple Diffuse Scattering in a Distorted Crystal System: An Exact Theory

Z.L. Wang

Philosophical Magazine Part B, 74 (1996) 6, 733-749.

简介：从具有点缺陷的晶体结构出发，本文首次给出了描述高能电子衍射中薛定谔方程的“严格解”，并对不同缺陷分布的晶体结构做了解析统计运算。这是发展利用电子衍射来进行点缺陷研究的一个里程碑性理论贡献。

ABSTRACT In electron scattering, diffuse scattering can be generated by atom vibrations, point vacancies and growth islands (or surface roughness). Most of the existing dynamical theories have been developed under the first-order diffuse scattering approximation; thus they are restricted to cases where the lattice distortion is small. In this paper, a formal dynamical theory is presented for calculating diffuse scattering with the inclusion of multiple diffuse scattering. By inclusion of a complex potential in dynamical calculation, a rigorous proof is given to show that the high-order diffuse scattering is fully recovered in the calculations using the equation derived under the distorted-wave Born approximation, and more importantly, the statistical time and structure averages over the distorted crystal lattices are evaluated analytically prior to numerical calculation. This conclusion establishes the basis for expanding the applications of the existing theories. The exact form of the optical potential is given using a general solution of the Green function, which can be computed numerically using existing dynamical diffraction theories. These conclusions are universal for both low-and high-energy electrons. For transmission electron diffraction, the final result is given in the Bloch wave representation for crystals containing distorted structure. The theory can also be applied to calculate electron images of diffusely scattered electrons in transmission and scanning transmission electron microscopy.

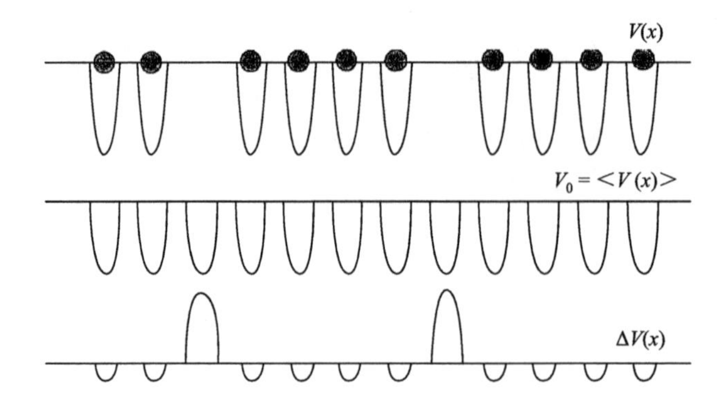

扫码阅全文

http://www.binn.cas.cn/book/202106/W020210609403004767204.pdf

6. Electron Statistical Dynamical Diffuse Scattering in Crystals Containing Short-Range-Order Point Defects

Z.L. Wang

Acta Crystallographica Section A, 52 (1996) 717-729.

简介：从具有短程有序的点缺陷的晶体结构出发，本文首次给出了描述高能电子衍射中薛定谔方程的严格解，并对不同缺陷分布的晶体结构所产生的衍射图做了解析统计运算。这是发展利用电子衍射来研究点缺陷短程有序的一个重要理论基石。

ABSTRACT A formal statistical dynamical theory is developed to calculate diffuse scattering produced by short-range order (SRO) in a distorted crystal structure with consideration of atomic thermal vibrations. Diffuse scattering not only produces fine details in diffraction patterns but also introduces a non-local imaginary potential function that reduces the intensities of the Bragg reflected beams. The distribution of the diffusely scattered electrons and the Fourier coefficients of the absorption potential are directly related to a dynamic form factor $S(\mathbf{Q},\mathbf{Q}')$, which has been calculated with consideration of SRO in the distorted lattices. The statistical structure average on imperfections is performed analytically and the final result is correlated to Cowley's short-range-order parameters. The theory is formulated in the Bloch-wave scheme (Bethe theory) for the convenience of numerical calculation in transmission electron diffraction. A rigorous theoretical proof is given to show that the inclusion of a complex potential in the dynamical calculation automatically recovers the contributions made by the high-order diffuse scattering, although the calculation is done using the equation derived for single diffuse scattering. This simply expands the capability of conventional single diffuse scattering theories. Therefore, the complex potential has a much richer meaning than the conventional interpretation of absorption effect.

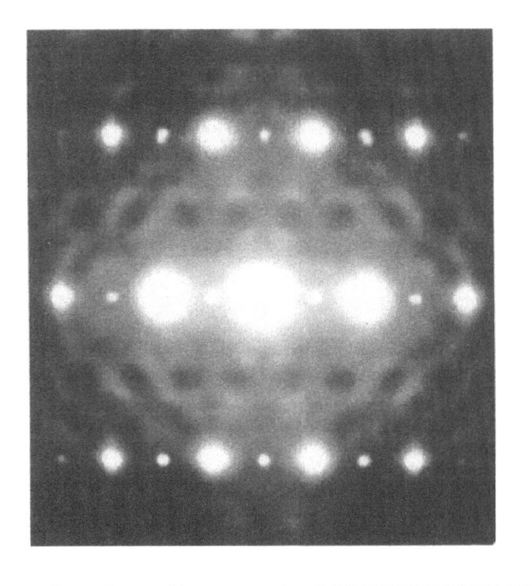

扫码阅全文

http://www.binn.cas.cn/book/202106/W020210609403005037143.pdf

7. The 'Frozen Lattice' Approach for Incoherent Phonon Excitations in Electron Scattering. How Accurate Is It?

Z.L. Wang

Acta Crystallographica Section A, 54 (1998) 460-467.

简介：在高能电子衍射中，由于电子的运动速度接近光速，晶体由于热振动所产生的晶格微小畸变对于入射电子就好像一个被瞬时冻结了的晶格，可以按静态的组态来处理。最后所得到的电子衍射图是所有这些组态所产生的衍射强度的非相干叠加。这个模型称之为晶格冻结模型（frozen lattice model）。而这个模型看起来是不同于电子和声子相互作用的量子力学散射模型。在这篇文章中，王中林从解析方程出发证明利用晶格冻结模型所计算的电子衍射图是和量子力学计算的模型全等的，从此奠定了这个模型半经典模型的量子力学基础。

ABSTRACT The 'frozen-lattice' model is a semi-classical approach for calculating electron diffuse scattering in crystals that has arisen from thermal vibration of crystal atoms. This quasi-elastic scattering approach is, however, unproven since its equivalence with the incoherent phonon-excitation model is not yet established. As quantitative electron microscopy is becoming a realistic method, it is necessary to examine the accuracy of the model. In this paper, based on a rigorous quantum-mechanical phonon excitation theory, it is proved that an identical result would be obtained using the frozen-lattice model and the formal phonon-excitation model if (i) the incoherence between different orders of thermal diffuse scattering is considered in the frozen-lattice-model calculation and (ii) the specimen thickness and the mean-free-path length for phonon excitation are both smaller than the distance travelled by the electron within the lifetime of the phonon ($\tau_0 v$, which is 5 μm for 100 kV electrons). Condition (ii) is usually absolutely satisfied and condition (i) can be precisely accounted for in the calculation with the introduction of the mixed dynamic form factor S(**Q**, **Q**'). The conclusion holds for each and all orders of diffuse scattering, thus, the quantum-mechanical basis of the frozen-lattice model is established, confirming the validity, reliability and accuracy of using this model in quantitative dynamical electron diffraction and imaging calculations. It has also been shown that the frozen-lattice model is suitable for low-energy electrons.

扫码阅全文

http://www.binn.cas.cn/book/202106/W020210609403005615773.pdf

8. An Optical Potential Approach to Incoherent Multiple Thermal Diffuse Scattering in Quantitative HRTEM

Z.L. Wang

Ultramicroscopy, 74 (1998) 7-26.

简介：本文首次给出了计算高能电子衍射和成像中包括声子散射情况下动力学方程的严格解。该文是进行定量显微学，特别是超高分辨电子成像的重要理论基石。

ABSTRACT The theory for the absorption potential (or optical potential) in electron scattering was first proposed by Yoshioka in 1957 based on an approximation that the Green's function is replaced by its form in free-space. This approximation has dramatically simplified the calculation, but the function of the optical potential has been partially lost. In this paper, a rigorous theoretical proof is given based on the quantum inelastic excitation theory to show that the inclusion of the optical potential in the dynamic calculation automatically recovers the contributions made by the high-order diffuse scattering although the calculation is done using the equation derived for single diffuse scattering. This conclusion generalizes the existing first-order diffuse scattering theories to cases in which the incoherent multiple diffuse scattering are important. It is suggested that multiple thermal diffuse scattering (or phonon excitations) is likely to be the dominant source for affecting image contrast in quantitative electron microscopy because of the channeling effect and the localized scattering nature of phonon excitation. Surprisingly, phonon scattering is anticipated to contribute fine atomic-scale contrast in the image, which is believed to be more pronounced than the white noise effect of conventional understanding. The optical potential is no longer a simple potential function, rather it is a non-local function strongly dependent on the dynamic diffraction in the crystal because of the involvement of Green's function. These characteristics can be properly taken into the computation using a proposed Bloch wave-multislice approach, in which the non-local effect is resolved in the Bloch wave matrix diagonalization and the dynamical effect is taken care of using Green's function calculated by the multislice theory under the small angle (or high-energy) scattering approximation. In the ground-state approximation, by which we mean that the crystal is in its ground state before each and every inelastic excitation, an introduction of the mixed dynamic form factor and the density matrix automatically produces the incoherence between different order and different phonon excitation processes, resolving a big problem encountered by many other theories. In the conventional dynamical calculation for elastic wave, the inclusion of the Debye-Waller factor and the optical potential accounts only for the effects of the phonon excitation on the elastic wave rather than the contribution made by the diffusely scattered electrons to the image/diffraction pattern.

$$\rho(\boldsymbol{r},\boldsymbol{r}') = \Phi_0(\boldsymbol{r})\Phi_0^*(\boldsymbol{r}') + e^2\xi^2 \int \mathrm{d}\boldsymbol{Q} \int \mathrm{d}\boldsymbol{Q}' S(\boldsymbol{Q},\boldsymbol{Q}')\psi(\boldsymbol{Q},\boldsymbol{b},z)\psi^*(\boldsymbol{Q}',\boldsymbol{b}',z)$$

9. Phonon Scattering: How Does It Affect the Image Contrast in High-Resolution Transmission Electron Microscopy?

Z.L. Wang

Philosophical Magazine Part B, 79 (1999) 1, 37-48.

简介：本文首次阐述了为什么在超高显微分辨率成像中必须考虑声子散射所做的贡献，这对定量显微学极其重要，可从理论计算高能电子衍射和成像中，包括声子散射情况下动力学方程的严格解。该文是进行定量显微学，特别是超高分辨电子成像的重要理论基石。

ABSTRACT In quantitative high-resolution transmission electron microscopy (HRTEM), the theoretically calculated images usually give better contrast than the experimentally observed images although all the factors have been accounted for. It is suggested that this discrepancy is due to thermal diffusely scattered electrons, which were not included in the image calculation. The question is: how do they affect the image contrast? In this paper, under the weak-phase object approximation. it is shown that the contribution of the thermal diffusely scattered electrons to the image is of the same order as the cross-interference terms for the Bragg reflected beams in the dark-field HRTEM imaging. Indirect experimental measurements showed that thermal diffuse scattering (TDS) is not a small effect; rather it is the dominant scattering at large angles. The TDS absorption is measured and the result indicates that about 12% of the incident electrons have been diffusely scattered to angles larger than 15.6° (the column angle of the transmission electron microscope) by a Si foil as thin as 15-20nm. The data clearly show the magnitude and importance of TDS in HRTEM. It is therefore mandatory to include this component in image calculation.

扫码阅全文

http://www.binn.cas.cn/book/202106/W020210609403006006092.pdf

10. Picoscale Science and Nanoscale Engineering by Electron Microscopy

Zhong Lin Wang

Journal of Electron Microscopy, 60 (Supplement 1) (2011) S269-S278 .

简介：本文预言了电子显微技术未来发展的方向和领域，必将是结合实空间成像、倒空间衍射、电子和化学谱仪学，再加上原位力-光-电微测量技术为一体的分析、表征和测量系统。

ABSTRACT A future scanning/transmission electron microscope is proposed to be a comprehensive machine that is capable of providing picoseconds time-resolved information at sub-nanometer scale and even at picometer scale, spatial resolution. At the same time, physical and chemical properties can be measured *in situ* from a region as small as a few nanometers by introducing local electric, mechanical, thermal, magnetic and/or optical stimulations/ excitations under vacuum or even in a quasi-ambient environment. It is anticipated that nanoscopy and picoscopy will be key tools for studying picoscale science and developing nanoscale technology related to materials science, biology, physics and chemistry.

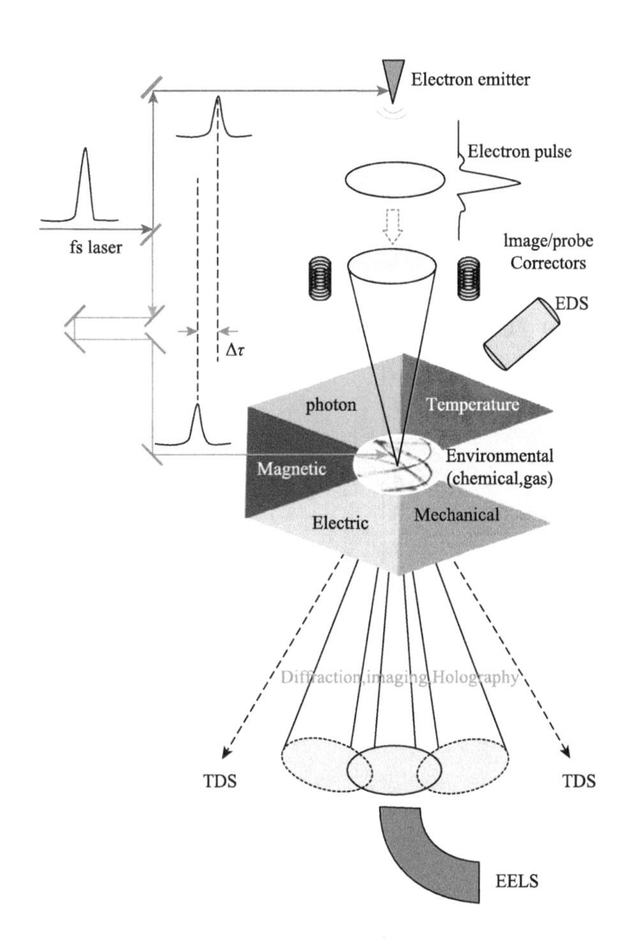

扫码阅全文

http://www.binn.cas.cn/book/202106/W020210609403006367719.pdf

I.4 透射显微镜中的原位测量

In-situ Nanomeasurements in TEM

1. Carbon Nanotube Quantum Resistors

Stefan Frank, Philippe Poncharal, Z. L. Wang, Walt A. de Heer*

Science, 280 (1998) 1744-1746.

简介：本文报道了结合原子力探针与透射显微镜为一体，首次原位观察到了碳纳米管的量子导电效应，开创了透射显微镜中实现电测量的先河。自从该文章发表后，王中林的研究转向了原位微测量阶段。

ABSTRACT The conductance of multiwalled carbon nanotubes (MWNTs) was found to be quantized. The experimental method involved measuring the conductance of nanotubes by replacing the tip of a scanning probe microscope with a nanotube fiber, which could be lowered into a liquid metal to establish a gentle electrical contact with a nanotube at the tip of the fiber. The conductance of arc-produced MWNTs is one unit of the conductance quantum $G_0 = 2e^2/h = (12.9\ \text{kilohms})^{-1}$. The nanotubes conduct current ballistically and do not dissipate heat. The nanotubes, which are typically 15 nanometers wide and 4 micrometers long, are several orders of magnitude greater in size and stability than other typical room-temperature quantum conductors. Extremely high stable current densities, $J > 10^7$ amperes per square centimeter, have been attained.

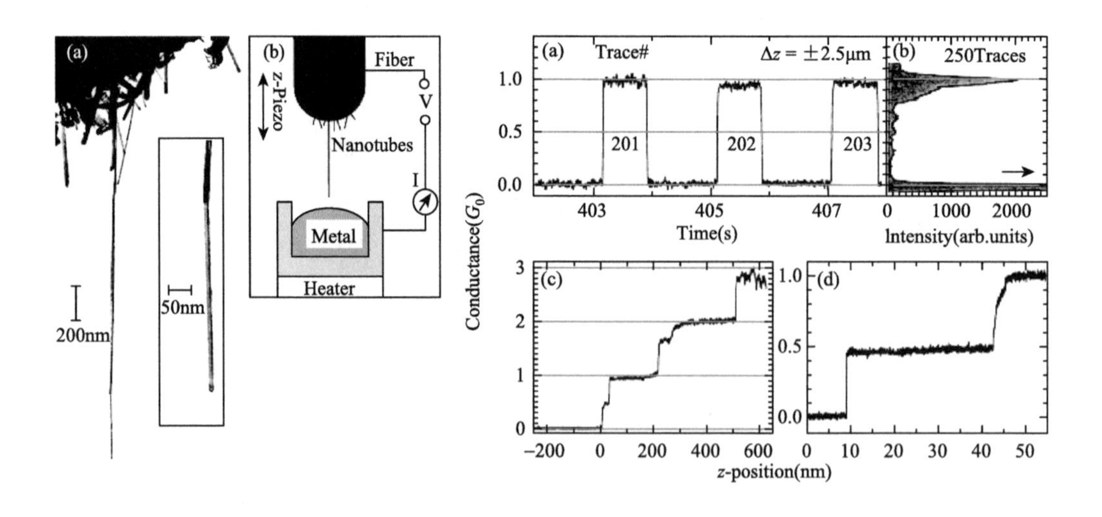

扫码阅全文

http://www.binn.cas.cn/book/202106/W020210609404880045160.pdf

2. Electrostatic Deflections and Electromechanical Resonances of Carbon Nanotubes

Philippe Poncharal, Z. L. Wang, Daniel Ugarte, Walt A. de Heer*

Science, 283 (1999) 1513-1516.

简介：本文首次报道了在透射显微镜中利用悬臂梁的共振效应，实现单根碳纳米管杨氏模量的原位测量，并提出了可以测量单个纳米颗粒的“纳米秤”。该文章开启了原位纳米力学的新领域。

ABSTRACT Static and dynamic mechanical deflections were electrically induced in cantilevered, multiwalled carbon nanotubes in a transmission electron microscope.The nanotubes were resonantly excited at the fundamental frequency and higher harmonics as revealed by their deflected contours, which correspond closely to those determined for cantilevered elastic beams. The elastic bending modulus as a function of diameter was found to decrease sharply (from about 1 to 0.1 terapascals) with increasing diameter (from 8 to 40 nanometers), which indicates a crossover from a uniform elastic mode to an elastic mode that involves wavelike distortions in the nanotube. The quality factors of the resonances are on the order of 500. The methods developed here have been applied to a nanobalance for nanoscopic particles and also to a Kelvin probe based on nanotubes.

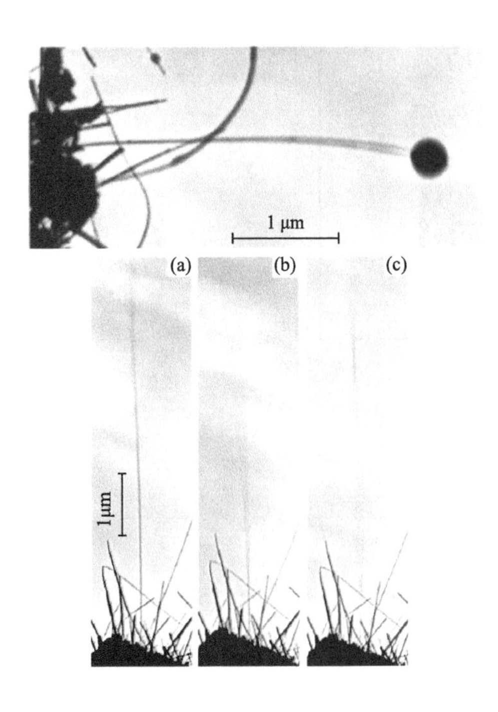

扫码阅全文

http://www.binn.cas.cn/book/202106/W020210609404880285328.pdf

3. Work Function at the Tips of Multiwalled Carbon Nanotubes

Ruiping Gao, Zhengwei Pan *and* **Zhong L. Wang***

Applied Physics Letters, 78 (2001) 1757-1759.

简介：本文首次提出了在透射电子显微镜中，可以原位测量单根碳纳米管尖端的功函数的方法。

ABSTRACT The work function at the tips of individual multiwalled carbon nanotubes has been measured by an *in situ* transmission electron microscopy technique. The tip work function shows no significant dependence on the diameter of the nanotubes in the range of 14-55 nm. Majority of the nanotubes have a work function of 4.6-4.8eV at the tips, which is 0.2-0.4eV lower than that of carbon. A small fraction of the nanotubes have a work function of~5.6eV, about 0.6 eV higher than that of carbon. This discrepancy is suggested due to the metallic and semiconductive characteristics of the nanotube.

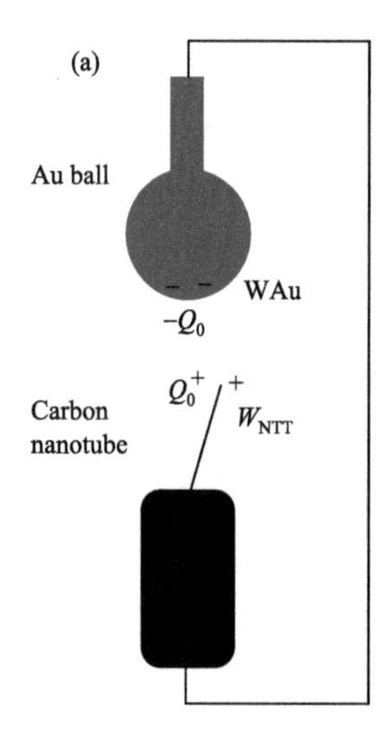

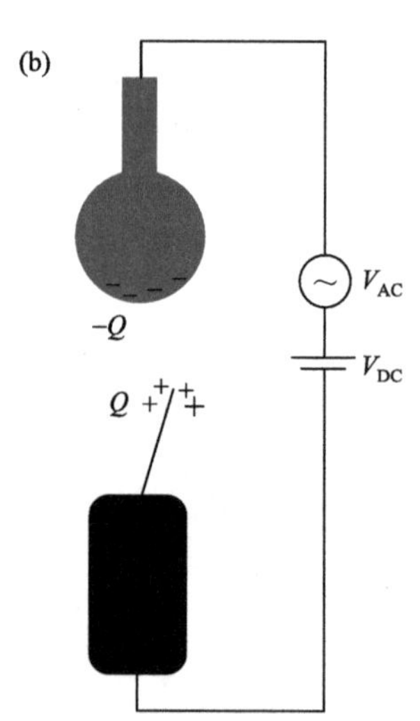

扫码阅全文

http://www.binn.cas.cn/book/202106/W020210609404880419214.pdf

4. Low-Temperature *in situ* Large Strain Plasticity of Ceramic SiC Nanowires and its Atomic-Scale Mechanism

X.D. Han, Y.F. Zhang, K. Zheng, X.N. Zhang, Z. Zhang*, Y.J. Hao, X.Y. Guo, J. Yuan *and* Z.L. Wang*

Nano Letters, 7 (2007) 452-457.

简介：本文首次报道了在透射显微镜中研究单根纳米线的塑性形变的方法和力学极限，拓展了原位纳米力学的领域。

ABSTRACT Large strain plasticity is phenomenologically defined as the ability of a material to exhibit an exceptionally large deformation rate during mechanical deformation. It is a property that is well established for metals and alloys but is rarely observed for ceramic materials especially at low temperature (~300 K). With the reduction in dimensionality, however, unusual mechanical properties are shown by ceramic nanomaterials. In this Letter, we demonstrated unusually large strain plasticity of ceramic SiC nanowires (NWs) at temperatures close to room temperature that was directly observed *in situ* by a novel high-resolution transmission electron microscopy technique. The continuous plasticity of the SiC NWs is accompanied by a process of increased dislocation density at an early stage, followed by an obvious lattice distortion, and finally reaches an entire structure amorphization at the most strained region of the NW. These unusual phenomena for the SiC NWs are fundamentally important for understanding the nanoscale fracture and strain-induced band structure variation for high-temperature semiconductors. Our result may also provide useful information for further studying of nanoscale elastic-plastic and brittle-ductile transitions of ceramic materials with superplasticity.

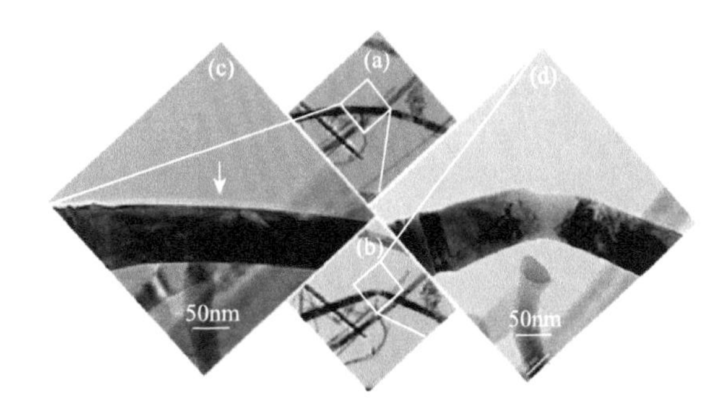

扫码阅全文

http://www.binn.cas.cn/book/202106/W020210609404880521681.pdf

5. Dynamic Fatigue Studies of ZnO Nanowires by *in-situ* Transmission Electron Microscopy

Zhiyuan Gao, Yong Ding, Shisheng Lin, Yue Hao *and* Zhong Lin Wang*

Physica Status Solidi RRL, 3 (2009) 260-262.

简介：本文首次报道了在透射显微镜中研究单根纳米线的疲劳。

ABSTRACT The fatigue behavior of ceramic ZnO nanowires (NWs) has been investigated under resonance cyclic loading conditions using *in-situ* transmission electron microscopy (TEM). After mechanical deformation at the resonance frequency at a vibration angle of 5.2° for 35 billion cycles, no failure or any defect generations have been found. We believe that the dislocation-free nature of NWs and the large surface-to-volume ratio contribute to the NWs' ability to undergo deformation without fatigue or fracture, proving their durability and toughness for nanogenerators and nanopiezotronics.

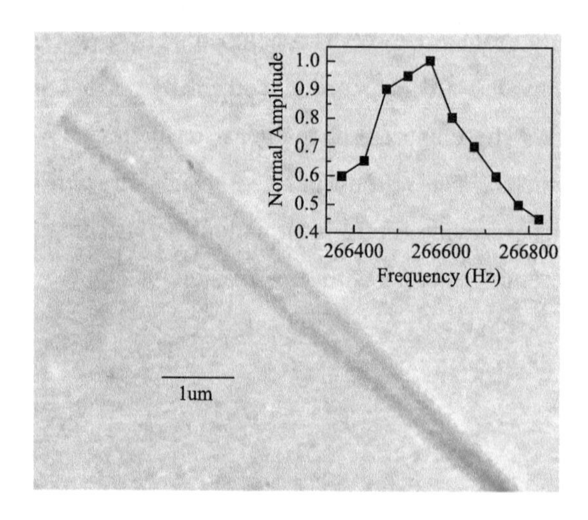

扫码阅全文

http://www.binn.cas.cn/book/202106/W020210609404880820709.pdf

6. Elastic Property of Vertically Aligned Nanowires

Jinhui Song, Xudong Wang, Elisa Riedo* *and* Zhong L. Wang*

Nano Letters, *5* (2005) 1954-1958.

简介：本文首次报道了利用原子力探针进行垂直阵列纳米线杨氏模量的测试，由此测量出纳米线压电系数，最后引发了压电式纳米发电机的发明。

ABSTRACT An atomic force microscopy (AFM) based technique is demonstrated for measuring the elastic modulus of individual nanowires/nanotubes aligned on a solid substrate without destructing or manipulating the sample. By simultaneously acquiring the topography and lateral force image of the aligned nanowires in the AFM contacting mode, the elastic modulus of the individual nanowires in the image has been derived. The measurement is based on quantifying the lateral force required to induce the maximal deflection of the nanowire where the AFM tip was scanning over the surface in contact mode. For the [0001] ZnO nanowires/nanorods grown on a sapphire surface with an average diameter of 45nm, the elastic modulus is measured to be 29±8 GPa.

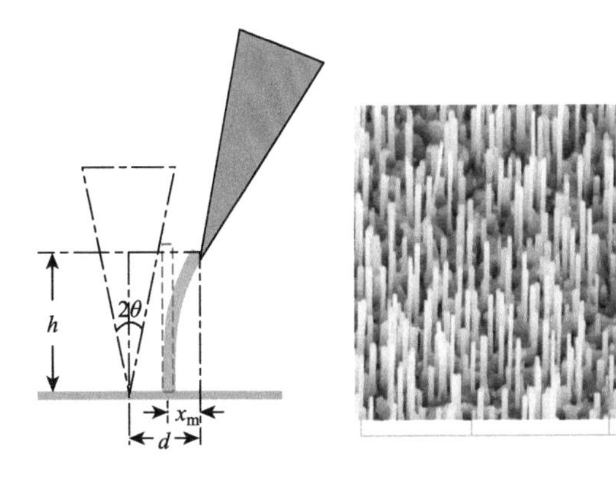

扫码阅全文

http://www.binn.cas.cn/book/202106/W020210609404880930661.pdf

7. Quantitative Nanoscale Tracking of Oxygen Vacancy Diffusion Inside Single Ceria Grains by *in situ* Transmission Electron Microscopy

Yong Ding*, Yong Man Choi, Yu Chen, Ken C. Pradel, Meilin Liu*, Zhong Lin Wang*

Materials Today, 38(2020), 24-34.

简介：研究固体样品中点缺陷，特别是无规则分布的点缺陷是透射显微镜的一大挑战。本文开创性地研究了 CeO_2 中氧空位的原位扩散行为，并推导出了扩散系数。

ABSTRACT Oxygen vacancy formation and migration in ceria is critical to its electrochemical and catalytic properties in systems for chemical and energy transformation, but its quantification is rather challenging especially at atomic-scale because of disordered distribution. Here we report a rational approach to track oxygen vacancy diffusion in single grains of pure and Sm-doped ceria at –20℃ to 160℃ using *in situ* (scanning) transmission electron microscopy ((S)TEM). To create a gradient in oxygen vacancy concentration, a small region (~30 nm in diameter) inside a ceria grain is reduced to the C-type $CeO_{1.68}$ phase by the ionization or radiolysis effect of a high-energy electron beam. The evolution in oxygen vacancy concentration is then mapped through lattice expansion measurement using scanning nano-beam diffraction or 4D STEM at a spatial resolution better than 2 nm; this allows direct determination of local oxygen vacancy diffusion coefficients in a very small domain inside pure and Sm-doped ceria at different temperatures. Further, the activation energies for oxygen transport are determined to be 0.59, 0.66, 1.12, and 1.27 eV for pure CeO_2, $Ce_{0.94}Sm_{0.06}O_{1.97}$, $Ce_{0.89}Sm_{0.11}O_{1.945}$, and $Ce_{0.8}Sm_{0.2}O_{1.9}$, respectively, implying that activation energy increases due to impurity scattering. The results are qualitatively supported by density functional theory (DFT) calculations. In addition, our *in situ* TEM investigation reveals that dislocations impede oxygen vacancy diffusion by absorbing oxygen vacancies from the surrounding areas and pinning them locally. With more oxygen vacancies absorbed, dislocations show extended strain fields with local tensile zone sandwiched between the compressed ones. Therefore, dislocation density should be reduced in order to minimize the resistance to oxygen vacancy diffusion at low temperatures.

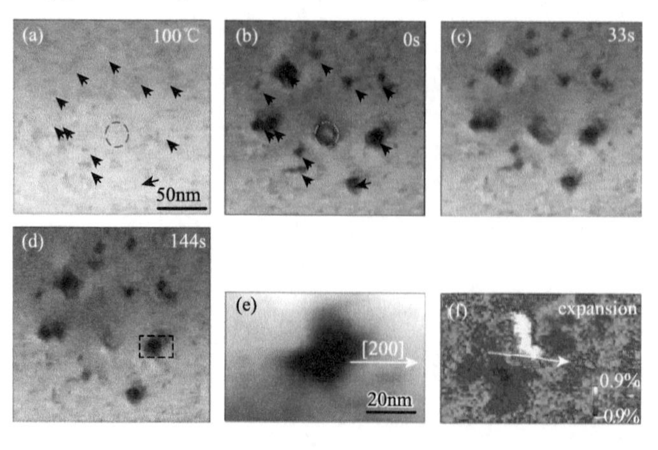

扫码阅全文

http://www.binn.cas.cn/book/202106/W020210609404881107849.pdf

I.5 纳米材料与 ZnO 纳米结构

Nanomaterials and ZnO Nanostructures

1. Nanobelts of Semiconducting Oxides

Zheng Wei Pan, Zu Rong Dai, Zhong Lin Wang*

Science, 291 (2001) 1947-1949.

简介：本文首次报道了氧化物纳米带结构，开创了氧化锌等氧化物纳米结构研究的先河，使对于氧化锌的研究等同于对碳纳米管和硅纳米线的研究。自从该文章发表后，王中林的研究转向了纳米材料合成、表征和测量的新阶段。该文章也是王中林最高引用的文章之一。

ABSTRACT　Ultralong beltlike (or ribbonlike) nanostructures (so-called nanobelts) were successfully synthesized for semiconducting oxides of zinc, tin, indium, cadmium, and gallium by simply evaporating the desired commercial metal oxide powders at high temperatures. The as-synthesized oxide nanobelts are pure, structurally uniform, and single crystalline, and most of them are free from defects and dislocations. They have a rectanglelike cross section with typical widths of 30 to 300 nanometers, width-to-thickness ratios of 5 to 10, and lengths of up to a few millimeters. The beltlike morphology appears to be a distinctive and common structural characteristic for the family of semiconducting oxides with cations of different valence states and materials of distinct crystallographic structures. The nanobelts could be an ideal system for fully understanding dimensionally confined transport phenomena in functional oxides and building functional devices along individual nanobelts.

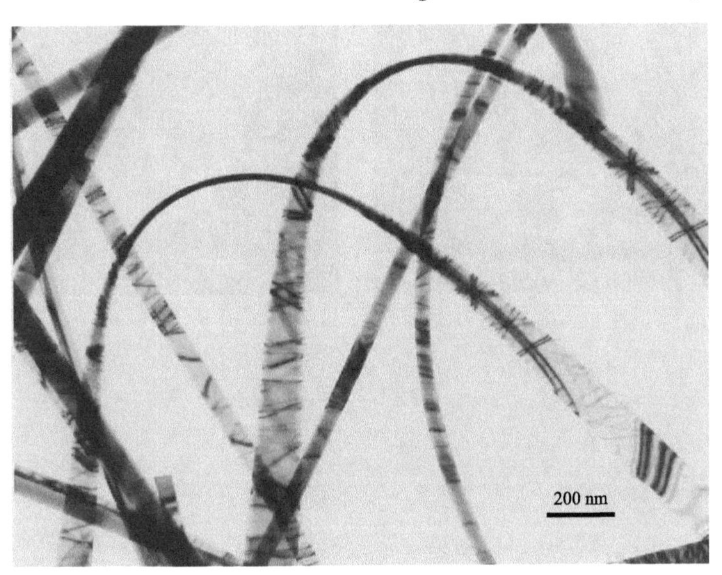

扫码阅全文

http://www.binn.cas.cn/book/202106/W020210609407279298886.pdf

2. Single-Crystal Nanorings Formed by Epitaxial Self-Coiling of Polar Nanobelts

Xiang Yang Kong, Yong Ding, Rusen Yang, Zhong Lin Wang*

Science, 303 (2004) 1348-1351.

简介：本文首次报道了单晶氧化锌纳米带由于极性面的作用而形成的环状结构，展现了氧化锌独特的生长机制。

ABSTRACT Freestanding single-crystal complete nanorings of zinc oxide were formed via a spontaneous self-coiling process during the growth of polar nanobelts. The nanoring appeared to be initiated by circular folding of a nanobelt, caused by long-range electrostatic interaction. Coaxial and uniradial loop-by-loop winding of the nanobelt formed a complete ring. Short-range chemical bonding among the loops resulted in a single-crystal structure. The self-coiling is likely to be driven by minimizing the energy contributed by polar charges, surface area, and elastic deformation. Zinc oxide nanorings formed by self-coiling of nanobelts may be useful for investigating polar surface-induced growth processes, fundamental physics phenomena, and nanoscale devices.

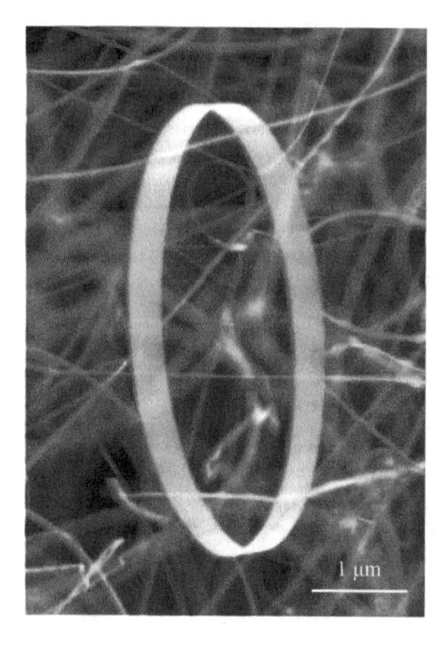

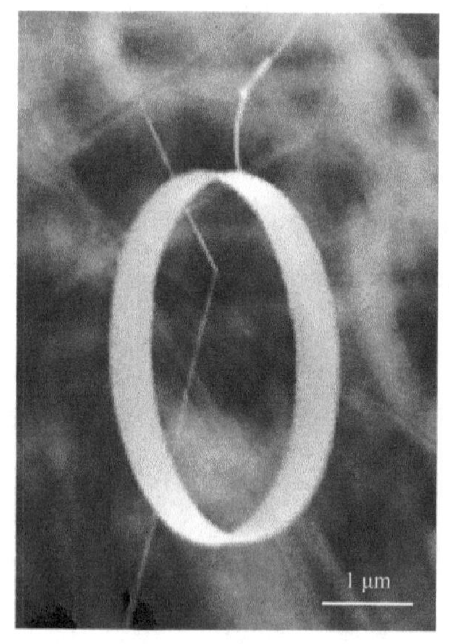

扫码阅全文

http://www.binn.cas.cn/book/202106/W020210609407279531808.pdf

3. Large-Scale Hexagonal-Patterned Growth of Aligned ZnO Nanorods for Nano-Optoelectronics and Nanosensor Arrays

Xudong Wang, Christopher J. Summers, ***and*** **Zhong Lin Wang***

Nano Letters, 4 (2004) 423-426.

简介：本文首次介绍了大规模生长六角形分布氧化锌纳米线阵列的新方法，为实现压电电子学芯片打下材料基础。

ABSTRACT An effective approach is demonstrated for growing large-area, hexagonally patterned, aligned ZnO nanorods. The synthesis uses a catalyst template produced by a self-assembled monolayer of submicron spheres and guided vapor-liquid-solid (VLS) growth on a single crystal alumina substrate. The ZnO nanorods have uniform shape and length, align vertically on the substrate, and are distributed according to the pattern defined by the catalyst template. The nanorods grow along [0001] with side surfaces defined by $\{2\bar{1}\bar{1}0\}$. This approach opens the possibility of creating patterned one-dimensional nanostructures for applications as sensor arrays, piezoelectric antenna arrays, optoelectronic devices, and interconnects.

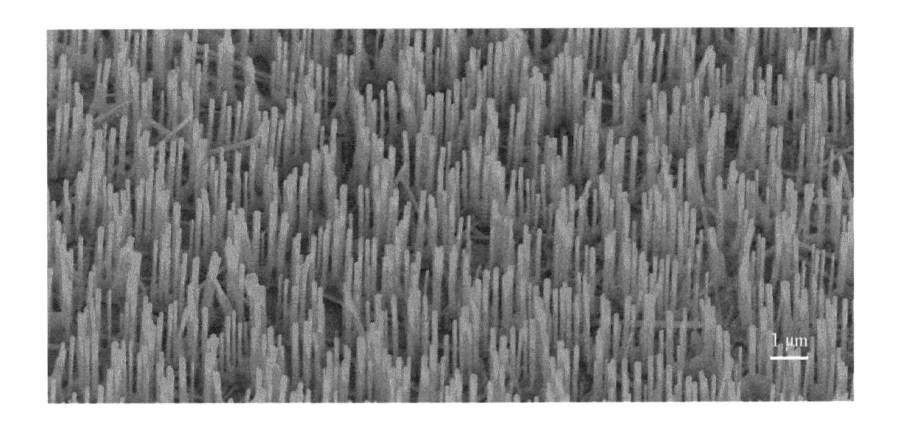

扫码阅全文

http://www.binn.cas.cn/book/202106/W020210609407279772877.pdf

4. Optimizing and Improving the Growth Quality of ZnO Nanowire Arrays Guided by Statistical Design of Experiments

Sheng Xu, Nagesh Adiga, Shan Ba, Tirthankar Dasgupta, C. F. Jeff Wu*, *and* Zhong Lin Wang*

ACS Nano, 3 (2009) 1803-1812.

简介：本文首次把统计设计的理念引入到纳米材料的可控优化生长中，希望像优化工业制造过程一样可以优化纳米材料的生长。

ABSTRACT Controlling the morphology of the as-synthesized nanostructures is usually challenging, and there lacks of a general theoretical guidance in experimental approach. In this study, a novel way of optimizing the aspect ratio of hydrothermally grown ZnO nanowire (NW) arrays is presented by utilizing a systematic statistical design and analysis method. In this work, we use pick-the-winner rule and one-pair-at-a-time main effect analysis to sequentially design the experiments and identify optimal reaction settings. By controlling the hydrothermal reaction parameters (reaction temperature, time, precursor concentration, and capping agent), we improved the aspect ratio of ZnO NWs from around 10 to nearly 23. The effect of noise on the experimental results was identified and successfully reduced, and the statistical design and analysis methods were very effective in reducing the number of experiments performed and in identifying the optimal experimental settings. In addition, the antireflection spectrum of the as-synthesized ZnO NWs clearly shows that higher aspect ratio of the ZnO NW arrays leads to about 30% stronger suppression in the UV vis range emission. This shows great potential applications as antireflective coating layers in photovoltaic devices.

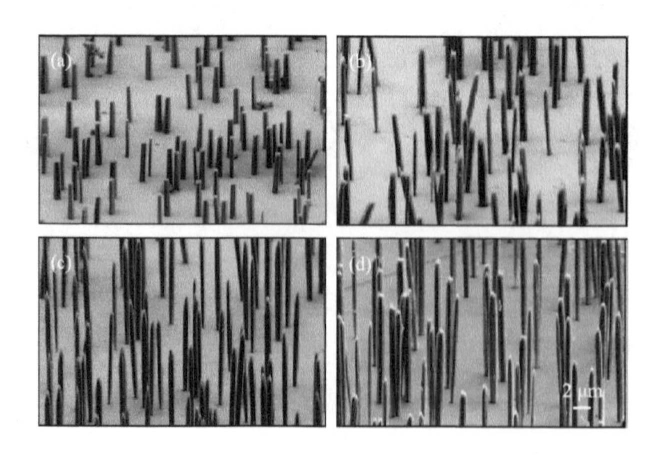

扫码阅全文

http://www.binn.cas.cn/book/202106/W020210609407279941041.pdf

5. Wafer-Scale High-Throughput Ordered Growth of Vertically Aligned ZnO Nanowire Arrays

Yaguang Wei, Wenzhuo Wu, Rui Guo, Dajun Yuan, Suman Das* *and* **Zhong Lin Wang***

Nano Letters, 10 (2010) 3414-3419.

简介：本文首次介绍了一种利用激光干涉制备衬底的方法来大规模生长氧化锌纳米线阵列，为实现压电电子学芯片打下材料基础。

ABSTRACT This article presents an effective approach for patterned growth of vertically aligned ZnO nanowire (NW) arrays with high throughput and low cost at wafer scale without using cleanroom technology. Periodic hole patterns are generated using laser interference lithography on substrates coated with the photoresist SU-8. ZnO NWs are selectively grown through the holes via a low-temperature hydrothermal method without using a catalyst and with a superior control over orientation, location/density, and as-synthesized morphology. The development of textured ZnO seed layers for replacing single crystalline GaN and ZnO substrates extends the large-scale fabrication of vertically aligned ZnO NW arrays on substrates of other materials, such as polymers, Si, and glass. This combined approach demonstrates a novel method of manufacturing large-scale patterned one-dimensional nanostructures on various substrates for applications in energy harvesting, sensing, optoelectronics, and electronic devices.

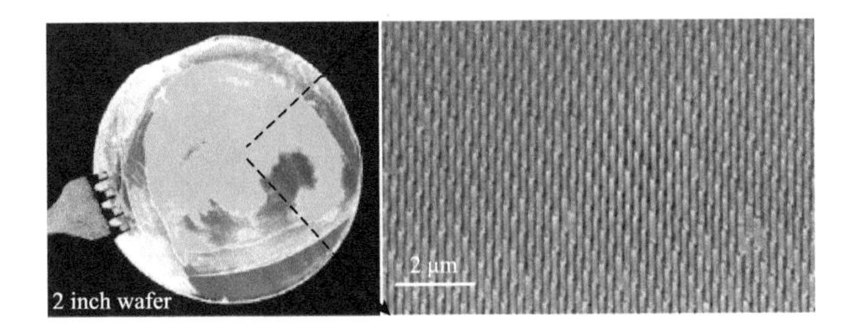

扫码阅全文

http://www.binn.cas.cn/book/202106/W020210609407280397609.pdf

6. Nanostructures of Zinc Oxide

Zhong Lin Wang

Materials Today, 7 (2004) 26-33.

简介：本文系统总结了极性面所导致的各种氧化锌纳米结构的形貌及其生长机制，是一篇经典文献。

ABSTRACT Zinc oxide (ZnO) is a unique material that exhibits semiconducting, piezoelectric, and pyroelectric multiple properties. Using a solid-vapor phase thermal sublimation technique, nanocombs, nanorings, nanohelixes/nanosprings, nanobows, nanobelts, nanowires, and nanocages of ZnO have been synthesized under specific growth conditions. These unique nanostructures unambiguously demonstrate that ZnO is probably the richest family of nanostructures among all materials, both in structures and properties. The nanostructures could have novel applications in optoelectronics, sensors, transducers, and biomedical science because it is bio-safe.

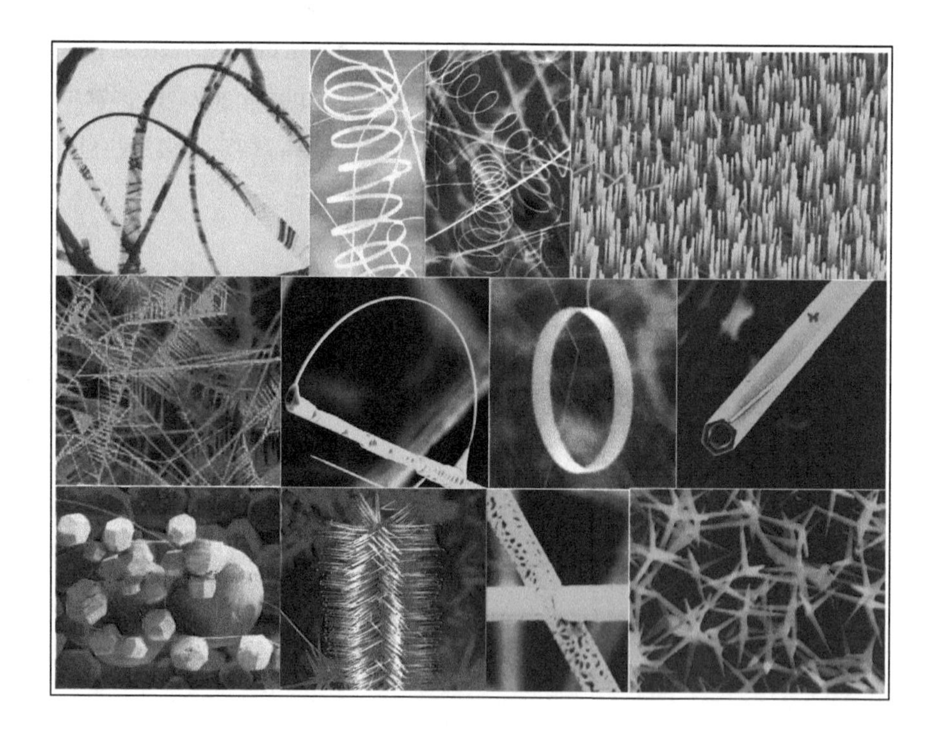

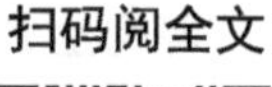

http://www.binn.cas.cn/book/202106/W020210609407280580673.pdf

7. Wafer-Level Patterned and Aligned Polymer Nanowire/Micro-and Nanotube Arrays on Any Substrate

Jenny Ruth Morber, Xudong Wang, Jin Liu, Robert L. Snyder *and* **Zhong Lin Wang***

Advanced Materials, 21 (2009) 2072-2076.

简介：本文首次介绍了一个普适的大规模生长高分子纳米线阵列的方法。

ABSTRACT A novel technique for fabrication of patterned and aligned polymer-nanowire/micro-and nanotube (PNW/PNT) arrays on a wafer-level substrate of any material is reported. By creating a designed pattern on a spin-coated polymer film using techniques, such as stamping or micro-tip writing, plasma etching results in the formation of aligned PNW arrays distributed according to the pattern.

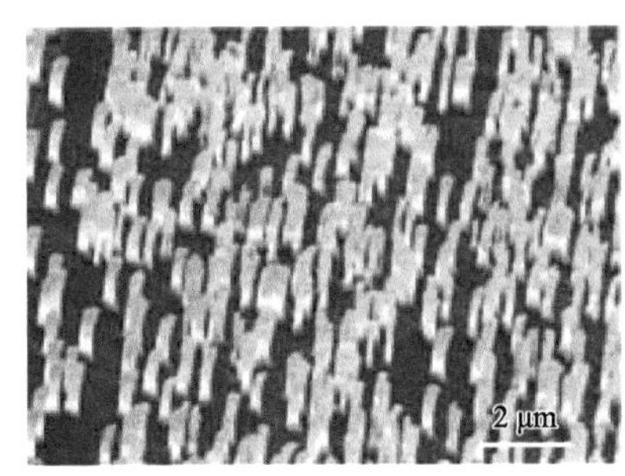

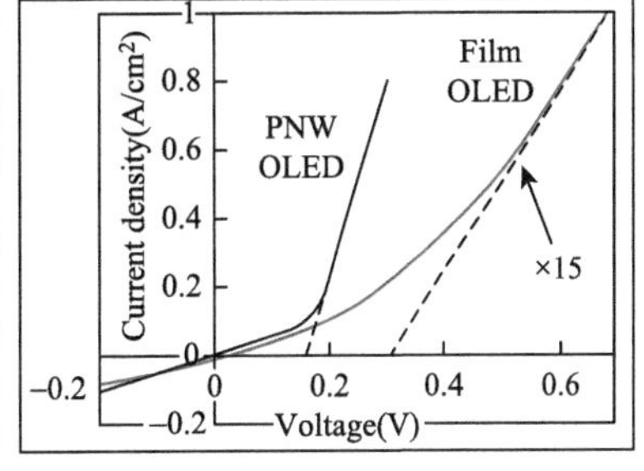

扫码阅全文

http://www.binn.cas.cn/book/202106/W020210609407281178492.pdf

8. Converting Ceria Polyhedral Nanoparticles into Single-Crystal Nanospheres

Xiangdong Feng*, Dean C. Sayle, Zhong Lin Wang*, M. Sharon Paras, Brian Santora, Anthony C. Sutorik, Thi X. T. Sayle, Yi Yang, Yong Ding, Xudong Wang, Yie-Shein Her

Science, 312 (2006)1504-1508.

简介：芯片抛光一般要用到 CeO_2 纳米颗粒。然而，由于受晶体面能量的限制，一般陶瓷纳米颗粒是多面体，这样的材料做芯片抛光难免有划痕。本文首次报道了如何合成单晶的球型 CeO_2 纳米颗粒的方法。这种纳米颗粒在芯片的化学抛光中不但抛光平，而且抛光速度快。

ABSTRACT Ceria nanoparticles are one of the key abrasive materials for chemical-mechanical planarization of advanced integrated circuits. However, ceria nanoparticles synthesized by existing techniques are irregularly faceted, and they scratch the silicon wafers and increase defect concentrations. We developed an approach for large-scale synthesis of single-crystal ceria nanospheres that can reduce the polishing defects by 80% and increase the silica removal rate by 50%, facilitating precise and reliable mass-manufacturing of chips for nanoelectronics. We doped the ceria system with titanium, using flame temperatures that facilitate crystallization of the ceria yet retain the titania in a molten state. In conjunction with molecular dynamics simulation, we show that under these conditions, the inner ceria core evolves in a single-crystal spherical shape without faceting, because throughout the crystallization it is completely encapsulated by a molten 1- to 2-nanometer shell of titania that, in liquid state, minimizes the surface energy. The principle demonstrated here could be applied to other oxide systems.

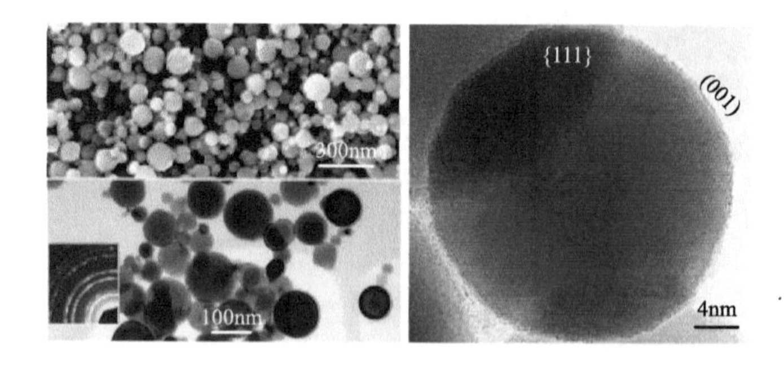

扫码阅全文

http://www.binn.cas.cn/book/202106/W020210609407281382065.pdf

9. Synthesis of Tetrahexahedral Platinum Nanocrystals with High-Index Facets and High Electro-Oxidation Activity

Na Tian, Zhi-You Zhou, Shi-Gang Sun*, Yong Ding, Zhong Lin Wang*

Science, 316 (2007) 732-735.

简介：金属纳米材料例如金、铂等的晶体面一般为由{100}和{111}组成的多面体。本文首次报道了铂颗粒可以具有更高指数的晶格面，打破了我们对纳米颗粒表面形成的常规认识。

ABSTRACT The shapes of noble metal nanocrystals (NCs) are usually defined by polyhedra that are enclosed by {111} and {100} facets, such as cubes, tetrahedra, and octahedra. Platinum NCs of unusual tetrahexahedral (THH) shape were prepared at high yield by an electrochemical treatment of Pt nanospheres supported on glassy carbon by a square-wave potential. The single-crystal THH NC is enclosed by 24 high-index facets such as {730}, {210}, and/or {520} surfaces that have a large density of atomic steps and dangling bonds. These high-energy surfaces are stable thermally (to 800 ℃) and chemically and exhibit much enhanced (up to 400%) catalytic activity for equivalent Pt surface areas for electro-oxidation of small organic fuels such as formic acid and ethanol.

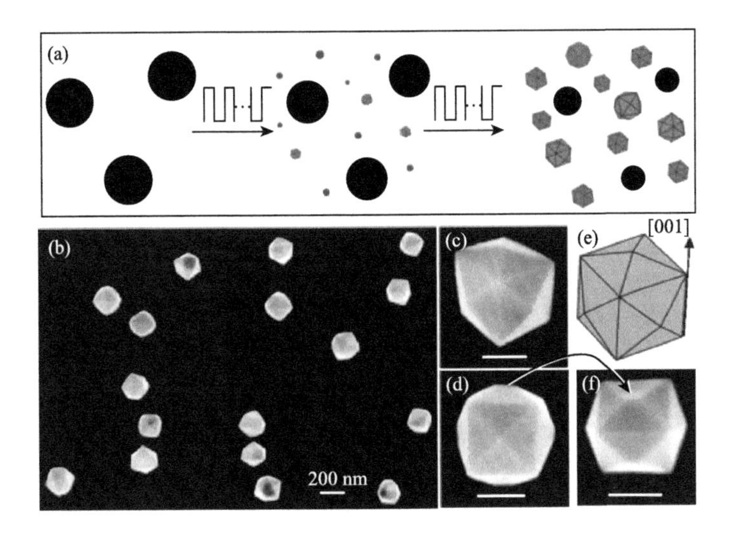

扫码阅全文

http://www.binn.cas.cn/book/202106/W020210609407281541426.pdf

1.6 纳米器件

Nanodevices

1. Stable and Highly Sensitive Gas Sensors Based on Semiconducting Oxide Nanobelts

E. Comini, G. Faglia *and* G. Sberveglieri; Zhengwei Pan *and* Zhong L. Wang

Applied Physics Letters, 81 (2002) 1869-1871.

简介：这是报道把氧化物纳米带首次用在气体传感器上的文章。

ABSTRACT Gas sensors have been fabricated using the single-crystalline SnO_2 nanobelts. Electrical characterization showed that the contacts were ohmic and the nanobelts were sensitive to environmental polluting species like CO and NO_2, as well as to ethanol for breath analyzers and food control applications. The sensor response, defined as the relative variation in conductance due to the introduction of the gas, is 4160% for 250 ppm of ethanol and −1550% for 0.5 ppm NO_2 at 400℃. The results demonstrate the potential of fabricating nanosized sensors using the integrity of a single nanobelt with a sensitivity at the level of a few ppb.

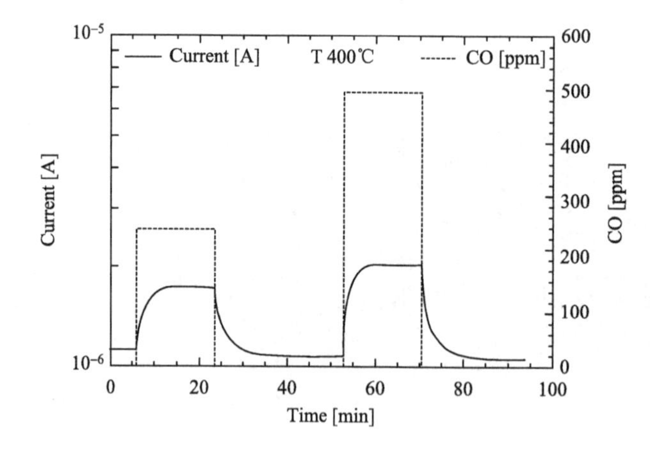

扫码阅全文

http://www.binn.cas.cn/book/202106/W020210609408770389920.pdf

2. Gigantic Enhancement in Sensitivity Using Schottky Contacted Nanowire Nanosensor

Te-Yu Wei, Ping-Hung Yeh, Shih-Yuan Lu* *and* **Zhong Lin Wang***

Journal of the American Chemical Society, 131 (2009) 17690-17695.

简介：在2009年之前，为了充分最大化由表面状态的变化而引起的电阻变化，一般制造纳米传感器件用的是欧姆型接触。然而，2009年王中林组发现利用肖特基接触的器件更灵敏，反应更快。该文章是根据肖特基接触做出的第一个气体传感器，比欧姆型接触器件的灵敏度高三到四个数量级。

ABSTRACT A new single nanowire based nanosensor is demonstrated for illustrating its ultrahigh sensitivity for gas sensing. The device is composed of a single ZnO nanowire mounted on Pt electrodes with one end in Ohmic contact and the other end in Schottky contact. The Schottky contact functions as a "gate" that controls the current flowing through the entire system. By tuning the Schottky barrier height through the responsive variation of the surface chemisorbed gases and the amplification role played by the nanowire to Schottky barrier effect, an ultrahigh sensitivity of 32 000% was achieved using the Schottky contacted device operated in reverse bias mode at 275 ℃ for detection of 400 ppm CO, which is 4 orders of magnitude higher than that obtained using an Ohmic contact device under the same conditions. In addition, the response time and reset time have been shortened by a factor of 7. The methodology and principle illustrated in the paper present a new sensing mechanism that can be readily and extensively applied to other gas sensing systems.

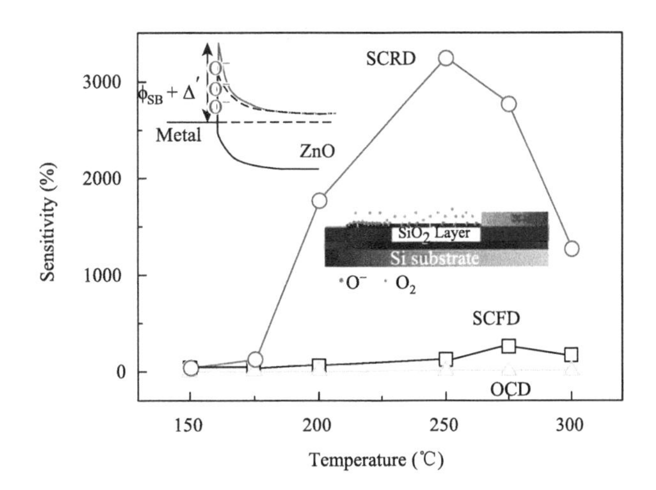

扫码阅全文

http://www.binn.cas.cn/book/202106/W020210609408770522772.pdf

3. Schottky-Gated Probe-Free ZnO Nanowire Biosensor

Ping-Hung Yeh, Zhou Li, *and* Zhong Lin Wang*

Advanced Materials, 21 (2009) 4975-4978.

简介：该文章是根据肖特基接触做出的第一个生物传感器，比欧姆型接触器件的灵敏度高十倍。

ABSTRACT A nanowire-based nanosensor for detecting biologically and chemically charged molecules that is probe-free and highly sensitive is demonstrated. The device relies on the nonsymmetrical Schottky contact under reverse bias (see figure), and is much more sensitive than the device based on the symmetric ohmic contact. This approach serves as a guideline for designing more practical chemical and biochemical sensors.

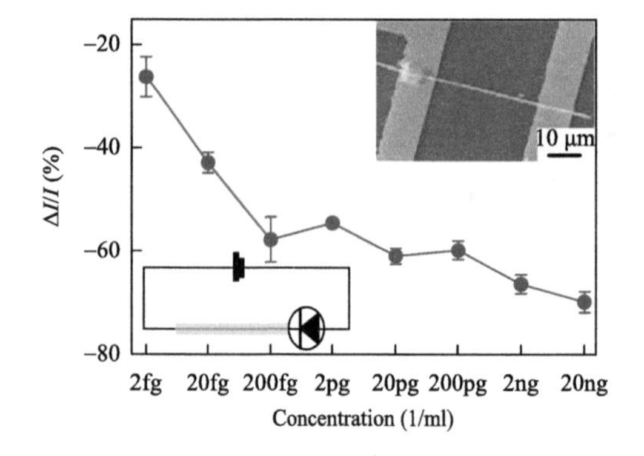

扫码阅全文

http://www.binn.cas.cn/book/202106/W020210609408770689159.pdf

4. Piezotronic Effect Enhanced Performance of Schottky-Contacted Optical, Gas, Chemical and Biological Nanosensors

Ruomeng Yu, Simiao Niu, Caofeng Pan, Zhong Lin Wang*

Nano Energy, 14 (2015) 312-339.

简介：该文章是根据肖特基接触，基于压电电子学效应做出的各种纳米传感器的首篇综述文章，其中包括光学、气体、化学和生物传感等。

ABSTRACT In this review, we systematically analyze the enhanced performances of Schottky-structured nanowire sensors by the piezotronic effect. Compared with traditional Ohmic-contact nanowire sensing systems, Schottky-contact provides ultra-high sensitivity and superfast response depending on the barrier height at local metal-semiconductor (M-S) interface. Utilizing the strain-induced piezoelectric polarization charges presented at the vicinity of M-S contact, piezotronic effect is applied to tune/control the charge carriers transport process through the interface by modulating the Schottky barrier height (SBH) at local contact, and thus hugely enhances the performances of Schottky-structured sensors. Our results indicate that piezotronic effect is a universal effect that provides an effective approach to improve the sensitivity, resolution, response time and other general properties of Schottky-contact nanowire sensors in different categories, including bio/chemical sensing, gas sensing, humidity sensing, temperature sensing and others.

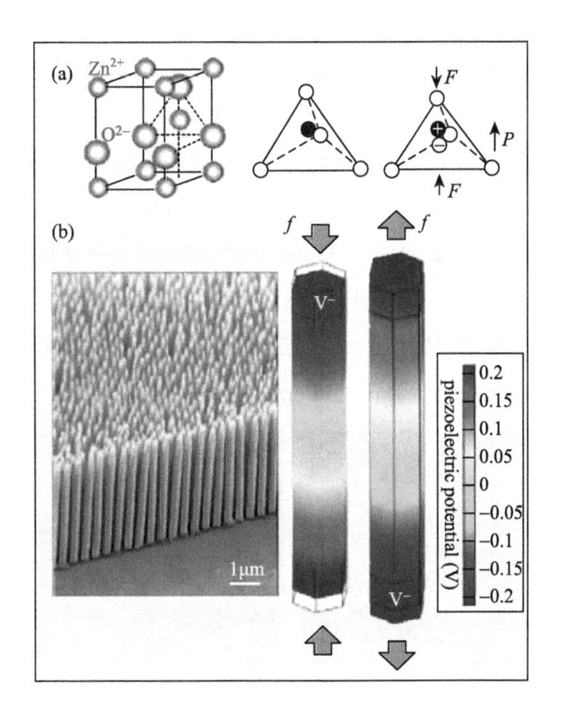

扫码阅全文

http://www.binn.cas.cn/book/202106/W020210609408770826926.pdf

5. Shell-Isolated Nanoparticle-Enhanced Raman Spectroscopy

Jian Feng Li, Yi Fan Huang, Yong Ding, Zhi Lin Yang, Song Bo Li, Xiao Shun Zhou, Feng Ru Fan, Wei Zhang, Zhi You Zhou, De Yin Wu, Bin Ren, Zhong Lin Wang,* Zhong Qun Tian*

Nature, 464 (2010) 392-395 (news and view highlights, *Nature*, 464 (2010) 357).

简介：单个探针型拉曼增强灵敏度高但信号很弱。该文章首次报道了用核壳结构做出的纳米颗粒进行同步拉曼成像，其原理是利用单层纳米颗粒的自组装所形成的平行阵列探针，大大提高了收集拉曼信号的强度，并表明拉曼检测可以应用到各种表面结构和形貌场景，有可能推动拉曼检测的工业应用。

ABSTRACT Surface-enhanced Raman scattering (SERS) is a powerful spectroscopy technique that can provide non-destructive and ultra-sensitive characterization down to single molecular level, comparable to single-molecule fluorescence spectroscopy. However, generally substrates based on metals such as Ag, Au and Cu, either with roughened surfaces or in the form of nanoparticles, are required to realise a substantial SERS effect, and this has severely limited the breadth of practical applications of SERS. A number of approaches have extended the technique to non-traditional substrates most notably tip-enhanced Raman spectroscopy (TERS) where the probed substance (molecule or material surface) can be on a generic substrate and where a nanoscale gold tip above the substrate acts as the Raman signal amplifier. The drawback is that the total Raman scattering signal from the tip area is rather weak, thus limiting TERS studies to molecules with large Raman cross-sections. Here, we report an approach, which we name shell-isolated nanoparticle-enhanced Raman spectroscopy, in which the Raman signal amplification is provided by gold nanoparticles with an ultrathin silica or alumina shell. A monolayer of such nanoparticles is spread as "smart dust" over the surface that is to be probed. The ultrathin coating keeps the nanoparticles from agglomerating, separates them from direct contact with the probed material and allows the nanoparticles to conform to different contours of substrates. High-quality Raman spectra were obtained on various molecules adsorbed at Pt and Au single-crystal surfaces and from Si surfaces with hydrogen monolayers. These measurements and our studies on yeast cells and citrus fruits with pesticide residues illustrate that our method significantly expands the flexibility of SERS for useful applications in the materials and life sciences, as well as for the inspection of food safety, drugs, explosives and environment pollutants.

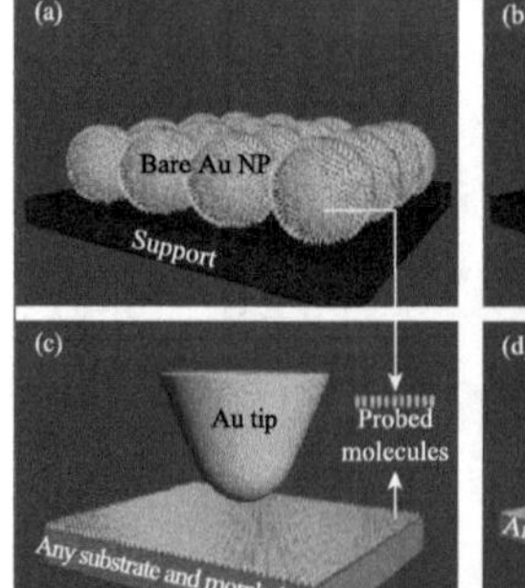

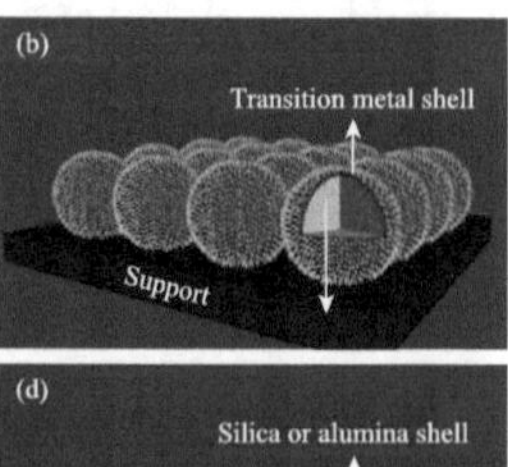

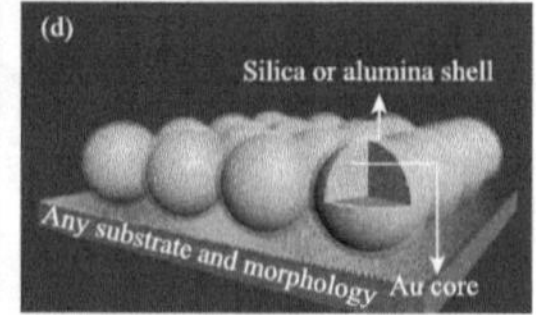

http://www.binn.cas.cn/book/202106/W020210609408773764633.pdf

1.7 压电纳米发电机

Piezoelectric Nanogenerators

1. Piezoelectric Nanogenerators Based on Zinc Oxide Nanowire Arrays

Zhong Lin Wang* *and* **Jinhui Song**

Science, 312 (2006) 242-246.

简介：该文章是报道基于扫描探针的压电式纳米发电机的开山之作。本文引出了三大领域：基于纳米发电机的微纳能源、自驱动系统与传感、基于压电势的压电电子学。

ABSTRACT We have converted nanoscale mechanical energy into electrical energy by means of piezoelectric zinc oxide nanowire (NW) arrays. The aligned NWs are deflected with a conductive atomic force microscope tip in contact mode. The coupling of piezoelectric and semiconducting properties in zinc oxide creates a strain field and charge separation across the NW as a result of its bending. The rectifying characteristic of the Schottky barrier formed between the metal tip and the NW leads to electrical current generation. The efficiency of the NW-based piezoelectric power generator is estimated to be 17% to 30%. This approach has the potential of converting mechanical, vibrational, and/or hydraulic energy into electricity for powering nanodevices.

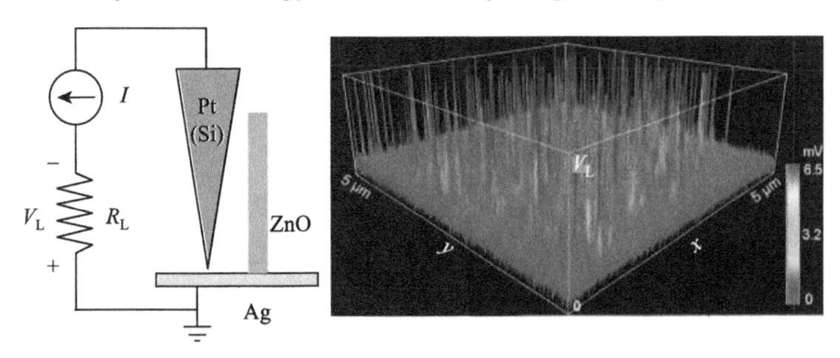

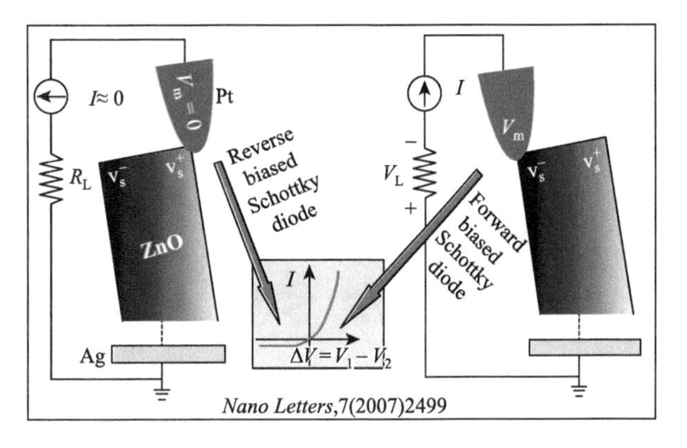

扫码阅全文

http://www.binn.cas.cn/book/202106/W020210609412254953177.pdf

2. Direct-Current Nanogenerator Driven by Ultrasonic Waves

Xudong Wang, Jinhui Song, Jin Liu, Zhong Lin Wang*

Science, 316 (2007) 102-105.

简介：该文章是基于平行探针的压电式纳米发电机的原作，是对单个探针结构的延伸。

ABSTRACT We have developed a nanowire nanogenerator that is driven by an ultrasonic wave to produce continuous direct-current output. The nanogenerator was fabricated with vertically aligned zinc oxide nanowire arrays that were placed beneath a zigzag metal electrode with a small gap. The wave drives the electrode up and down to bend and/or vibrate the nanowires. A piezoelectric-semiconducting coupling process converts mechanical energy into electricity. The zigzag electrode acts as an array of parallel integrated metal tips that simultaneously and continuously create, collect, and output electricity from all of the nanowires. The approach presents an adaptable, mobile, and cost-effective technology for harvesting energy from the environment, and it offers a potential solution for powering nanodevices and nanosystems.

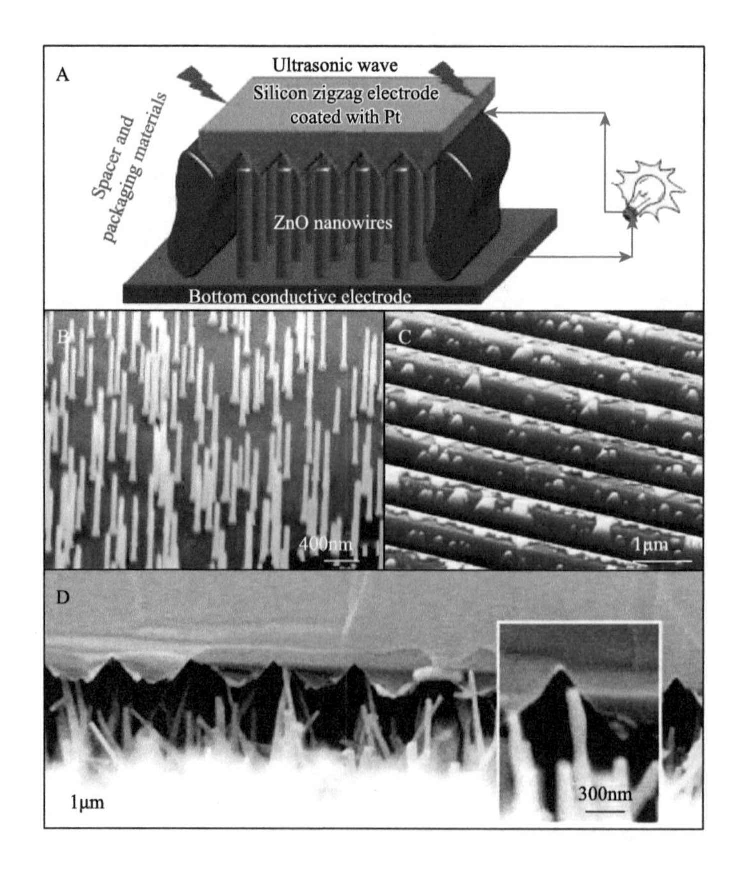

扫码阅全文

http://www.binn.cas.cn/book/202106/W020210609412255360869.pdf

3. Microfibre-Nanowire Hybrid Structure for Energy Scavenging

Yong Qin, Xudong Wang *and* Zhong Lin Wang*

Nature, 451 (2008) 809-813.

简介：该文章是首篇利用纳米线与纤维相结合的结构进行能量回收的开山之作，从此诞生了很多纤维基的能量回收与存储器件。

ABSTRACT A self-powering nanosystem that harvests its operating energy from the environment is an attractive proposition for sensing, personal electronics and defence technologies. This is in principle feasible for nanodevices owing to their extremely low power consumption. Solar, thermal and mechanical (wind, friction, body movement) energies are common and may be scavenged from the environment, but the type of energy source to be chosen has to be decided on the basis of specific applications. Military sensing/surveillance node placement, for example, may involve difficult-to-reach locations, may need to be hidden, and may be in environments that are dusty, rainy, dark and/or in deep forest. In a moving vehicle or aeroplane, harvesting energy from a rotating tyre or wind blowing on the body is a possible choice to power wireless devices implanted in the surface of the vehicle. Nanowire nanogenerators built on hard substrates were demonstrated for harvesting local mechanical energy produced by high-frequency ultrasonic waves. To harvest the energy from vibration or disturbance originating from footsteps, heartbeats, ambient noise and air flow, it is important to explore innovative technologies that work at low frequencies (such as <10 Hz) and that are based on flexible soft materials. Here we present a simple, low-cost approach that converts low-frequency vibration/friction energy into electricity using piezoelectric zinc oxide nanowires grown radially around textile fibres. By entangling two fibres and brushing the nanowires rooted on them with respect to each other, mechanical energy is converted into electricity owing to a coupled piezoelectric-semiconductor process. This work establishes a methodology for scavenging light-wind energy and body-movement energy using fabrics.

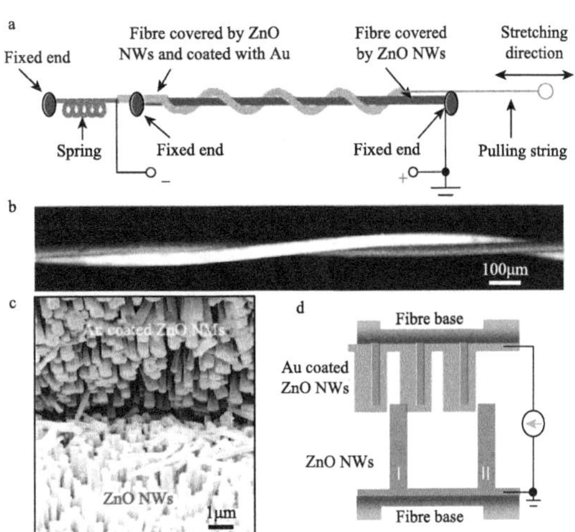

扫码阅全文

http://www.binn.cas.cn/book/202106/W020210609412255545717.pdf

4. Power Generation with Laterally Packaged Piezoelectric Fine Wires

Rusen Yang, Yong Qin, Liming Dai *and* **Zhong Lin Wang** *

Nature Nanotechnology, 4 (2009) 34-39.

简介：该文是首篇利用平躺的单根纳米线进行能量回收的文章，并首次明确阐述了压电发电的基本原理。文中发展出的操纵纳米线结构的方法对后来发展压电电子学与压电光电子学起到了关键作用。

ABSTRACT Converting mechanical energy into electricity could have applications in sensing, medical science, defence technology and personal electronics, and the ability of nanowires to "scavenge" energy from ambient and environmental sources could prove useful for powering nanodevices Previously reported nanowire generators were based on vertically aligned piezoelectric nanowires that were attached to a substrate at one end and free to move at the other. However, there were problems with the output stability, mechanical robustness, lifetime and environmental adaptability of such devices. Here we report a flexible power generator that is based on cyclic stretching-releasing of a piezoelectric fine wire that is firmly attached to metal electrodes at both ends, is packaged on a flexible substrate, and does not involve sliding contacts. Repeatedly stretching and releasing a single wire with a strain of 0.05%-0.1% creates an oscillating output voltage of up to ~50mV, and the energy conversion efficiency of the wire can be as high as 6.8%.

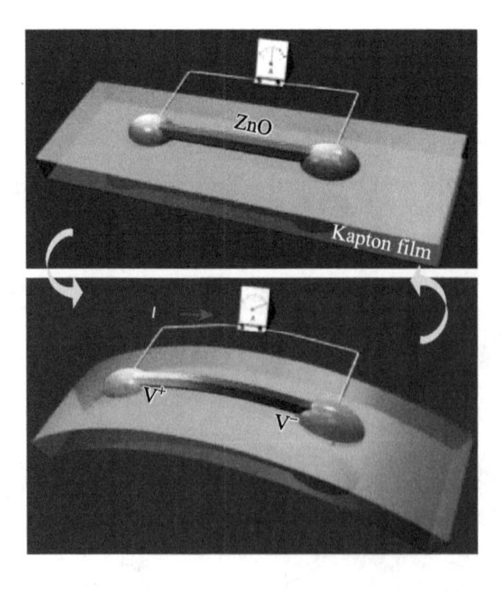

扫码阅全文

http://www.binn.cas.cn/book/202106/W020210609412255934167.pdf

5. Converting Biomechanical Energy into Electricity by a Muscle-Movement-Driven Nanogenerator

Rusen Yang, Yong Qin, Cheng Li, Guang Zhu *and* **Zhong Lin Wang***

Nano Letters, 9 (2009) 1201-1205.

简介：该文是首篇利用压电式纳米发电机收集生物机械能的文献。

ABSTRACT A living species has numerous sources of mechanical energy, such as muscle stretching, arm/leg swings, walking/running, heart beats, and blood flow. We demonstrate a piezoelectric nanowire based nanogenerator that converts biomechanical energy, such as the movement of a human finger and the body motion of a live hamster (Campbell's dwarf), into electricity. A single wire generator (SWG) consists of a flexible substrate with a ZnO nanowire affixed laterally at its two ends on the substrate surface. Muscle stretching results in the back and forth stretching of the substrate and the nanowire. The piezoelectric potential created inside the wire leads to the flow of electrons in the external circuit. The output voltage has been increased by integrating multiple SWGs. A series connection of four SWGs produced an output voltage of up to ~0.1-0.15V. The success of energy harvesting from a tapping finger and a running hamster reveals the potential of using the nanogenerators for scavenging low-frequency energy from regular and irregular biomotion.

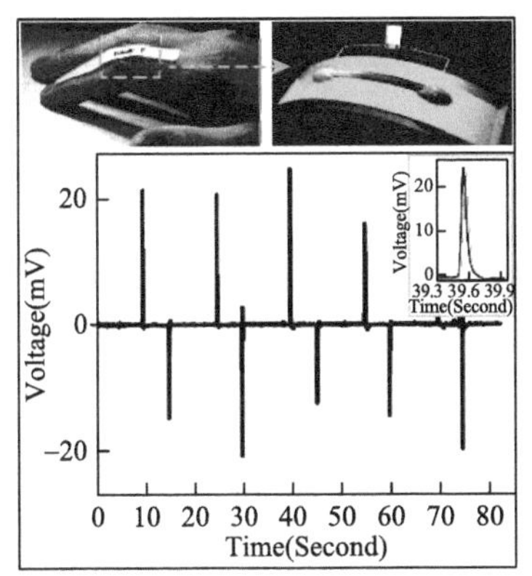

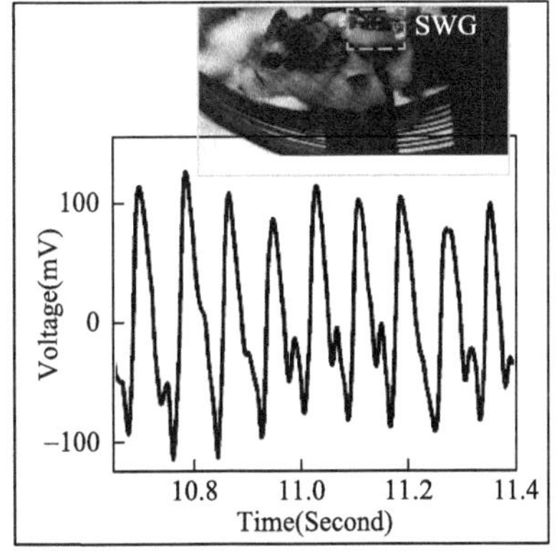

扫码阅全文

http://www.binn.cas.cn/book/202106/W020210609412256105280.pdf

6. Electrostatic Potential in a Bent Piezoelectric Nanowire. The Fundamental Theory of Nanogenerator and Nanopiezotronics

Yifan Gao *and* **Zhong Lin Wang***

Nano Letters, 7 (2007) 2499-2505.

简介：该文是首篇计算纳米线中压电势分布的文章，奠定了建立压电式纳米发电机与压电电子学的基础。

ABSTRACT We have applied the perturbation theory for calculating the piezoelectric potential distribution in a nanowire (NW) as pushed by a lateral force at the tip. The analytical solution given under the first-order approximation produces a result that is within 6% from the full numerically calculated result using the finite element method. The calculation shows that the piezoelectric potential in the NW almost does not depend on the z-coordinate along the NW unless very close to the two ends, meaning that the NW can be approximately taken as a "parallel plated capacitor". This is entirely consistent to the model established for nanopiezotronics, in which the potential drop across the nanowire serves as the gate voltage for the piezoelectric field effect transistor. The maximum potential at the surface of the NW is directly proportional to the lateral displacement of the NW and inversely proportional to the cube of its length-to-diameter aspect ratio. The magnitude of piezoelectric potential for a NW of diameter 50 nm and length 600nm is ~0.3 V. This voltage is much larger than the thermal voltage (~25mV) and is high enough to drive the metal-semiconductor Schottky diode at the interface between atomic force microscope tip and the ZnO NW, as assumed in our original mechanism for the nanogenerators.

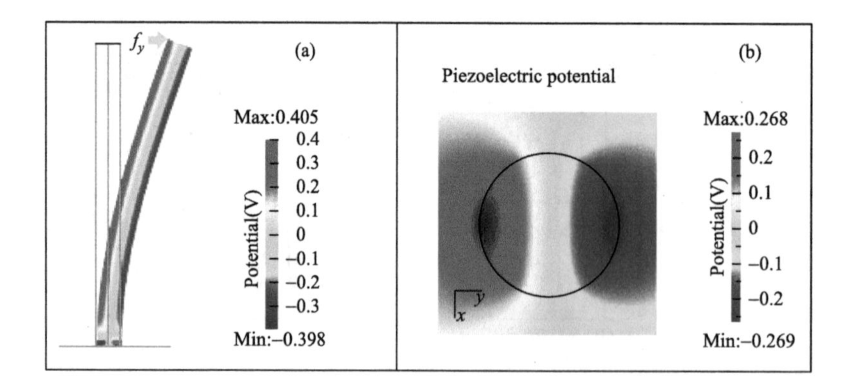

扫码阅全文

http://www.binn.cas.cn/book/202106/W020210609412256553615.pdf

7. Characteristics of Output Voltage and Current of Integrated Nanogenerators

Rusen Yang, Yong Qin, Cheng Li, Liming Dai *and* **Zhong Lin Wang***

Applied Physics Letters, 94 (2009) 022905.

简介：为了判别实验中所测到的微小电信号是否真实而非随机噪声，该文首次提出了进行信号判断的“三大纪律、八项注意”的标准，对于发展新的技术很有指导意义。该文章是王中林为了回应德国一个小组质疑纳米发电机信号的真实性而发展出来的方法。

ABSTRACT Owing to the anisotropic property and small output signals of the piezoelectric nanogenerators (NGs) and the influence of the measurement system and environment, identification of the true signal generated by the NG is critical. We have developed three criteria: Schottky behavior test, switching-polarity tests, and linear superposition of current and voltage tests. The 11 tests can effectively rule out the system artifacts, whose sign does not change with the switching measurement polarity, and random signals, which might change signs but cannot consistently add up or cancel out under designed connection configurations. This study establishes the standards for designing and scale up of integrated nanogenerators.

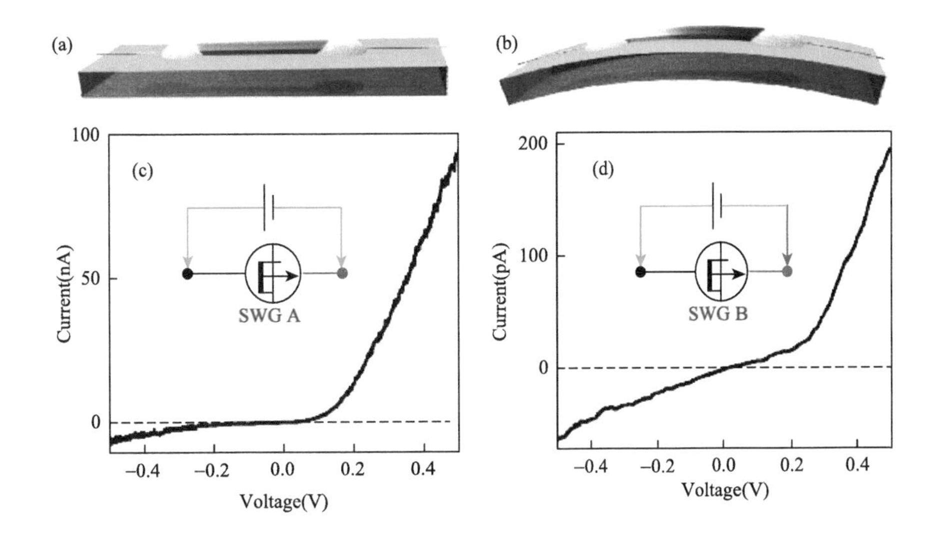

扫码阅全文

http://www.binn.cas.cn/book/202106/W020210609412256663157.pdf

8. Muscle-Driven *in vivo* Nanogenerator

Zhou Li, Guang Zhu, Rusen Yang, Aurelia C. Wang *and* Zhong Lin Wang*

Advanced Materials, 22 (2010) 2534-2537.

简介：该文是首篇利用活体器官来带动基于单根纳米线的纳米发电机的文章，从此开始了研发自驱动心脏起搏器的漫长历程。历经十年直到 2019 年，商用的心脏起搏器可以直接用纳米发电机来驱动，实现了里程碑式的飞跃。

ABSTRACT A nanogenerator based on a single piezoelectric fine wire producing an alternating current (AC) is successfully used for the harvesting of biomechanical energy under *in vivo* conditions. We demonstrate the implanting and working of such a nanogenerator in a live rat where it harvests energy generated by its breathing or heart beating. This study shows the potential of applying these nanogenerators for driving in vivo nanodevices.

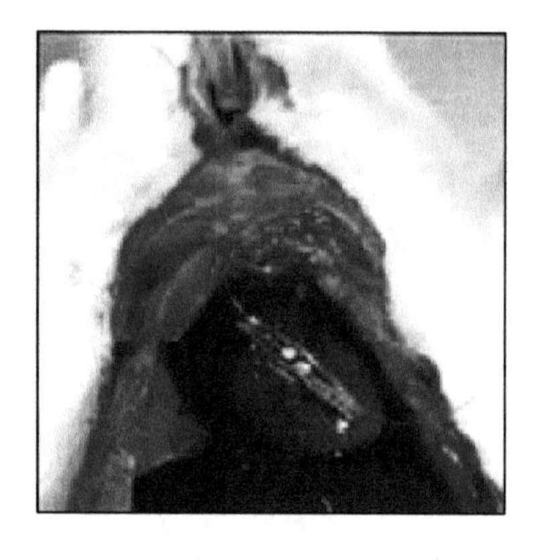

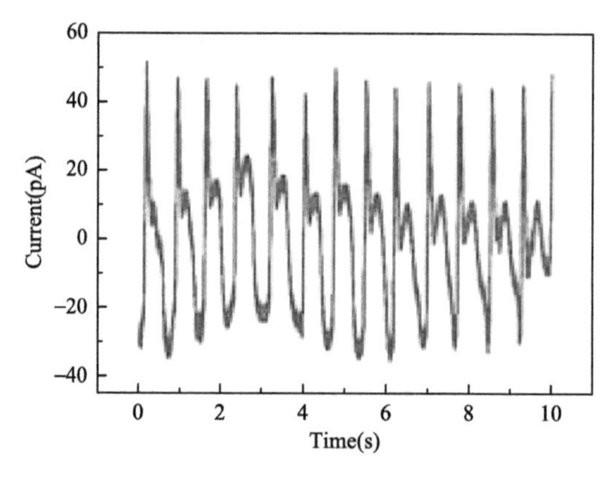

扫码阅全文

http://www.binn.cas.cn/book/202106/W020210609412256842195.pdf

9. Flexible High-Output Nanogenerator Based on Lateral ZnO Nanowire Array

Guang Zhu, Rusen Yang, Sihong Wang *and* Zhong Lin Wang*

Nano Letters, 10 (2010) 3151-3155.

简介：该文是首篇基于多级平行组装的纳米线阵列的结构来提升压电式纳米发电机的输出电压和电流，首次利用输出 2 伏特的纳米发电机点亮了一盏 LED 灯，是纳米发电机发展历史上一个质的飞跃。

ABSTRACT We report here a simple and effective approach, named scalable sweeping-printing-method, for fabricating flexible high-output nanogenerator (HONG) that can effectively harvesting mechanical energy for driving a small commercial electronic component. The technique consists of two main steps. In the first step, the vertically aligned ZnO nanowires (NWs) are transferred to a receiving substrate to form horizontally aligned arrays. Then, parallel stripe type of electrodes are deposited to connect all of the NWs together. Using a single layer of HONG structure, an open-circuit voltage of up to 2.03 V and a peak output power density of ~11 mW/cm^3 have been achieved. The generated electric energy was effectively stored by utilizing capacitors, and it was successfully used to light up a commercial light-emitting diode (LED), which is a landmark progress toward building self-powered devices by harvesting energy from the environment. This research opens up the path for practical applications of nanowire-based piezoelectric nanogeneragtors for self-powered nanosystems.

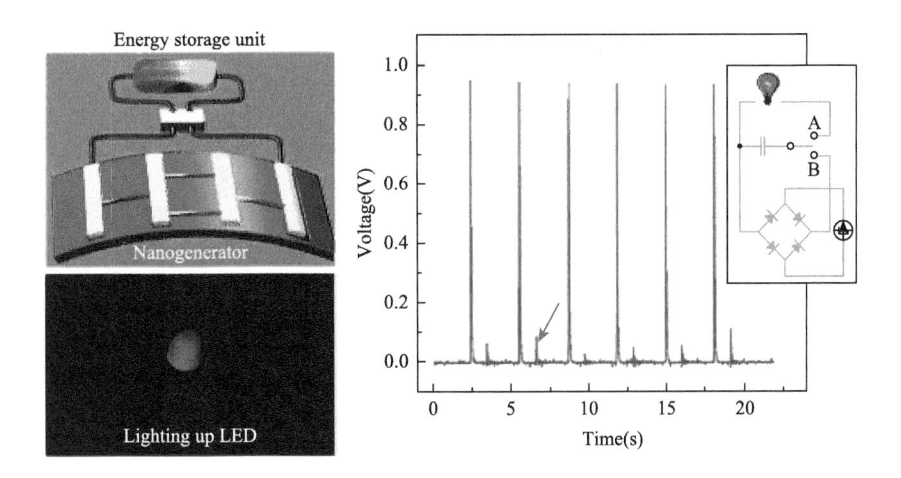

扫码阅全文

http://www.binn.cas.cn/book/202106/W020210609412257001303.pdf

10. Replacing a Battery by a Nanogenerator with 20 V Output

Youfan Hu, Long Lin, Yan Zhang *and* Zhong Lin Wang*

Advanced Materials, 24 (2012) 110-114.

简介：该文是首篇报道利用阵列式纳米线制造的压电式纳米发电机来驱动一个电子手表，是压电式纳米发电机发展历史上第一个可以驱动一个小电子产品的例证，也是迈向应用的第一步。

ABSTRACT Replacing batteries by nanogenerators (NGs) in small consumer electronics is one of the goals in the emerging field of self-powered nanotechnology. We show that the maximum measured output voltage of an NG optimized with pretreatments on the as-grown ZnO nanowire films reaches 20 V and the output current exceeds 6 μA, which corresponds to a power density of 0.2 W·cm^{-3}. The NG is also demonstrated to replace a battery for driving a electronic watch.

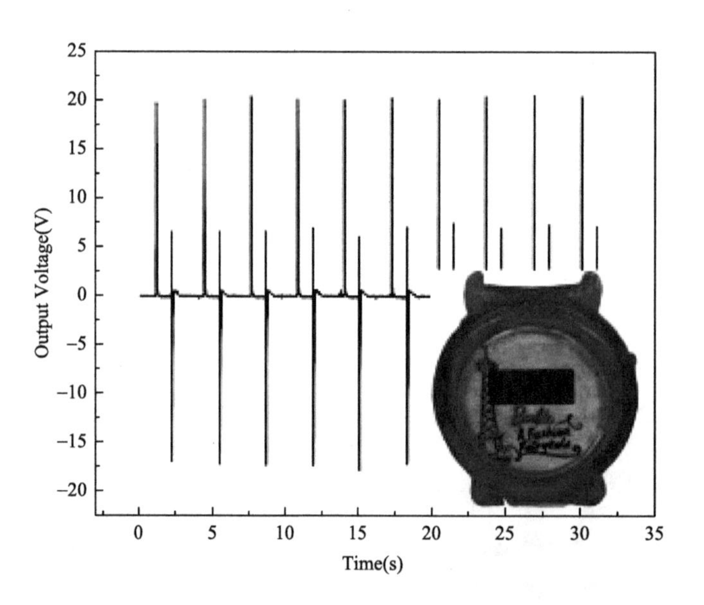

扫码阅全文

http://www.binn.cas.cn/book/202106/W020210609412257154129.pdf

11. Functional Electrical Stimulation by Nanogenerator with 58 V Output Voltage

Guang Zhu, Aurelia C. Wang, Ying Liu, Yusheng Zhou & Zhong Lin Wang*

Nano Letters, 12 (2012) 3086-3090.

简介：该文是首篇利用阵列式纳米线制造的多个压电式纳米发电机系统来实现 58 V 和 134 μA 输出的文章，进而电刺激一个生物器官。这是王中林团队利用氧化锌纳米线做出的最大输出的纳米发电机。

ABSTRACT We demonstrate a new type of integrated nanogenerator based on arrays of vertically aligned piezoelectric ZnO nanowires. The peak open-circuit voltage and short-circuit current reach a record high level of 58 V and 134 μA, respectively, with a maximum power density of 0.78 W/cm^3. The electric output was directly applied to a sciatic nerve of a frog, inducing innervation of the nerve. Vibrant contraction of the frog's gastrocnemius muscle is observed as a result of the instantaneous electric input from the nanogenerator.

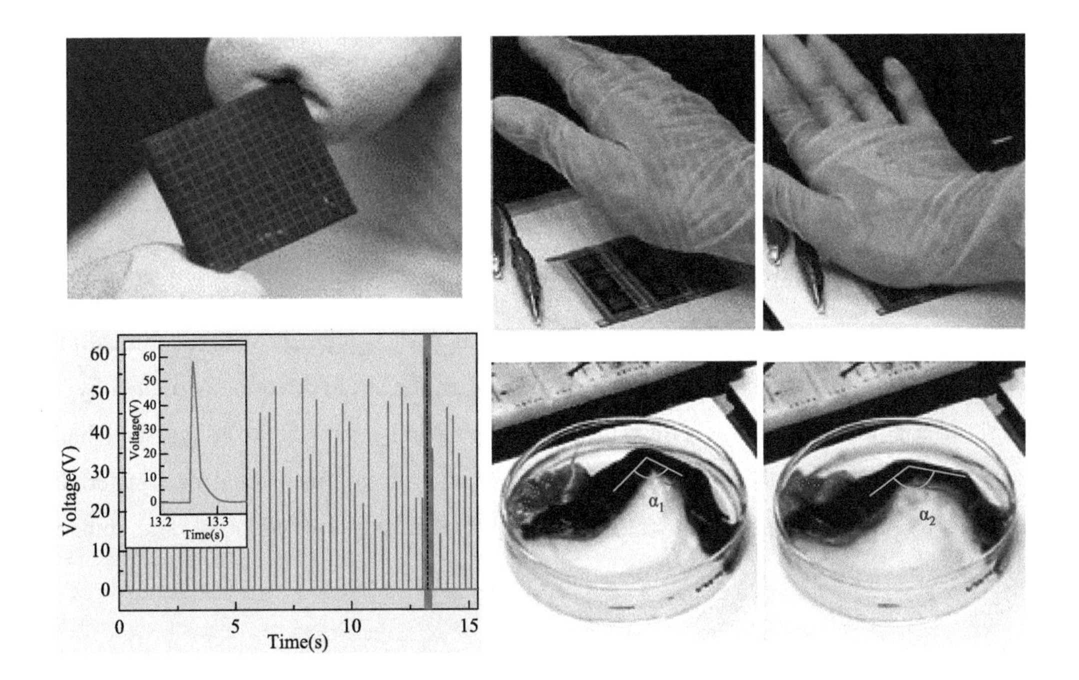

扫码阅全文

http://www.binn.cas.cn/book/202106/W020210609412257320280.pdf

12. Piezoelectricity of Single-Atomic-Layer MoS_2 for Energy Conversion and Piezotronics

Wenzhuo Wu, Lei Wang*, Yilei Li, Fan Zhang, Long Lin, Simiao Niu, Daniel Chenet, Xian Zhang, Yufeng Hao, Tony F. Heinz, James Hone* *and* Zhong Lin Wang*

Nature, 514 (2014) 470-474.

简介：块状 MoS_2 是不具有压电效应的，但当材料的原子层数小为一层或三层原子层时，由于非中心结构的出现就导致了压电效应。本文是二维材料中首篇报道具有压电效应和压电电子学效应的文献，从此开辟了研究二维材料力电光耦合的新方向。

ABSTRACT The piezoelectric characteristics of nanowires, thin films and bulk crystals have been closely studied for potential applications in sensors, transducers, energy conversion and electronics. With their high crystallinity and ability to withstand enormous strain, two-dimensional materials are of great interest as high-performance piezoelectric materials. Monolayer MoS_2 is predicted to be strongly piezoelectric, an effect that disappears in the bulk owing to the opposite orientations of adjacent atomic layers. Here we report the first experimental study of the piezoelectric properties of two-dimensional MoS_2 and show that cyclic stretching and releasing of thin MoS_2 flakes with an odd number of atomic layers produces oscillating piezoelectric voltage and current outputs, whereas no output is observed for flakes with an even number of layers. A single monolayer flake strained by 0.53% generates a peak output of 15 mV and 20 pA, corresponding to a power density of 2 $mW \cdot m^{-2}$ and a 5.08% mechanical-to-electrical energy conversion efficiency. In agreement with theoretical predictions, the output increases with decreasing thickness and reverses sign when the strain direction is rotated by 90°. Transport measurements show a strong piezotronic effect in single-layer MoS_2, but not in bilayer and bulk MoS_2. The coupling between piezoelectricity and semiconducting properties in two-dimensional nanomaterials may enable the development of applications in powering nanodevices, adaptive bioprobes and tunable/stretchable electronics/optoelectronics.

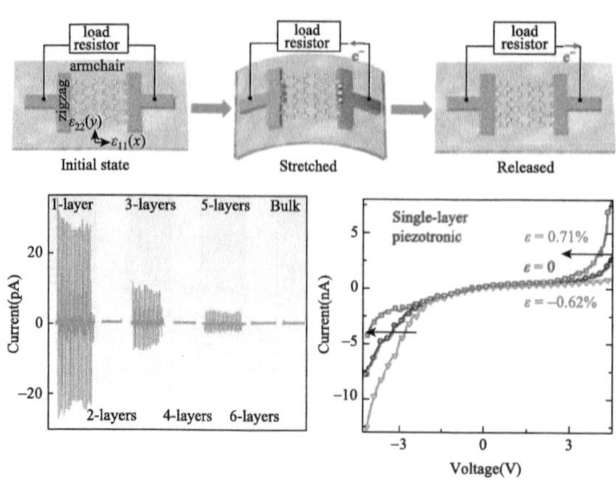

扫码阅全文

http://www.binn.cas.cn/book/202106/W020210609412257684105.pdf

13. Lateral Nanowire/Nanobelt Based Nanogenerators, Piezotronics and Piezo-Phototronics

Zhong Lin Wang*, Rusen Yang, Jun Zhou, Yong Qin, Chen Xu, Youfan Hu, Sheng Xu

Materials Science and Engineering R, 70 (2010) 320-329.

简介：这是一篇关于压电式纳米发电机、压电电子学与压电光电子学的综合性综述论文。

ABSTRACT Relying on the piezopotential created in ZnO under straining, nanogenerators, piezotronics and piezo-phototronics developed based on laterally bonded nanowires on a polymer substrate have been reviewed. The principle of the nanogenerator is a transient flow of electrons in external load as driven by the piezopotential created by dynamic straining. By integrating the contribution made by millions of nanowires, the output voltage has been raised to 1.2 V. Consequently, self-powered nanodevices have been demonstrated. This is an important platform technology for the future sensor network and the internet of things. Alternatively, the piezopotential can act as a gate voltage that can tune/gate the transport process of the charge carriers in the nanowire, which is a gate-electrode free field effect transistor (FET). The device fabricated based on this principle is called the piezotronic device. Piezo-phototronic effect is about the tuning and controlling of electro-optical processes by strain induced piezopotential. The piezotronic, piezophotonic and pieozo-phototronic devices are focused on low frequency applications in areas involving mechanical actions, such as MEMS/NEMS, nanorobotics, sensors, actuators and triggers.

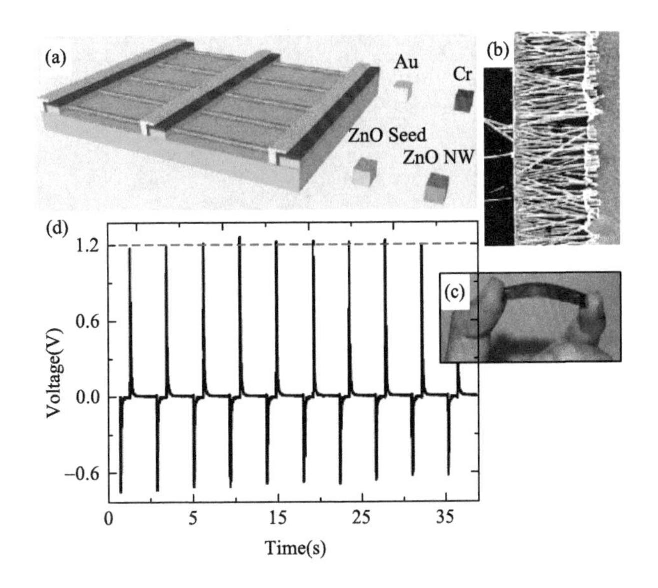

扫码阅全文

http://www.binn.cas.cn/book/202106/W020210609412258171630.pdf

14. Nanotechnology-Enabled Energy Harvesting for Self-Powered Micro-/Nanosystems

Zhong Lin Wang *and* **Wenzhuo Wu**

Angewandte Chemie, 51 (2012) 11700-11721.

简介：这是一篇关于利用纳米技术进行能量回收和自驱动系统的一篇综述，其中包括太阳能、热电发电、纳米发电机等，并对各种技术进行了详细比较。

ABSTRACT Health, infrastructure, and environmental monitoring as well as networking and defense technologies are only some of the potential areas of application of micro-/nanosystems (MNSs). It is highly desirable that these MNSs operate without an external electricity source and instead draw the energy they require from the environment in which they are used. This Review covers various approaches for energy harvesting to meet the future demand for self-powered MNSs.

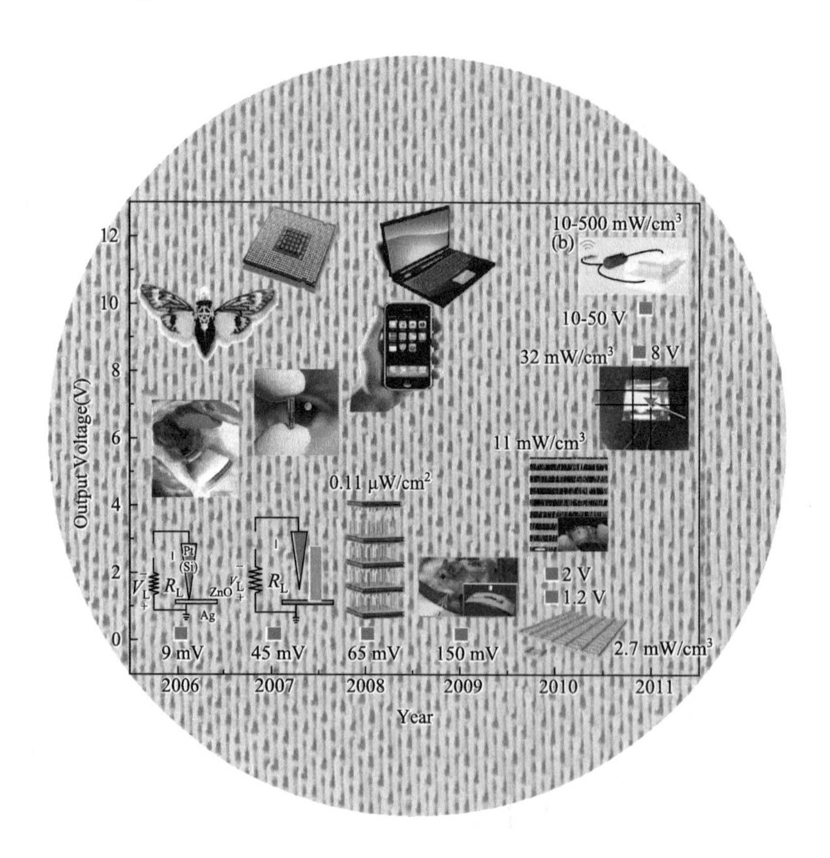

扫码阅全文

http://www.binn.cas.cn/book/202106/W020210609412258472874.pdf

15. Recent Progress in Piezoelectric Nanogenerators as a Sustainable Power Source in Self-Powered Systems and Active Sensors

Youfan Hu, Zhong Lin Wang*

Nano Energy, 14 (2015) 3-14.

简介：这是一篇聚焦在压电式纳米发电机的综述论文，可能是介绍本领域最全面的一篇综述。

ABSTRACT Mechanical energy sources are abundant in our living environment, such as body motion, vehicle transportation, engine vibrations and breezy wind, which have been underestimated in many cases. They could be converted into electrical energy and utilized for many purposes, including driving small electronic devices or even constructing an integrated system operated without bulky batteries and power cables. Many progresses have been made recently in the mechanical energy harvesting technology based on piezoelectric nanogenerators (PENGs). By introducing a new sandwich structure design, high performance PENGs can be achieved through very simple fabrication process with good mechanical stability by utilizing ZnO nanowires (NWs). By further optimizing the nanomaterials' properties and device structure, the PENG's open circuit (OC) voltage can be elevated to over 37 V. Two important applications of this technology are that the nanogenerator can be used as a sustainable power source for self-powered system and can worked as active sensors. Several demonstrations are reviewed here. Finally, perspectives of this mechanical energy harvesting technology are discussed. Co-operation with power management circuit, capability of integrating with a system, and low cost large-scale manufacturing processing are suggested to be the key points toward commercialization of PENGs.

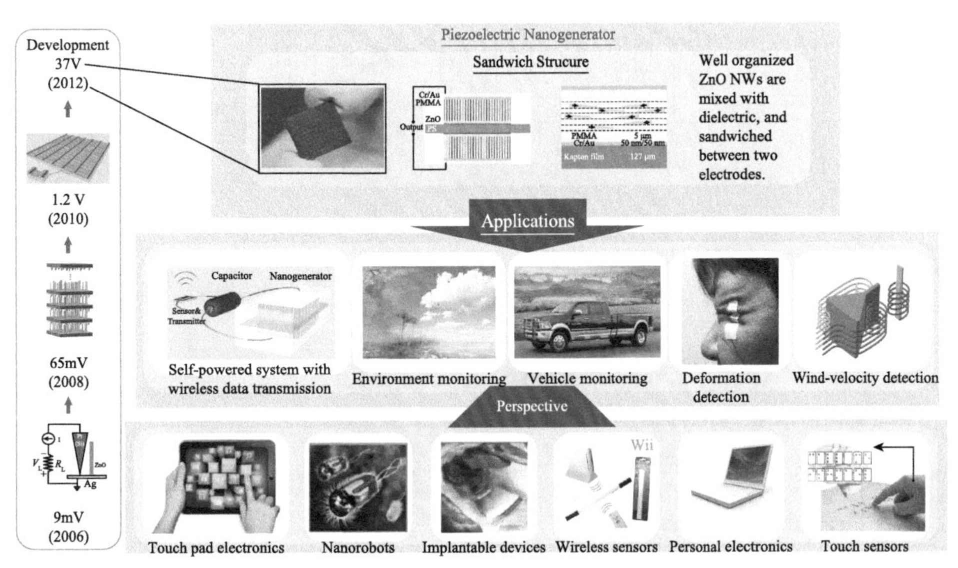

扫码阅全文

http://www.binn.cas.cn/book/202106/W020210609412259408906.pdf

16. Energy Harvesting Using Piezoelectric Nanowires—A Correspondence on "Energy Harvesting Using Nanowires?" by Alexe et al.

Zhong Lin Wang

Advanced Materials, 21 (2009) 1311-1315.

简介：自从纳米发电机第一篇文章发表后，德国一个小组在 2007～2008 年间不断质疑纳米发电机信号的真实性。该文章是王中林为了回应这个质疑而发表的正式回复。文中详细分析了德国组所犯的技术错误和不科学的论述。自从该文发表后，这个质疑被彻底打消了。建议从事科学研究的同仁参考阅读此文。

ABSTRACT A response to the questions raised by Alexe et al. concerning nanowire-based nanogenerators is presented. Evidence is given about the existence and detection of a piezoelectric potential in ZnO nanowires. The role played by the piezoelectric potential is to overcome the threshold voltage at the Pt-ZnO junction, while the observed output signal of ～10 mV is the difference in Fermi levels between the two electrodes. The measurement system used by Alexe et al. is questioned, as is their model.

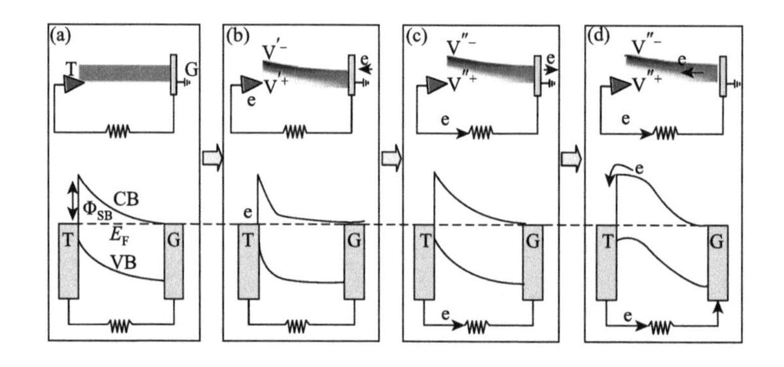

扫码阅全文

http://www.binn.cas.cn/book/202106/W020210609412260424101.pdf

I.8 压电催化

Piezocatalysis

1. Piezo-Potential Enhanced Photocatalytic Degradation of Organic Dye Using ZnO Nanowires

Xinyu Xue, Weili Zang, Ping Deng, Qi Wang, Lili Xing, Yan Zhang*, Zhong Lin Wang*

Nano Energy, 13 (2015) 414-422.

简介：该文利用压电所产生的分布在纳米颗粒两端的正负极化电荷进行氧化还原反应，提升有机分子的分解速率。

ABSTRACT Piezoelectric semiconductors, such as wurtzite structure ZnO, GaN, and InN, have novel properties of coupling of piezoelectric and semiconductor. The piezoelectric field is created inside ZnO nanowires by applying strain. The photo-generated electrons and holes will be separated under the driving of piezoelectric field. The photocatalytic activity of ZnO nanowires for degrading methylene blue has been enhanced by the piezoelectric-driven separation of photo-generated carriers. Coupling the piezoelectric and photocatalytic properties of ZnO nanowires, a new fundamental mechanism for the degradation of organic dye has been demonstrated.

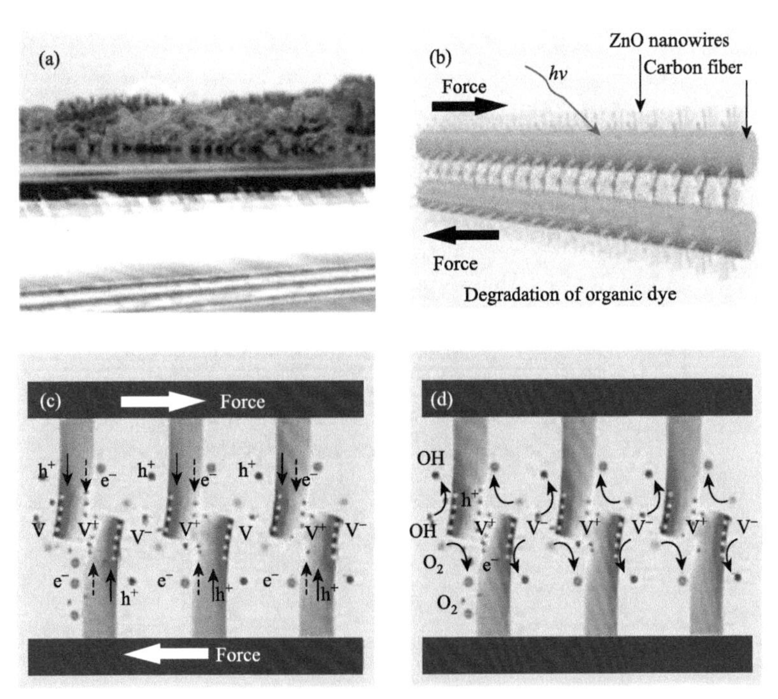

2. Enhanced Ferroelectric-Nanocrystal-Based Hybrid Photocatalysis by Ultrasonic-Wave-Generated Piezophototronic Effect

Haidong Li, Yuanhua Sang, Sujie Chang, Xin Huang, Yan Zhang, Rusen Yang, Huaidong Jiang, Hong Liu* *and* Zhong Lin Wang*

Nano Letters, 15 (2015) 2372-2379.

简介：该文利用超声波所产生的铁电纳米颗粒所具有的压电光电子学效应进行催化反应，展示了一个重要的研究方向。

ABSTRACT An electric field built inside a crystal was proposed to enhance photoinduced carrier separation for improving photocatalytic property of semiconductor photocatalysts. However, a static built-in electric field can easily be saturated by the free carriers due to electrostatic screening, and the enhancement of photocatalysis, thus, is halted. To overcome this problem, here, we propose sonophotocatalysis based on a new hybrid photocatalyst, which combines ferroelectric nanocrystals ($BaTiO_3$) and semiconductor nanoparticles (Ag_2O) to form an Ag_2O-$BaTiO_3$ hybrid photocatalyst. Under periodic ultrasonic excitation, a spontaneous polarization potential of $BaTiO_3$ nanocrystals in responding to ultrasonic wave can act as alternating built-in electric field to separate photoinduced carriers incessantly, which can significantly enhance the photocatalytic activity and cyclic performance of the Ag_2O-$BaTiO_3$ hybrid structure. The piezoelectric effect combined with photoelectric conversion realizes an ultrasonic-wave-driven piezophototronic process in the hybrid photocatalyst, which is the fundamental of sonophotocatalysis.

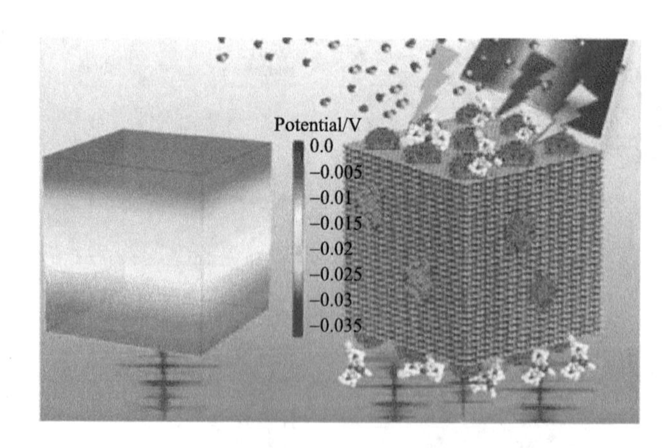

扫码阅全文

http://www.binn.cas.cn/book/202106/W020210609413498891983.pdf

3. Advances in Piezo-Phototronic Effect Enhanced Photocatalysis and Photoelectrocatalysis

Lun Pan, Shangcong Sun, Ying Chen, Peihong Wang, Jiyu Wang, Xiangwen Zhang, Ji-Jun Zou* *and* **Zhong Lin Wang***

Advanced Energy Materials, 10 (2020) 2000214.

简介：这是一篇全面介绍利用压电光电子学效应增强光催化和光电催化反应的综述文章。

ABSTRACT Direct conversion of solar light into chemical energy by means of photocatalysis or photoelectrocatalysis is currently a point of focus for sustainable energy development and environmental remediation. However, its current efficiency is still far from satisfying, suffering especially from severe charge recombination. To solve this problem, the piezo-phototronic effect has emerged as one of the most effective strategies for photo(electro)catalysis. Through the integration of piezoelectricity, photoexcitation, and semiconductor properties, the built-in electric field by mechanical stimulation induced polarization can serve as a flexible autovalve to modulate the charge-transfer pathway and facilitate carrier separation both in the bulk phase and at the surfaces of semiconductors. This review focuses on illustrating the trends and impacts of research based on piezo-enhanced photocatalytic reactions. The fundamental mechanisms of piezo-phototronics modulated band bending and charge migration are highlighted. Through comparing and classifying different categories of piezo-photocatalysts (like the typical ZnO, MoS_2, and $BaTiO_3$), the recent advances in polarization-promoted photo(electro)catalytic processes involving water splitting and pollutant degradation are overviewed. Meanwhile the optimization methods to promote their catalytic activities are described. Finally, the outlook for future development of polarization-enhanced strategies is presented.

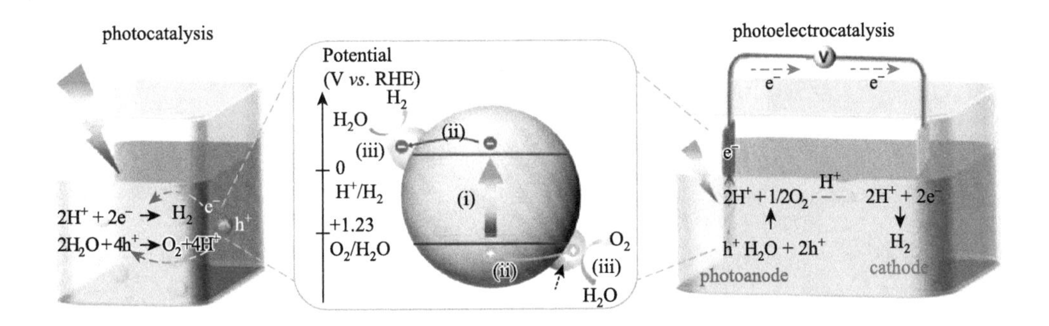

扫码阅全文

http://www.binn.cas.cn/book/202106/W020210609413499740377.pdf

第二部分

Part Ⅱ

Ⅱ.1 摩擦纳米发电机的基本模式
The Working Modes of Triboelectric Nanogenerators

1. Flexible Triboelectric Generator!

Feng-Ru Fan, Zhong-Qun Tian, Zhong Lin Wang*

Nano Energy, 1 (2012) 328-334.

简介：该文章是摩擦纳米发电机的开山之作，开辟了能源回收中的新领域。本文也首次介绍了接触分离式摩擦纳米发电机-TENG 的第一个模式。这种交流式发电机利用的是基于接触起电与静电感应相结合的效应，而电子是在外电路来回流动。这种交流电摩擦纳米发电机被认为是常规型摩擦纳米发电机。

ABSTRACT Charges induced in triboelectric process are usually referred as a negative effect either in scientific research or technological applications, and they are wasted energy in many cases. Here, we demonstrate a simple, low cost and effective approach of using the charging process in friction to convert mechanical energy into electric power for driving small electronics. The triboelectric generator (TEG) is fabricated by stacking two polymer sheets made of materials having distinctly different triboelectric characteristics, with metal films deposited on the top and bottom of the assembled structure. Once subjected to mechanical deformation, a friction between the two films, owing to the nano-scale surface roughness, generates equal amount but opposite signs of charges at two sides. Thus, a triboelectric potential layer is formed at the interface region, which serves as a charge "pump" for driving the flow of electrons in the external load if there is a variation in the capacitance of the system. Such a flexible polymer TEG gives an output voltage of up to 3.3 V at a power density of ～10.4 mW/cm^3. TEGs have the potential of harvesting energy from human activities, rotating tires, ocean waves, mechanical vibration and more, with great applications in self-powered systems for personal electronics, environmental monitoring, medical science and even large-scale power.

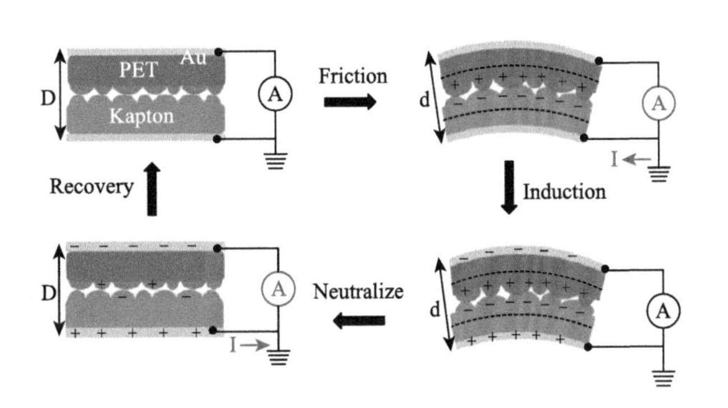

扫码阅全文

http://www.binn.cas.cn/book/202106/W020210609420164085370.pdf

2. Sliding-Triboelectric Nanogenerators Based on In-Plane Charge-Separation Mechanism

Sihong Wang, Long Lin, Yannan Xie, Qingshen Jing, Simiao Niu *and* **Zhong Lin Wang***

Nano Letters, 13 (2013) 2226-2233.

简介：该文章首次介绍了滑动式摩擦纳米发电机的原理，是 TENG 的第二个模式。

ABSTRACT Aiming at harvesting ambient mechanical energy for self-powered systems, triboelectric nanogenerators (TENGs) have been recently developed as a highly efficient, cost-effective and robust approach to generate electricity from mechanical movements and vibrations on the basis of the coupling between triboelectrification and electrostatic induction. However, all of the previously demonstrated TENGs are based on vertical separation of triboelectric-charged planes, which requires sophisticated device structures to ensure enough resilience for the charge separation, otherwise there is no output current. In this paper, we demonstrated a newly designed TENG based on an in-plane charge separation process using the relative sliding between two contacting surfaces. Using Polyamide 6,6 (Nylon) and polytetrafluoroethylene (PTFE) films with surface etched nanowires, the two polymers at the opposite ends of the triboelectric series, the newly invented TENG produces an open-circuit voltage up to 1300V and a short-circuit current density of 4.1 mA/m^2 with a peak power density of 5.3 W/m^2, which can be used as a direct power source for instantaneously driving hundreds of serially connected light-emitting diodes (LEDs). The working principle and the relationships between electrical outputs and the sliding motion are fully elaborated and systematically studied, providing a new mode of TENGs with diverse applications. Compared to the existing vertical-touching based TENGs, this planar-sliding TENG has a high efficiency, easy fabrication, and suitability for many types of mechanical triggering. Furthermore, with the relationship between the electrical output and the sliding motion being calibrated, the sliding-based TENG could potentially be used as a self-powered displacement/speed/acceleration sensor.

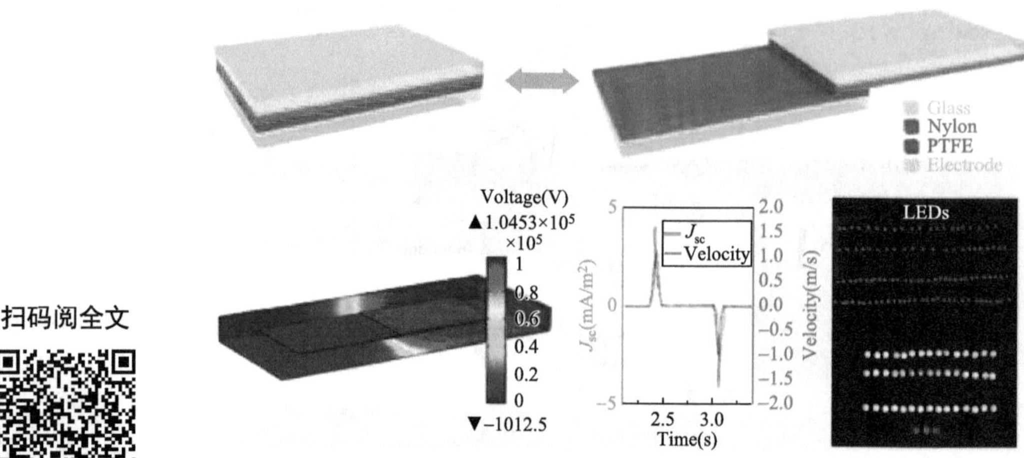

扫码阅全文

http://www.binn.cas.cn/book/202106/W020210609420164566202.pdf

3. Linear-Grating Triboelectric Generator Based on Sliding Electrification

Guang Zhu, Jun Chen, Ying Liu, Peng Bai, Yu Sheng Zhou, Qingshen Jing,
Caofeng Pan *and* **Zhong Lin Wang***

Nano Letters, 13 (2013) 2282-2289.

简介：该文章首次报道了线状格栅式摩擦纳米发电机，是TENG第二个模式的延伸。

ABSTRACT　The triboelectric effect is known for many centuries and it is the cause of many charging phenomena. However, it has not been utilized for energy harvesting until very recently. Here we developed a new principle of triboelectric generator (TEG) based on a fully contacted, sliding electrification process, which lays a fundamentally new mechanism for designing universal, high-performance TEGs to harvest diverse forms of mechanical energy in our daily life. Relative displacement between two sliding surfaces of opposite triboelectric polarities generates uncompensated surface triboelectric charges; the corresponding polarization created a voltage drop that results in a flow of induced electrons between electrodes. Grating of linear rows on the sliding surfaces enables substantial enhancements of total charges, output current, and current frequency. The TEG was demonstrated to be an efficient power source for simultaneously driving a number of small electronics. The principle established in this work can be applied to TEGs of different configurations that accommodate the needs of harvesting energy and/or sensing from diverse mechanical motions, such as contacted sliding, lateral translation, and rotation/rolling.

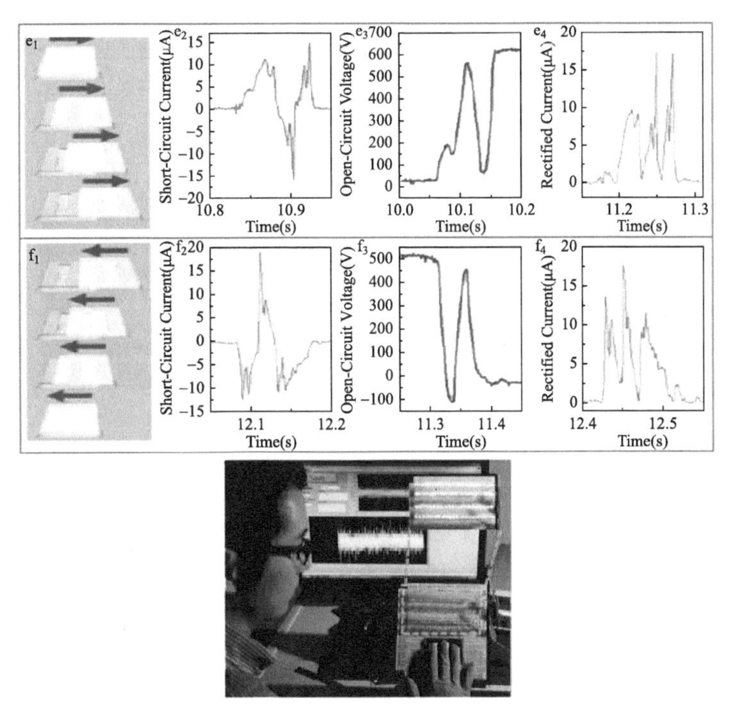

扫码阅全文

http://www.binn.cas.cn/book/202106/W020210609420164768541.pdf

4. Segmentally Structured Disk Triboelectric Nanogenerator for Harvesting Rotational Mechanical Energy

Long Lin, Sihong Wang, Yannan Xie, Qingshen Jing, Simiao Niu, Youfan Hu *and* **Zhong Lin Wang***

Nano Letters, 13 (2013) 2916-2923.

简介：该文章首次报道了转盘格栅式摩擦纳米发电机，是 TENG 第二个模式的扩展。

ABSTRACT We introduce an innovative design of a disk triboelectric nanogenerator (TENG) with segmental structures for harvesting rotational mechanical energy. Based on a cyclic in-plane charge separation between the segments that have distinct triboelectric polarities, the disk TENG generates electricity with unique characteristics, which have been studied by conjunction of experimental results with finite element calculations. The role played by the segmentation number is studied for maximizing output. A distinct relationship between the rotation speed and the electrical output has been thoroughly investigated, which not only shows power enhancement at high speed but also illuminates its potential application as a self-powered angular speed sensor. Owing to the nonintermittent and ultrafast rotation-induced charge transfer, the disk TENG has been demonstrated as an efficient power source for instantaneously or even continuously driving electronic devices and/or charging an energy storage unit. This work presents a novel working mode of TENGs and opens up many potential applications of nanogenerators for harvesting even large-scale energy.

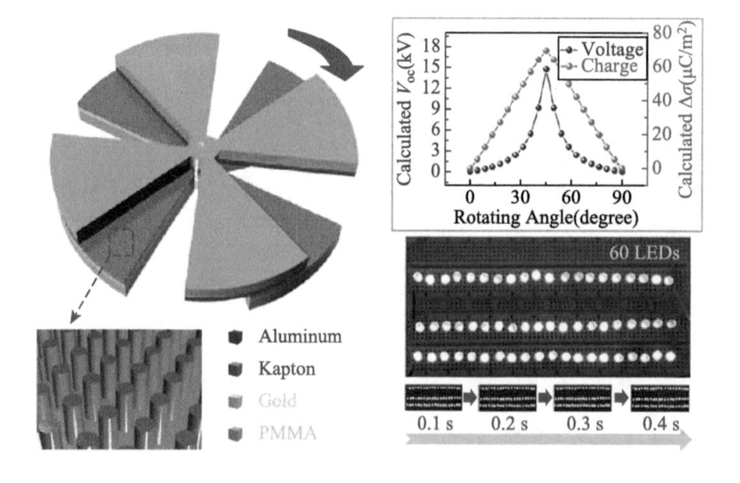

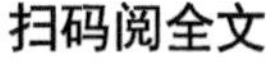
扫码阅全文

http://www.binn.cas.cn/book/202106/W020210609420165320726.pdf

5. Single-Electrode-Based Sliding Triboelectric Nanogenerator for Self-Powered Displacement Vector Sensor System

Ya Yang, Hulin Zhang, Jun Chen, Qingshen Jing, Yu Sheng Zhou, Xiaonan Wen, *and* **Zhong Lin Wang***

ACS Nano, 7 (2013) 7342-7351.

简介：该文首次报道了单电极式摩擦纳米发电机，是 TENG 的第三个模式。

ABSTRACT We report a single-electrode-based sliding-mode triboelectric nanogenerator (TENG) that not only can harvest mechanical energy but also is a self-powered displacement vector sensor system for touching pad technology. By utilizing the relative sliding between an electrodeless polytetrafluoroethylene (PTFE) patch with surface-etched nanoparticles and an Al electrode that is grounded, the fabricated TENG can produce an open-circuit voltage up to 1100 V, a short-circuit current density of 6 mA/m^2, and a maximum power density of 350 mW/m^2 on a load of 100MΩ, which can be used to instantaneously drive 100 green-light-emitting diodes (LEDs). The working mechanism of the TENG is based on the charge transfer between the Al electrode and the ground by modulating the relative sliding distance between the tribo-charged PTFE patch and the Al plate. Grating of linear rows on the Al electrode enables the detection of the sliding speed of the PTFE patch along one direction. Moreover, we demonstrated that 16 Al electrode channels arranged along four directions were used to monitor the displacement (the direction and the location) of the PTFE patch at the center, where the output voltage signals in the 16 channels were recorded in real-time to form a mapping figure. The advantage of this design is that it only requires the bottom Al electrode to be grounded and the top PTFE patch needs no electrical contact, which is beneficial for energy harvesting in automobile rotation mode and touch pad applications.

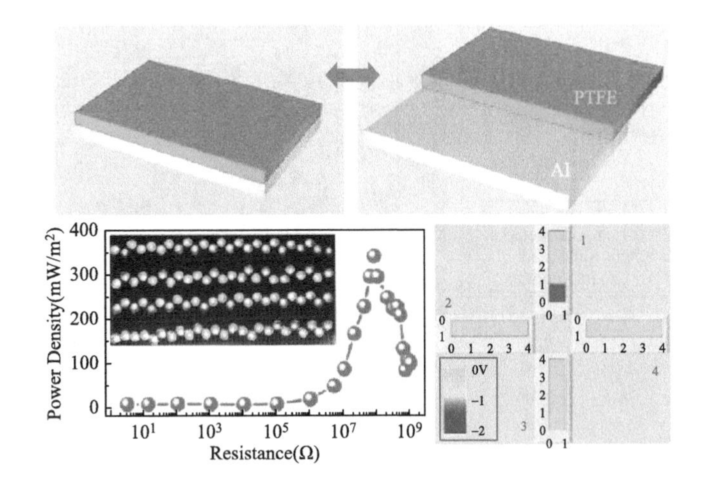

扫码阅全文

http://www.binn.cas.cn/book/202106/W020210609420165721956.pdf

6. Freestanding Triboelectric-Layer-Based Nanogenerators for Harvesting Energy from a Moving Object or Human Motion in Contact and Non-contact Modes

Sihong Wang, Yannan Xie, Simiao Niu, Long Lin *and* Zhong Lin Wang*

Advanced Materials, 26 (2014) 2818-2824.

简介：该文首次报道了自由悬浮式摩擦纳米发电机，是TENG的第四个模式。

ABSTRACT For versatile mechanical energy harvesting from arbitrary moving objects such as humans, a new mode of triboelectric nanogenerator is developed based on the sliding of a freestanding triboelectric-layer between two stationary electrodes on the same plane. With two electrodes alternatively approached by the tribo-charges on the sliding layer, electricity is effectively generated due to electrostatic induction. A unique feature of this nanogenerator is that it can operate in non-contact sliding mode, which greatly increases the lifetime and the efficiency of such devices.

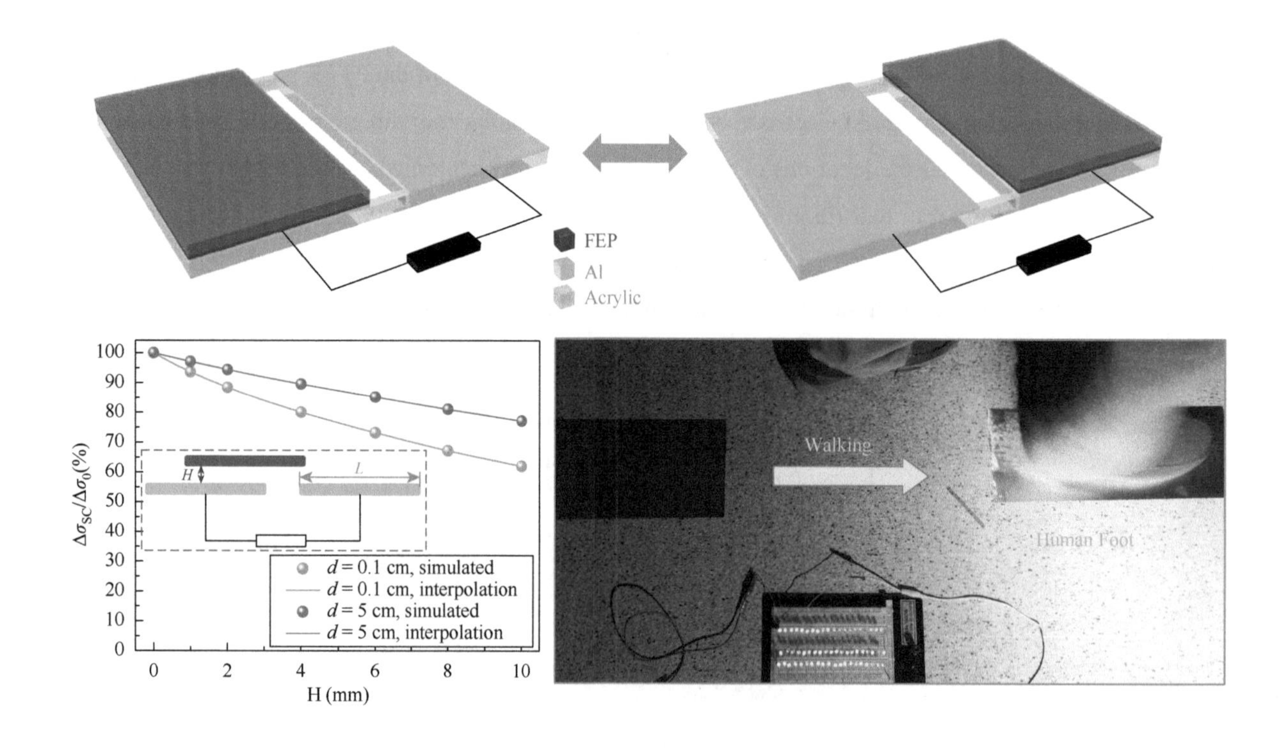

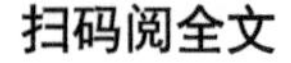
扫码阅全文

http://www.binn.cas.cn/book/202106/W020210609420166137099.pdf

7. Robust Triboelectric Nanogenerator Based on Rolling Electrification and Electrostatic Induction at an Instantaneous Energy Conversion Efficiency of～55%

Long Lin, Yannan Xie, Simiao Niu, Sihong Wang, Po-Kang Yang *and* **Zhong Lin Wang***

ACS Nano, 9 (2015) 922-930.

简介：该文首次报道了基于球或柱滚动式摩擦纳米发电机，是 TENG 的第五个模式，但可以分解成接触分离式和滑动式两种模式的叠加。

ABSTRACT In comparison to in-pane sliding friction, rolling friction not only is likely to consume less mechanical energy but also presents high robustness with minimized wearing of materials. In this work, we introduce a highly efficient approach for harvesting mechanical energy based on rolling electrification and electrostatic induction, aiming at improving the energy conversion efficiency and device durability. The rolling triboelectric nanogenerator is composed of multiple steel rods sandwiched by two fluorinated ethylene propylene (FEP) thin films. The rolling motion of the steel rods between the FEP thin films introduces triboelectric charges on both surfaces and leads to the change of potential difference between each pair of electrodes on back of the FEP layer, which drives the electrons to flow in the external load. As power generators, each pair of output terminals works independently and delivers an open-circuit voltage of 425 V, and a short-circuit current density of 5 mA/m^2. The two output terminals can also be integrated to achieve an overall power density of up to 1.6 W/m^2. The impacts of variable structural factors were investigated for optimization of the output performance, and other prototypes based on rolling balls were developed to accommodate different types of mechanical energy sources. Owing to the low frictional coefficient of the rolling motion, an instantaneous energy conversion efficiency of up to 55% was demonstrated and the high durability of the device was confirmed. This work presents a substantial advancement of the triboelectric nanogenerators toward large-scope energy harvesting and self-powered systems.

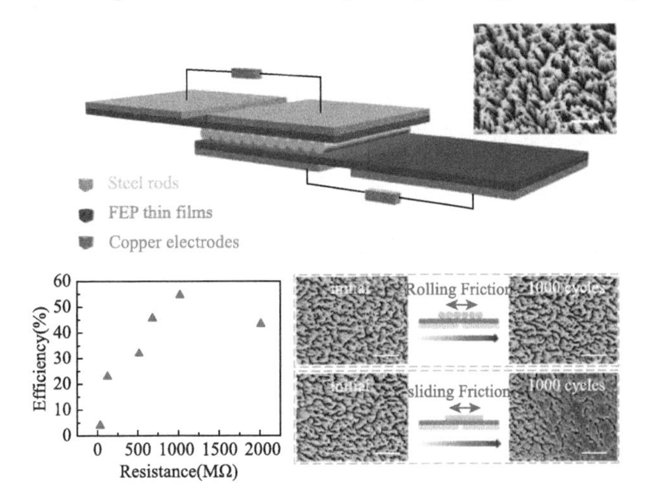

扫码阅全文

http://www.binn.cas.cn/book/202106/W020210609420166618273.pdf

8. Cylindrical Rotating Triboelectric Nanogenerator

Peng Bai, Guang Zhu, Ying Liu, Jun Chen, Qingshen Jing, Weiqing Yang, Jusheng Ma, Gong Zhang *and* **Zhong Lin Wang***

ACS Nano, 7 (2013) 6361-6366.

简介：该文首次报道了圆柱滚筒式摩擦纳米发电机。

ABSTRACT We demonstrate a cylindrical rotating triboelectric nanogenerator (TENG) based on sliding electrification for harvesting mechanical energy from rotational motion. The rotating TENG is based on a core-shell structure that is made of distinctly different triboelectric materials with alternative strip structures on the surface. The charge transfer is strengthened with the formation of polymer nanoparticles on surfaces. During coaxial rotation, a contact-induced electrification and the relative sliding between the contact surfaces of the core and the shell result in an "in-plane" lateral polarization, which drives the flow of electrons in the external load. A power density of 36.9 W/m^2 (short-circuit current of 90 μA and opencircuit voltage of 410 V) has been achieved by a rotating TENG with 8 strip units at a linear rotational velocity of 1.33 m/s (a rotation rate of 1000r/min). The output can be further enhanced by integrating more strip units and/or applying larger linear rotational velocity. This rotating TENG can be used as a direct power source to drive small electronics, such as LED bulbs. This study proves the possibility to harvest mechanical energy by TENGs from rotational motion, demonstrating its potential for harvesting the flow energy of air or water for applications such as self-powered environmental sensors and wildlife tracking devices.

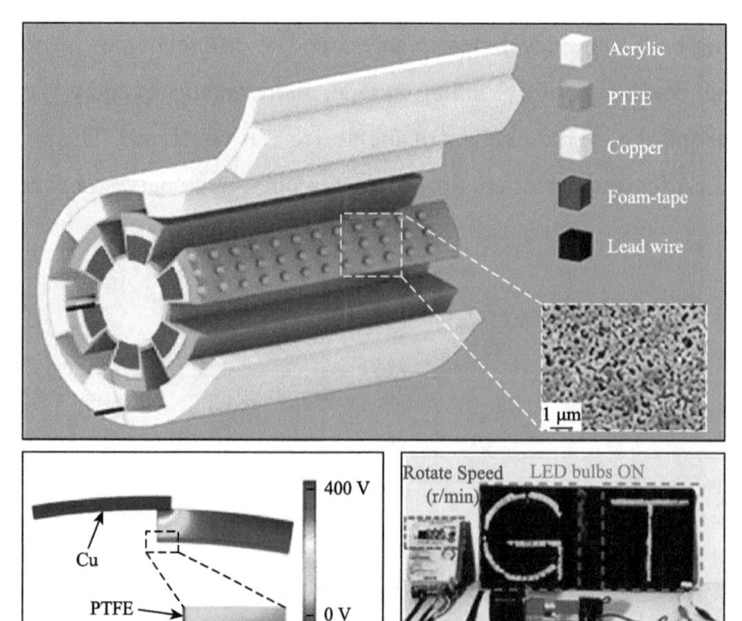

http://www.binn.cas.cn/book/202106/W020210609420167532317.pdf

9. A Constant Current Triboelectric Nanogenerator Arising from Electrostatic Breakdown

Di Liu, Xing Yin, Hengyu Guo, Linglin Zhou, Xinyuan Li, Chunlei Zhang, Jie Wang†, Zhong Lin Wang†

Science Advances, 5 (2019) eaav6437

简介：该文首次报道了直流电摩擦纳米发电机，其原理是基于接触起电所导致的内电路击穿放电效应。

ABSTRACT *In situ* conversion of mechanical energy into electricity is a feasible solution to satisfy the increasing power demand of the Internet of Things (IoTs). A triboelectric nanogenerator (TENG) is considered as a potential solution via building self-powered systems. Based on the triboelectrification effect and electrostatic induction, a conventional TENG with pulsed AC output characteristics always needs rectification and energy storage units to obtain a constant DC output to drive electronic devices. Here, we report a next-generation TENG, which realizes constant current (crest factor, ~1) output by coupling the triboelectrification effect and electrostatic breakdown. Meanwhile, a triboelectric charge density of 430 μC·m^{-2} is attained, which is much higher than that of a conventional TENG limited by electrostatic breakdown. The novel DC-TENG is demonstrated to power electronics directly. Our findings not only promote the miniaturization of self-powered systems used in IoTs but also provide a paradigm-shifting technique to harvest mechanical energy.

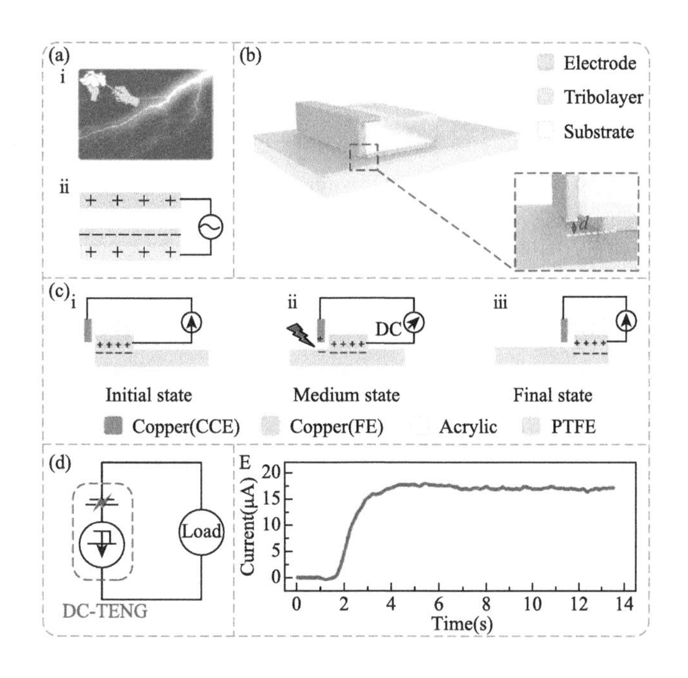

扫码阅全文

http://www.binn.cas.cn/book/202106/W020210609420167712317.pdf

10. Rationally Patterned Electrode of Direct-Current Triboelectric Nanogenerators for Ultrahigh Effective Surface Charge Density

Zhihao Zhao, Yejing Dai, Di Liu, Linglin Zhou, Shaoxin Li, Jie Wang* *and* **Zhong Lin Wang***

Nature Communications, 11 (2020) 6186.

简介：该文首次报道了如何提高直流电摩擦纳米发电机的输出性能。

ABSTRACT As a new-era of energy harvesting technology, the enhancement of triboelectric charge density of triboelectric nanogenerator (TENG) is always crucial for its large-scale application on Internet of Things (IoTs) and artificial intelligence (AI). Here, a microstructure-designed direct-current TENG (MDC-TENG) with rationally patterned electrode structure is presented to enhance its effective surface charge density by increasing the efficiency of contact electrification. Thus, the MDC-TENG achieves a record high charge density of ~5.4 $mC\cdot m^{-2}$, which is over 2-fold the state-of-art of AC-TENGs and over 10-fold compared to previous DCTENGs. The MDC-TENG realizes both the miniaturized device and high output performance. Meanwhile, its effective charge density can be further improved as the device size increases. Our work not only provides a miniaturization strategy of TENG for the application in IoTs and Al as energy supply or self-powered sensor, but also presents a paradigm shift for large-scale energy harvesting by TENGs.

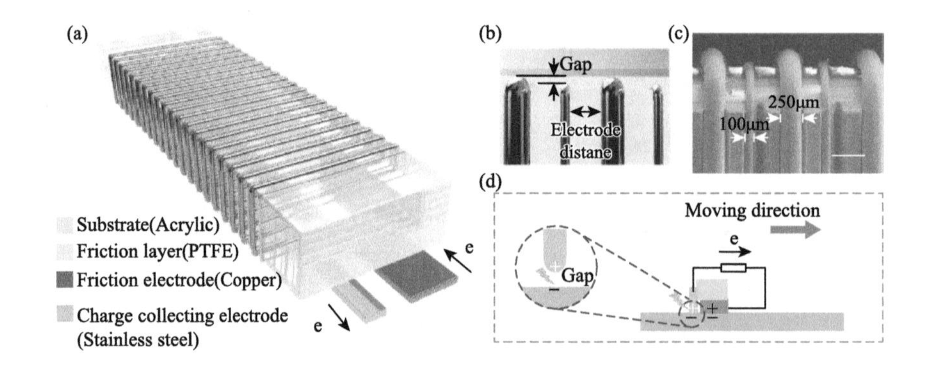

扫码阅全文

http://www.binn.cas.cn/book/202106/W020210609420168647443.pdf

II.2 无线传输型摩擦纳米发电机

The Triboelectric Nanogenerators with Wireless Power Transmission

1. Inductor-Free Wireless Energy Delivery via Maxwell's Displacement Current from an Electrodeless Triboelectric Nanogenerator

Xia Cao, Meng Zhang, Jinrong Huang, Tao Jiang, Jingdian Zou, Ning Wang* *and* **Zhong Lin Wang***

Advanced Materials, 30 (2018) 1704077.

简介：该文首次报道了不利用感应线圈而是基于麦克斯韦位移电流进行摩擦纳米发电机电力无线输出的新技术。

ABSTRACT Wireless power delivery has been a dream technology for applications in medical science, security, radio frequency identification (RFID), and the internet of things, and is usually based on induction coils and/or antenna. Here, a new approach is demonstrated for wireless power delivery by using the Maxwell's displacement current generated by an electrodeless triboelectric nanogenerator (TENG) that directly harvests ambient mechanical energy. A rotary electrodeless TENG is fabricated using the contact and sliding mode with a segmented structure. Due to the leakage of electric field between the segments during relative rotation, the generated Maxwell s displacement current in free space is collected by metal collectors. At a gap distance of 3 cm, the output wireless current density and voltage can reach 7 μA·cm^{-2} and 65 V, respectively. A larger rotary electrodeless TENG and flexible wearable electrodeless TENG are demonstrated to power light-emitting diodes (LEDs) through wireless energy delivery. This innovative discovery opens a new avenue for noncontact, wireless energy transmission for applications in portable and wearable electronics.

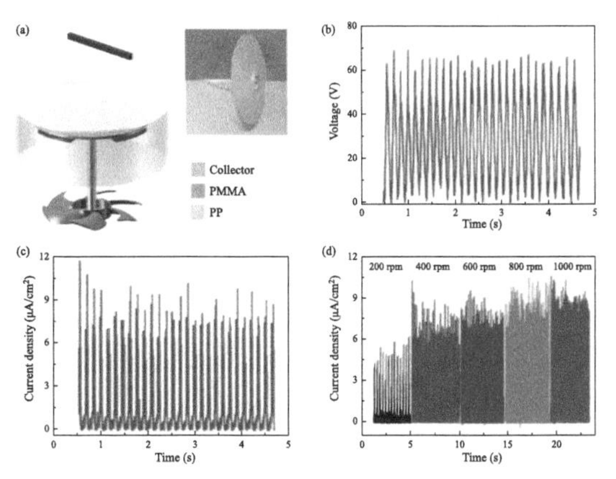

扫码阅全文

http://www.binn.cas.cn/book/202106/W020210609510195969396.pdf

2. Novel Wireless Power Transmission Based on Maxwell Displacement Current

Yan dong Chen, Yang Jie, Ning Wang*, Zhong Lin Wang*, Xia Cao*

Nano Energy, 76 (2020) 105051.

简介：该文报道了利用麦克斯韦位移电流进行摩擦纳米发电机电力无线输出的新技术。

ABSTRACT Wireless power transmission (WPT) is vitally important for portable electronics. Recently, wireless energy delivery via Maxwell displacement current from triboelectric nanogenerator may open new routes to develop novel technologies especially for implantable medical devices and sensor network due to its flexibility, adaption, convenience and safety. In this paper, the propagation characteristics of WPT are explored, which are based on displacement current from a spherical high frequency electric field. The results showed that the spherical surfaces with the same radius from the center have equal value of electric field intensity. When the receiver has a larger area or workload, the output has a higher value, which is subject to the electric field intensity. However, the current is almost unchanged with increasing output power below 15 kΩ of load resistance in series, which may have potential applications for sensors. Furthermore, the receiver circuit of wireless energy is in accordance with the traditional serial and parallel circuit regularity. Based on these, we are going to assume that the electric field source is a reservoir, and water flow is analogous to current. The assumption is suitable for studying propagation characteristics. This new discovery may also offer a new research direction for WPT based on Maxwell displacement electric field.

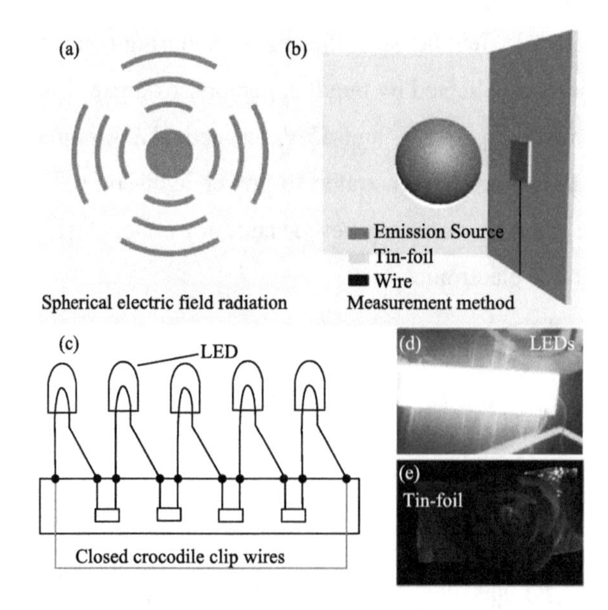

扫码阅全文

http://www.binn.cas.cn/book/202106/W020210609510196623198.pdf

II.3 摩擦纳米发电机的效率和特性

The Efficiency and Characteristics of Triboelectric Nanogenerators

1. Applicability of Triboelectric Generator over a Wide Range of Temperature

Xiaonan Wen, Yuanjie Su, Ya Yang, Hulin Zhang, Zhong Lin Wang*

Nano Energy, 4 (2014) 150-156.

简介：本文首次探索了摩擦纳米发电机在不同温度下的工作情况，证明 TENG 在 77～400 K 都可以工作。

ABSTRACT We studied the influence of temperature on the output performance of triboelectric generators (TEGs). PTFE film and aluminum foil are used as the contact materials for the TEG. A high temperature system and a low temperature system were used to conduct measurements of TEG output voltage and current from 300 K to 500 K and from 77 K to 300 K, respectively. Dependence of output performance on temperature was subsequently obtained by statistically analyzing the data as a function of temperature. The performance of the TEG is maximized at around 260 K and degrades at both higher and lower temperatures. A possible mechanism is proposed for explaining the observed phenomenon. Applicability of TEG over a wide range of temperature was confirmed, from as low as 77 K to as high as 500 K, spanning a range of 423 K. Several LEDs connected in series were successfully lit up by TEG as the sole power source at temperatures of 77 K, 300 K and 500 K.

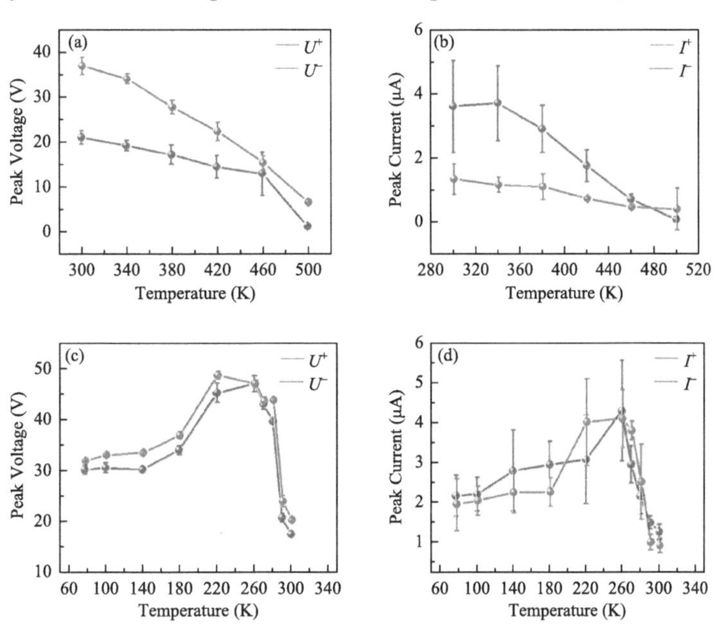

扫码阅全文

http://www.binn.cas.cn/book/202106/W020210609512020297511.pdf

2. A Shape-Adaptive Thin-Film-Based Approach for 50% High-Efficiency Energy Generation Through Micro-Grating Sliding Electrification

Guang Zhu, Yu Sheng Zhou, Peng Bai, Xiansong Meng, Qingshen Jing, Jun Chen *and* Zhong Lin Wang*

Advanced Materials, 26 (2014) 3788-3796.

简介：本文首次实现了摩擦纳米发电机在滑动模式下可以给出50%的能量转换效率和500W/m^{2}的面输出功率密度。

ABSTRACT Effectively harvesting ambient mechanical energy is the key for realizing self-powered and autonomous electronics, which addresses limitations of batteries and thus has tremendous applications in sensor networks, wireless devices, and wearable/implantable electronics, etc. Here, a thin-film-based micro-grating triboelectric nanogenerator (MG-TENG) is developed for high-efficiency power generation through conversion of mechanical energy. The shape-adaptive MG-TENG relies on sliding electrification between complementary micro-sized arrays of linear grating, which offers a unique and straightforward solution in harnessing energy from relative sliding motion between surfaces. Operating at a sliding velocity of 10 m/s, a MG-TENG of 60 cm^2 in overall area, 0.2 cm^3 in volume and 0.6 g in weight can deliver an average output power of 3 W (power density of 50 mW·cm^{-2} and 15 W·cm^{-3}) at an overall conversion efficiency of～50%, making it a sufficient power supply to regular electronics, such as light bulbs. The scalable and cost-effective MG-TENG is practically applicable in not only harvesting various mechanical motions but also possibly power generation at a large scale.

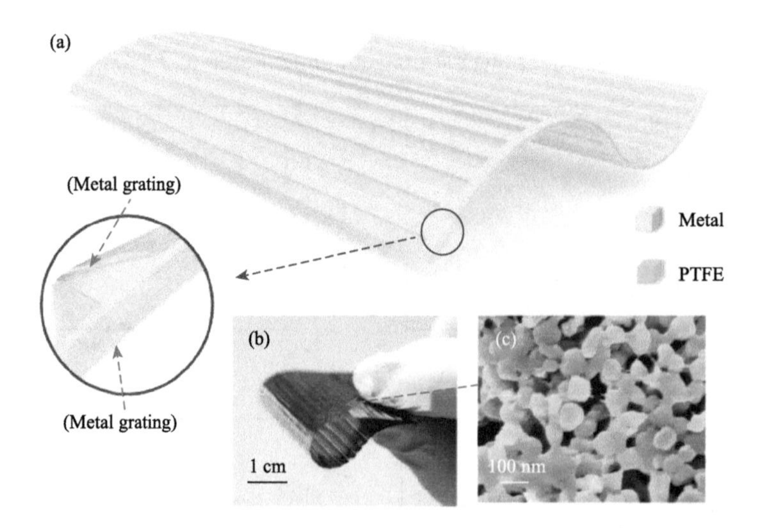

扫码阅全文

http://www.binn.cas.cn/book/202106/W020210609512021171374.pdf

3. Grating-Structured Freestanding Triboelectric-Layer Nanogenerator for Harvesting Mechanical Energy at 85% Total Conversion Efficiency

Yannan Xie, Sihong Wang, Simiao Niu, Long Lin, Qingshen Jing, Jin Yang, Zhengyun Wu *and* **Zhong Lin Wang***

Advanced Materials, 26 (2014) 6599-6607.

简介：本文首次实现了摩擦纳米发电机在自由悬浮式下高达85%的能量转换总效率。

ABSTRACT A newly-designed triboelectric nanogenerator is demonstrated which is composed of a grating-segmented freestanding triboelectric layer and two groups of interdigitated electrodes with the same periodicity. The sliding motion of the grating units across the electrode fingers can be converted into multiple alternating currents through the external load due to the contact electrification and electrostatic induction. Working in non-contact mode, the device shows excellent stability and the total conversion efficiency can reach up to 85% at low operation frequency.

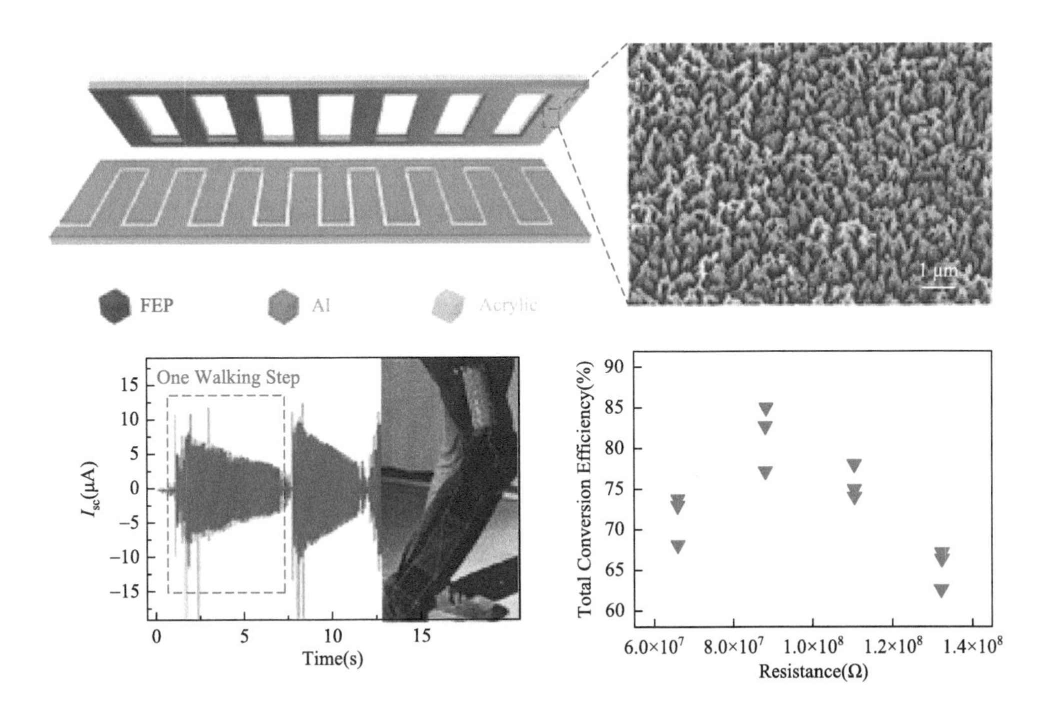

扫码阅全文

http://www.binn.cas.cn/book/202106/W020210609512021465310.pdf

4. Liquid-Metal Electrode for High-Performance Triboelectric Nanogenerator at an Instantaneous Energy Conversion Efficiency of 70.6%

Wei Tang, Tao Jiang, Feng Ru Fan, Ai Fang Yu, Chi Zhang, Xia Cao* *and* Zhong Lin Wang*

Advanced Functional Materials, 25 (2015) 3718-3725.

简介：本文首次利用液态金属作为一个电极的摩擦纳米发电机，并实现了转换效率高达 70%。这种结构的摩擦纳米发电机后来被发展成测试固体表面摩擦电荷密度的新方法和新标准。

ABSTRACT Harvesting ambient mechanical energy is a key technology for realizing self-powered electronics, which has tremendous applications in wireless sensing networks, implantable devices, portable electronics, etc. The currently reported triboelectric nanogenerator (TENG) mainly uses solid materials, so that the contact between the two layers cannot be 100% with considering the roughness of the surfaces, which greatly reduces the total charge density that can be transferred and thus the total energy conversion efficiency. In this work, a liquid-metal-based triboelectric nanogenerator (LM-TENG) is developed for high power generation through conversion of mechanical energy, which allows a total contact between the metal and the dielectric. Due to that the liquid-solid contact induces large contacting surface and its shape adaptive with the polymer thin fims, the LM-TENG exhibits a high output charge density of 430 $\mu C \cdot m^{-2}$, which is four to five times of that using a solid thin film electrode. And its power density reaches 6.7 $W \cdot m^{-2}$ and 133 $kW \cdot m^{-3}$ .More importantly, the instantaneous energy conversion efficiency is demonstrated to be as high as 70.6%. This provides a new approach for improving the performance of the TENG for special applications. Furthermore, the liquid easily fluctuates, which makes the LM-TENG inherently suitable for vibration energy harvesting.

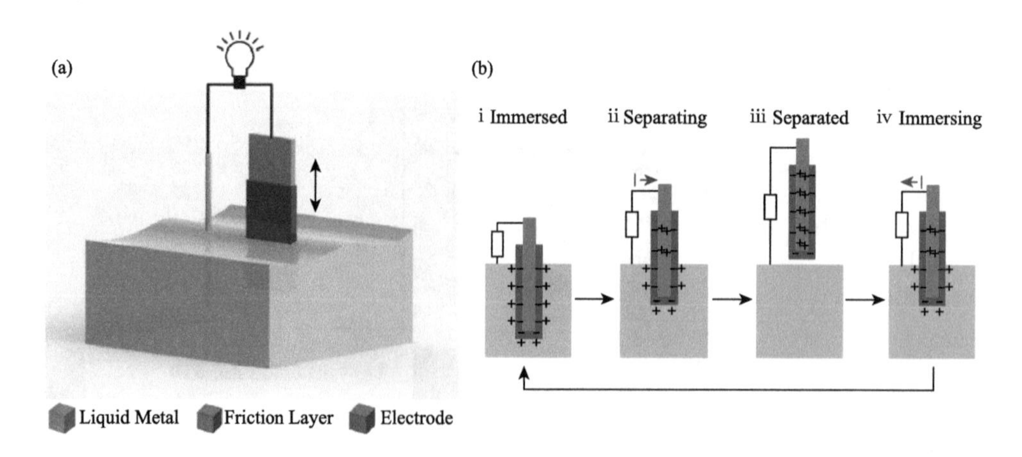

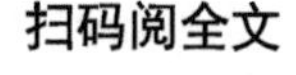
扫码阅全文

http://www.binn.cas.cn/book/202106/W020210609512021840464.pdf

5. Maximum Surface Charge Density for Triboelectric Nanogenerators Achieved by Ionized-Air Injection: Methodology and Theoretical Understanding

Sihong Wang, Yannan Xie, Simiao Niu, Long Lin, Chang Liu, Yu Sheng Zhou, *and* **Zhong Lin Wang***

Advanced Materials, 26 (2014) 6720-6728.

简介：本文首次介绍了利用空气离子化注入的方法来提高表面电荷密度的方法。

ABSTRACT For the maximization of the surface charge density in triboelectric nanogenerators, a new method of injecting single-polarity ions onto surfaces is introduced for the generation of surface charges. The triboelectric nanogenerator's output power gets greatly enhanced and its maximum surface charge density is systematically studied, which shows a huge room for the improvement of the output of triboelectric nanogenerators by surface modification.

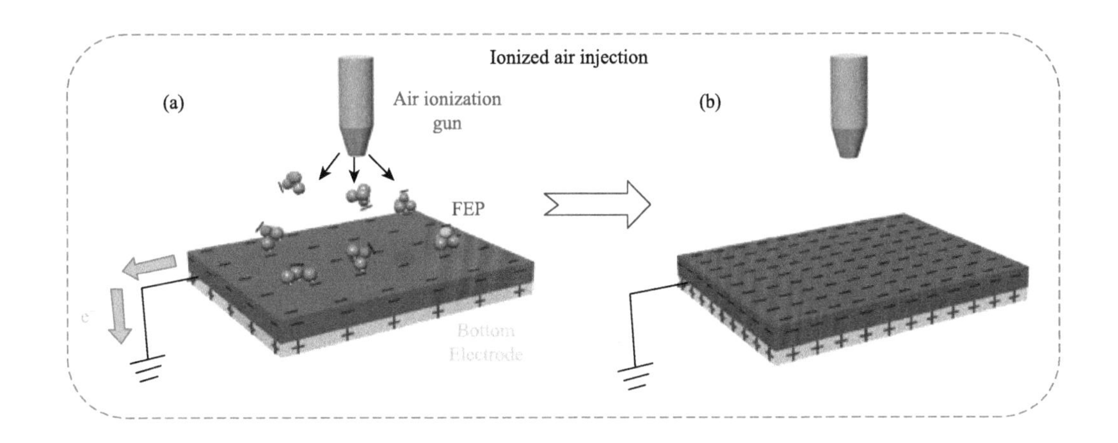

扫码阅全文

http://www.binn.cas.cn/book/202106/W020210609512022589372.pdf

II.4 摩擦纳米发电机与电磁发电机的性能比较

A Comparison of Triboelectric Nanogenerator and Electromagnetic Generator

1. Theoretical Comparison, Equivalent Transformation, and Conjunction Operations of Electromagnetic Induction Generator and Triboelectric Nanogenerator for Harvesting Mechanical Energy

Chi Zhang, Wei Tang, Changbao Han, Fengru Fan *and* Zhong Lin Wang*

Advanced Materials, 26 (2014) 3580-3591.

简介：本文首次把电磁发电机与摩擦纳米发电机的输出做了理论和性能方面的比较，充分展现了TENG的单位质量与单位体积的高输出功率密度。

ABSTRACT Triboelectric nanogenerator (TENG) is a newly invented technology that is effective using conventional organic materials with functionalized surfaces for converting mechanical energy into electricity, which is light weight, cost-effective and easy scalable. Here, we present the first systematic analysis and comparison of EMIG and TENG from their working mechanisms, governing equations and output characteristics, aiming at establishing complementary applications of the two technologies for harvesting various mechanical energies. The equivalent transformation and conjunction operations of the two power sources for the external circuit are also explored, which provide appropriate evidences that the TENG can be considered as a current source with a large internal resistance, while the EMIG is equivalent to a voltage source with a small internal resistance. The theoretical comparison and experimental validations presented in this paper establish the basis of using the TENG as a new energy technology that could be parallel or possibly equivalently important as the EMIG for general power application at large-scale. It opens a field of organic nanogenerator for chemists and materials scientists who can be first time using conventional organic materials for converting mechanical energy into electricity at a high efficiency.

(a)

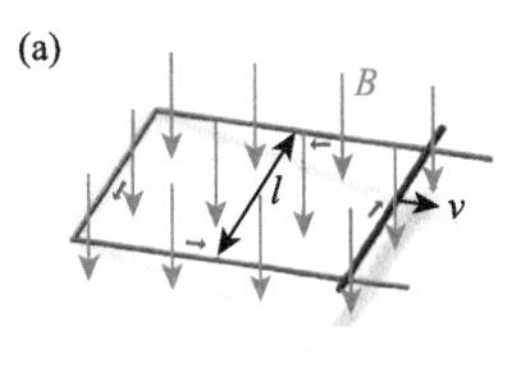

$E = B \cdot l \cdot v$

(b)

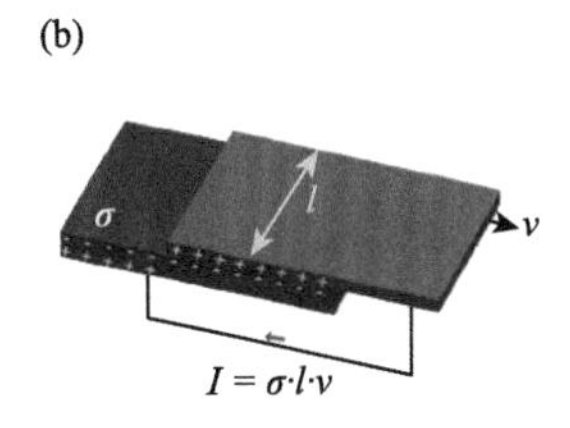

$I = \sigma \cdot l \cdot v$

(c)

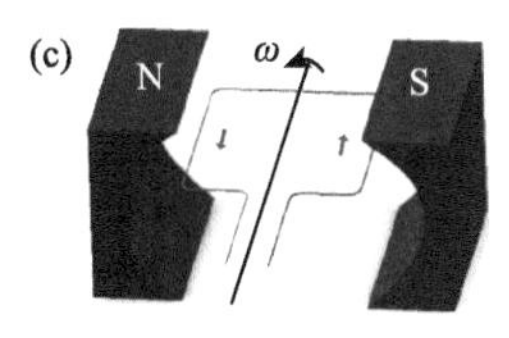

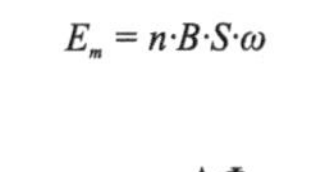

$E_m = n \cdot B \cdot S \cdot \omega$

(d)

ω

$|I| = \frac{n}{\pi} \cdot \sigma \cdot S \cdot \omega$

(e)

$$E = n \cdot \frac{\Delta \Phi}{\Delta t}$$

(In general)

(f)

$$I = n \cdot \frac{\Delta Q}{\Delta t}$$

(In general)

2. Harvesting Low-Frequency (<5 Hz) Irregular Mechanical Energy: A Possible Killer Application of Triboelectric Nanogenerator

Yunlong Zi, Hengyu Guo, Zhen Wen, Min-Hsin Yeh, Chenguo Hu *and* **Zhong Lin Wang***

ACS Nano, 10 (2016) 4797-4805.

简介：本文首次展现了在低频率运转的情况下，摩擦纳米发电机的输出功率远高于电磁发电机的输出。特别是，TENG 的输出电压与频率没有明显的关系。因此在低频下，摩擦纳米发电机有电磁发电机所不具有的优势，也是 TENG 的撒手锏应用。根据这个发现，后来设计出来高效收集水波能量的 TENG，奠定了蓝色能源的基础。

ABSTRACT Electromagnetic generators (EMGs) and triboelectric nanogenerators (TENGs) are the two most powerful approaches for harvesting ambient mechanical energy, but the effectiveness of each depends on the triggering frequency. Here, after systematically comparing the performances of EMGs and TENGs under low-frequency motion (<5 Hz), we demonstrated that the output performance of EMGs is proportional to the square of the frequency, while that of TENGs is approximately in proportion to the frequency. Therefore, the TENG has a much better performance than that of the EMG at low frequency (typically 0.1-3 Hz). Importantly, the extremely small output voltage of the EMG at low frequency makes it almost inapplicable to drive any electronic unit that requires a certain threshold voltage (0.2-4 V), so that most of the harvested energy is wasted. In contrast, a TENG has an output voltage that is usually high enough (>10-100 V) and independent of frequency so that most of the generated power can be effectively used to power the devices. Furthermore, a TENG also has advantages of light weight, low cost, and easy scale up through advanced structure designs. All these merits verify the possible killer application of a TENG for harvesting energy at low frequency from motions such as human motions for powering small electronics and possibly ocean waves for large-scale blue energy.

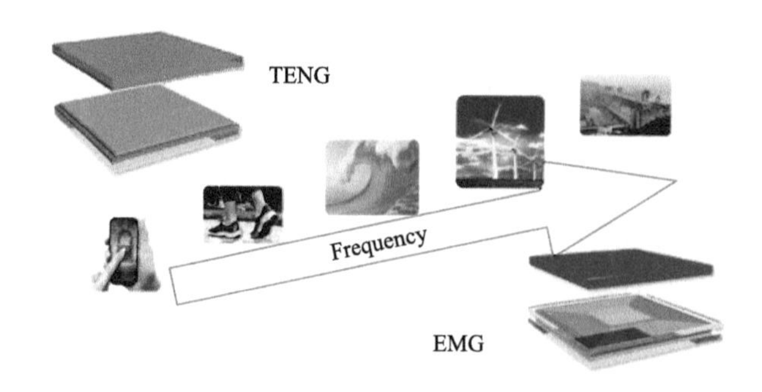

扫码阅全文

http://www.binn.cas.cn/book/202106/W020210609515662619791.pdf

3. Remarkable Merits of Triboelectric Nanogenerator than Electromagnetic Generator for Harvesting Small-Amplitude Mechanical Energy

Junqing Zhao, Gaowei Zhen, Guoxu Liu, Tianzhao Bu, Wenbo Liu, Xianpeng Fu, Ping Zhang*, Chi Zhang*, ZhongLin Wang*

Nano Energy, 61 (2019) 111-118.

简介：本文首次把电磁发电机与摩擦纳米发电机的输出功率做了定量比较，充分展现了TENG对小幅度、低频率的巨大响应和输出，正好弥补了电磁发电机的短板。

ABSTRACT Triboelectric nanogenerator (TENG) is a new energy technology that is as important as traditional electromagnetic generator (EMG) for converting mechanical energy into electricity, which shows great advantages of simple structure, high power density and low-frequency. Here, the effects of motion amplitude for both generators are first taken into consideration. Our result demonstrates that the TENG has a much better performance than that of the EMG at small amplitude. Under fixed operation frequency, the maximum output power of the TENG rapidly grows to saturation with the increase of the amplitude, while that of the EMG grows slowly and gradually. This contrastive characteristic is verified at different frequencies, which has demonstrated that the TENG has dominant scope over the EMG, in not only low-frequency but also small-amplitude. Moreover, electronics powered by the TENG in small-amplitude has been exhibited and an overall comparison of the TENG and EMG is summarized. Beyond low-frequency, this work has verified the small-amplitude is also a remarkable merit of the TENG for harvesting micro-mechanical energy, which has guided the development prospects of TENG as a foundation of the energy for the new era for internet of things, wearable electronics, robotics and artificial intelligence.

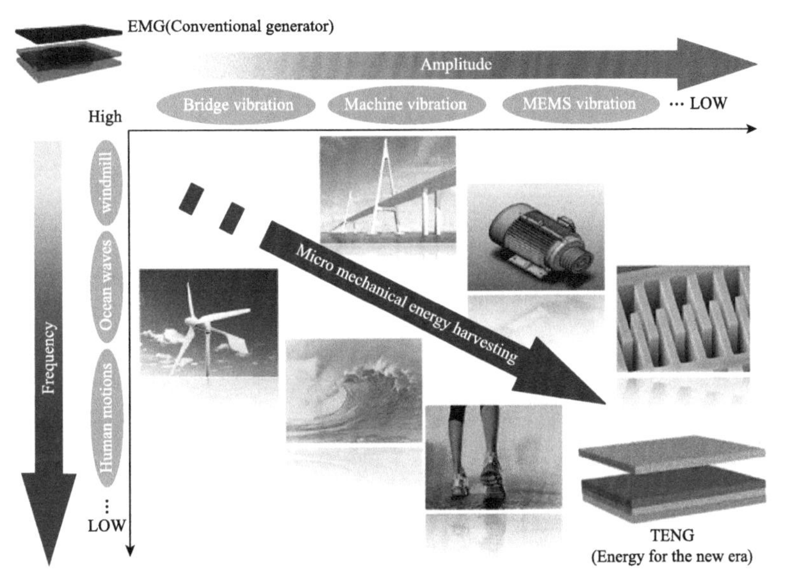

扫码阅全文

http://www.binn.cas.cn/book/202106/W020210609515663350051.pdf

II.5 提高摩擦纳米发电机的寿命

Improving the Durability of Triboelectric Nanogenerator

1. Largely Improving the Robustness and Lifetime of Triboelectric Nanogenerators through Automatic Transition between Contact and Noncontact Working States

Shengming Li, Sihong Wang, Yunlong Zi, Zhen Wen, Long Lin, Gong Zhang *and* Zhong Lin Wang*

ACS Nano, 9 (2015) 7479-7487.

简介：摩擦纳米发电机的寿命是一个极其重要的核心问题，如果解决不好就会大大限制它的应用。本文首次提出了摩擦纳米发电机运用接触式和非接触式自动互换的机制来大大延伸其寿命。

ABSTRACT Although a triboelectric nanogenerator (TENG) has been developed to be an efficient approach to harvest mechanical energy, its robustness and lifetime are still to be improved through an effective and widely applicable way. Here, we show a rational designing methodology for achieving a significant improvement of the long-term stability of TENGs through automatic transition between contact and noncontact working states. This is realized by structurally creating two opposite forces in the moving part of the TENG, in which the pulling-away force is controlled by external mechanical motions. In this way, TENGs can work in the noncontact state with minimum surface wear and also transit into contact state intermittently to maintain high triboelectric charge density. A wind-driven disk-based TENG and a rotary barrel-based TENG that can realize automatic state transition under different wind speeds and rotation speeds, respectively, have been demonstrated as two examples, in which their robustness has been largely improved through this automatic transition. This methodology will further expand the practical application of TENGs for long-time usage and for harvesting mechanical energies with fluctuating intensities.

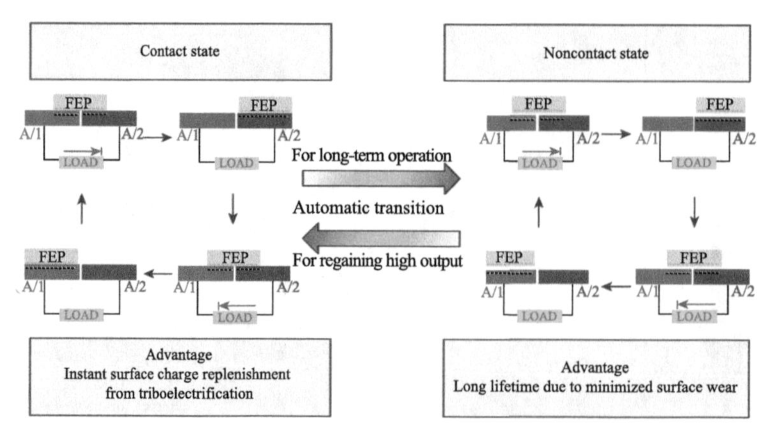

扫码阅全文

http://www.binn.cas.cn/book/202106/W020210610346735814045.pdf

2. A Spring-Based Resonance Coupling for Hugely Enhancing the Performance of Triboelectric Nanogenerators for Harvesting Low-Frequency Vibration Energy

Changsheng Wu, Ruiyuan Liu, Jie Wang, Yunlong Zi, Long Lin, Zhong Lin Wang*

Nano Energy, 32 (2017) 287-293.

简介：本文首次利用弹簧共振的原理来延长摩擦纳米发电机的寿命并提高其输出功率。

ABSTRACT　Low-frequency vibration is a ubiquitous energy that exists almost everywhere, but a high efficient harvesting of which remains challenging. Recently developed triboelectric nanogenerator (TENG) provides a promising alternative approach to conventional electromagnetic and piezoelectric generators, with the advantage of low cost and high output voltage. In this work, a mechanical spring-based amplifier with the ability of amplifying both the vibration frequency and amplitude is integrated with TENG to improve its low-frequency performance by up to 10 times. A new scheme for evaluating TENG using the average output power is proposed and the process of choosing an appropriate time interval for analysis is demonstrated. It takes into account the temporal variation in electrical output and offers a more accurate and convincing evaluation of TENG's performance in practical working environment compared to previously used instantaneous power. This work serves as an important progress for the future development and standardization of TENG, especially for harvesting low-frequency vibration energy as well as a great prospect of blue energy.

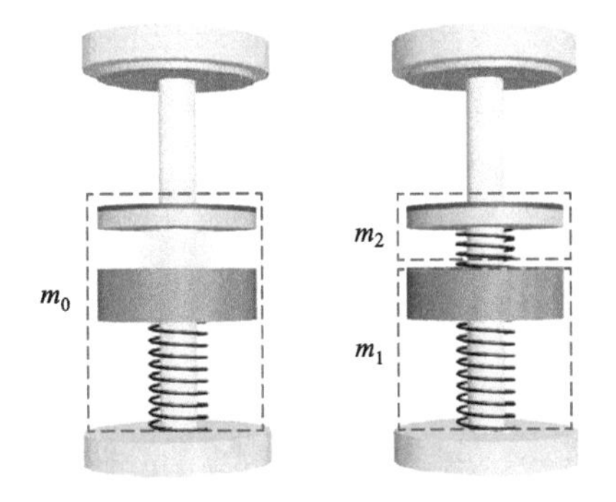

扫码阅全文

http://www.binn.cas.cn/book/202106/W020210610346737482055.pdf

3. Rationally Designed Rotation Triboelectric Nanogenerators with Much Extended Lifetime and Durability

Zhiming Lin, Binbin Zhang, Haiyang Zou, Zhiyi Wu, Hengyu Guo, Ying Zhang, Jin Yang, Zhong Lin Wang*

Nano Energy, 68 (2020) 104378.

简介：本文首次展示了利用摆钟结构设计的TENG，不但大大延长了摩擦纳米发电机的寿命，而且也大大提高了它收集低频能量的效率。设计的结构不但可以收集水波能量，也能收集微风能量。

ABSTRACT Triboelectric nanogenerators (TENG) is a promising route for harvesting ambient mechanical energy to obtain clean and renewable electricity. However, the durability of materials still limits its practical applications for continuous operation. Here, a radial-arrayed TENG was integrated with a transmission mechanism, which successfully maximized the sustainable operation with less triboelectric material wear and damage. Through the rational design of the transmission mechanism, triboelectric layers were first charged by the contact electrification due to the wave impact from the sides, and charges were transferred by the electrostatic induction continuously because of the tumbler's design. In this case, frictional resistance between the triboelectric layers were successfully eliminated, which largely enhances its robustness and durability. After running for 500 k cycles, there is only 1.8% degradation. The TENG could convert one single impact into multiple cycles of rotations, which improves the energy harvesting efficiency for sustainably driving and charging commercial electronics, immediately demonstrating the feasibility of the TENG as a practical power source with extended lifetime and durability. Given exceptional durability and unprecedented applicability resulting from novel designs, this work provides an attractive strategy for the study of triboelectric nanogenerator's durability and facilitates the development of the self-powered systems.

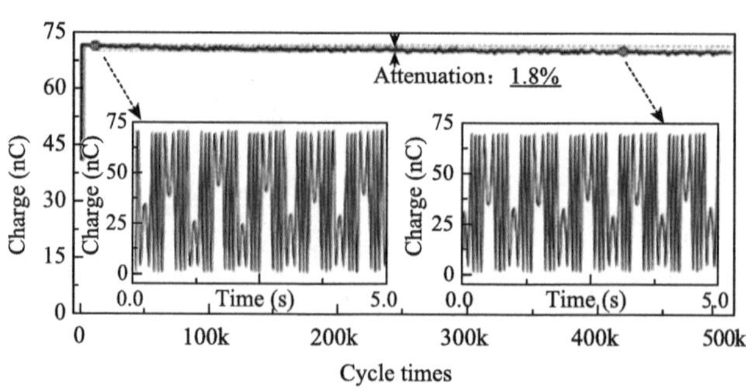

扫码阅全文

http://www.binn.cas.cn/book/202106/W020210610346737772716.pdf

Ⅱ.6 摩擦纳米发电机的综述介绍
Review Papers on Triboelectric Nanogenerator

1. Triboelectric Nanogenerators as New Energy Technology and Self-Powered Sensors-Principles, Problems and Perspectives

Zhong Lin Wang

Faraday Discussions, 176 (2014) 447-451.

简介：本文首次总结了摩擦纳米发电机的四种模式，率先提出了基于摩擦纳米发电机网格式结构的蓝色能源大概念，并提出了在迈向产业化的方向上要解决的十大问题。

ABSTRACT Triboelectrification is one of the most common effects in our daily life, but it is usually taken as a negative effect with very limited positive applications. Here, we invented a triboelectric nanogenerator (TENG) based on organic materials that is used to convert mechanical energy into electricity. The TENG is based on the conjunction of triboelectrification and electrostatic induction, and it utilizes the most common materials available in our daily life, such as papers, fabrics, PTFE, PDMS, Al, PVC etc. In this short review, we first introduce the four most fundamental modes of TENG, based on which a range of applications have been demonstrated. The area power density reaches 1200 $W \cdot m^{-2}$, volume density reaches 490 $kW \cdot m^{-3}$, and an energy conversion efficiency of ～50%-85% has been demonstrated. The TENG can be applied to harvest all kinds of mechanical energy that is available in our daily life, such as human motion, walking, vibration, mechanical triggering, rotation energy, wind, a moving automobile, flowing water, rain drops, tide and ocean waves. Therefore, it is a new paradigm for energy harvesting. Furthermore, TENG can be a sensor that directly converts a mechanical triggering into a self-generated electric signal for detection of motion, vibration, mechanical stimuli, physical touching, and biological movement. After a summary of TENG for micro-scale energy harvesting, mega-scale energy harvesting, and self-powered systems, we will present a set of questions that need to be discussed and explored for applications of the TENG. Lastly, since the energy conversion efficiencies for each mode can be different although the materials are the same, depending on the triggering conditions and design geometry. But one common factor that determines the performance of all the TENGs is the charge density on the two surfaces, the saturation value of which may independent of the triggering configurations of the TENG. Therefore, the triboelectric charge density or the relative charge density in reference to a standard material (such as polytetrafluoroethylene (PTFE)) can be taken as a measuring matrix for characterizing the performance of the material for the TENG.

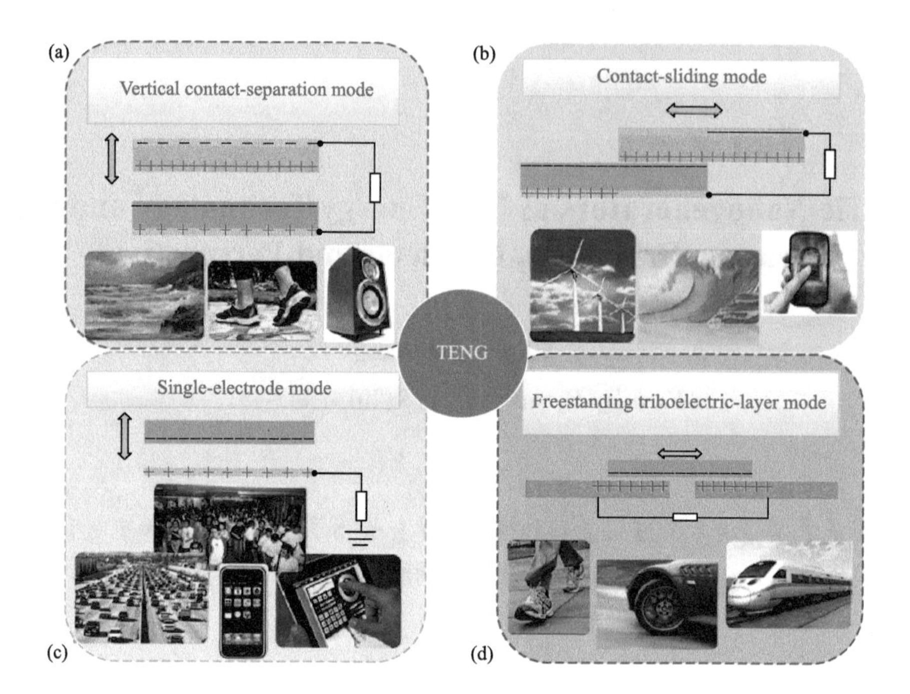

扫码阅全文

http://www.binn.cas.cn/book/202106/W020210610505818857667.pdf

2. Triboelectric Nanogenerators as New Energy Technology for Self-Powered Systems and as Active Mechanical and Chemical Sensors

Zhong Lin Wang

ACS Nano, 7 (2013) 9533-9557.

简介：这是第一篇系统总结摩擦纳米发电机的综述文章。

ABSTRACT Triboelectrification is an effect that is known to each and every one probably since ancient Greek time, but it is usually taken as a negative effect and is avoided in many technologies. We have recently invented a triboelectric nanogenerator (TENG) that is used to convert mechanical energy into electricity by a conjunction of triboelectrification and electrostatic induction. As for this power generation unit, in the inner circuit, a potential is created by the triboelectric effect due to the charge transfer between two thin organic/inorganic films that exhibit opposite tribo-polarity; in the outer circuit, electrons are driven to flow between two electrodes attached on the back sides of the films in order to balance the potential. Since the most useful materials for TENG are organic, it is also named organic nanogenerator, which is the first using organic materials for harvesting mechanical energy. In this paper, we review the fundamentals of the TENG in the three basic operation modes: vertical contact-separation mode, in-plane sliding mode, and single-electrode mode. Ever since the first report of the TENG in January 2012, the output power density of TENG has been improved 5 orders of magnitude within 12 months. The area power density reaches 313 W/m^2, volume density reaches 490 kW/m^3, and a conversion efficiency of ~60% has been demonstrated. The TENG can be applied to harvest all kinds of mechanical energy that is available but wasted in our daily life, such as human motion, walking, vibration, mechanical triggering, rotating tire, wind, flowing water, and more. Alternatively, TENG can also be used as a self-powered sensor for actively detecting the static and dynamic processes arising from mechanical agitation using the voltage and current output signals of the TENG, respectively, with potential applications for touch pad and smart skin technologies. To enhance the performance of the TENG, besides the vast choices of materials in the triboelectric series, from polymer to metal and to fabric, the morphologies of their surfaces can be modified by physical techniques with the creation of pyramid-, square-, or hemisphere-based micro- or nanopatterns, which are effective for enhancing the contact area and possibly the triboelectrification. The surfaces of the materials can be functionalized chemically using various molecules, nanotubes, nanowires, or nanoparticles, in order to enhance the triboelectric effect. The contact materials can be composites, such as embedding nanoparticles in a polymer matrix, which may change not only the surface electrification but also the permittivity of the materials so that they can be effective for electrostatic induction. Therefore, there are numerous ways to enhance the performance of the TENG from the materials point of view. This gives an excellent opportunity for chemists and materials scientists to do extensive study both in the basic science and in practical applications. We anticipate that a better enhancement of the output power density will be achieved in the next few years. The TENG is possible not only for self-powered portable electronics but also as a new energy technology with potential to contribute to the world energy in the near future.

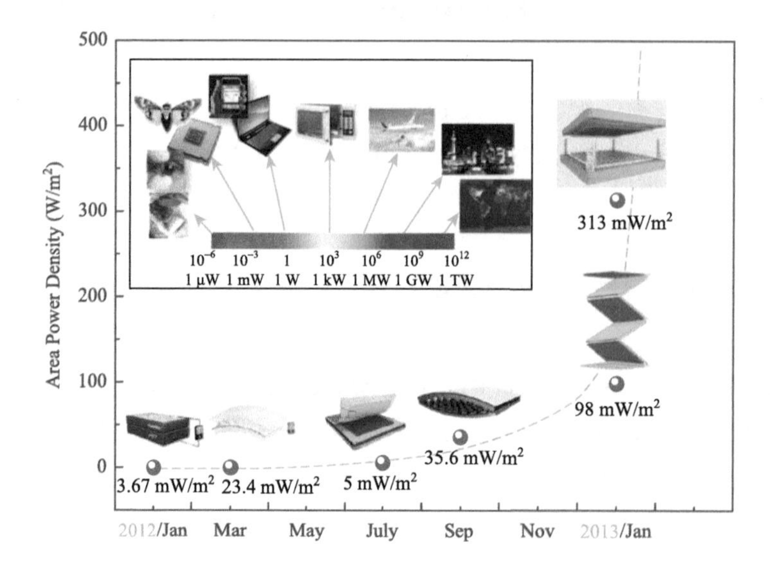

扫码阅全文

http://www.binn.cas.cn/book/202106/W020210610505819220188.pdf

3. Progress in Triboelectric Nanogenertors as a New Energy Technology and Self-Powered Sensors

Zhong Lin Wang,* Jun Chen *and* Long Lin

Energy & Environmental Science, 8 (2015) 2250-2282.

简介：这是第二篇系统总结摩擦纳米发电机的综述文章。

ABSTRACT Ever since the first report of the triboelectric nanogenerator (TENG) in January 2012, its output area power density has reached 500 $W·m^{-2}$, and an instantaneous conversion efficiency of ~70% and a total energy conversion efficiency of up to 85% have been demonstrated. We provide a comprehensive review of the four modes, their theoretical modelling, and the applications of TENGs for harvesting energy from human motion, walking, vibration, mechanical triggering, rotating tire, wind, flowing water and more as well as self-powered sensors.

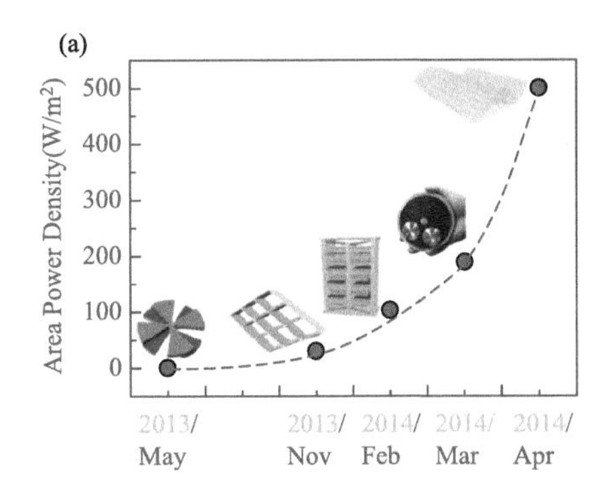

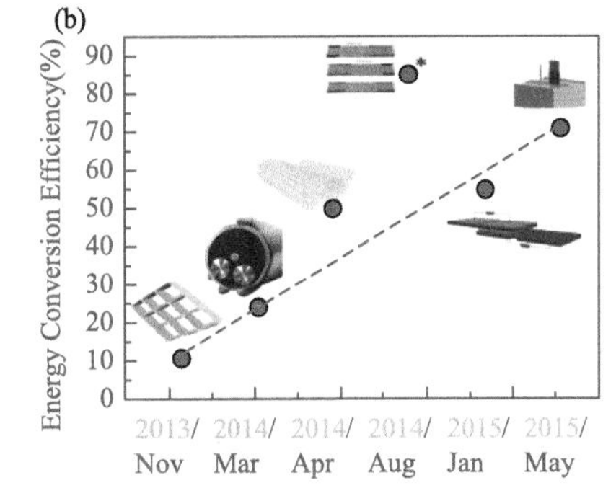

扫码阅全文

http://www.binn.cas.cn/book/202106/W020210610505820160608.pdf

4. Flexible Nanogenerators for Energy Harvesting and Self-Powered Electronics

Fengru Fan, Wei Tang *and* **Zhong Lin Wang***

Advanced Materials, 28 (2016) 4283-4305.

简介：这是一篇摩擦纳米发电机的综述文章。

ABSTRACT Flexible nanogenerators that efficiently convert mechanical energy into electrical energy have been extensively studied because of their great potential for driving low-power personal electronics and self-powered sensors. Integration of flexibility and stretchability to nanogenerator has important research significance that enables applications in flexible/stretchable electronics, organic optoelectronics, and wearable electronics. Progress in nanogenerators for mechanical energy harvesting is reviewed, mainly including two key technologies: flexible piezoelectric nanogenerators (PENGs) and flexible triboelectric nanogenerators (TENGs). By means of material classification, various approaches of PENGs based on ZnO nanowires, lead zirconate titanate (PZT), poly(vinylidene fluoride) (PVDF), 2D materials, and composite materials are introduced. For flexible TENG, its structural designs and factors determining its output performance are discussed, as well as its integration, fabrication and applications. The latest representative achievements regarding the hybrid nanogenerator are also summarized. Finally, some perspectives and challenges in this field are discussed.

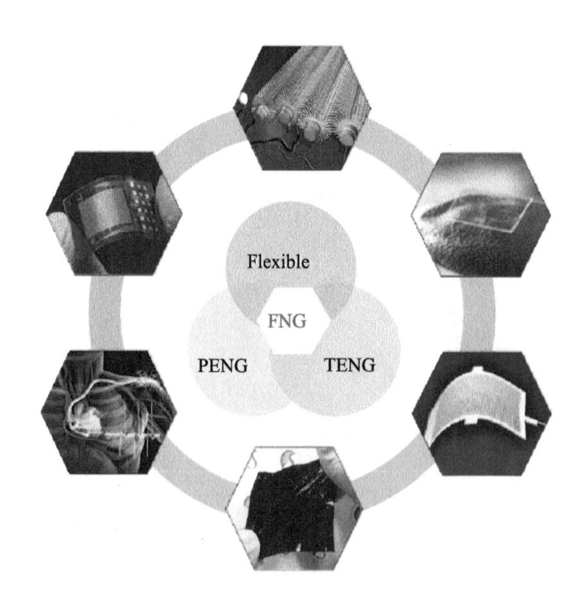

扫码阅全文

http://www.binn.cas.cn/book/202106/W020210610505821613097.pdf

5. Triboelectric Nanogenerator: A Foundation of the Energy for the New Era

Changsheng Wu, Aurelia C. Wang, Wenbo Ding, Hengyu Guo *and* Zhong Lin Wang*

Advanced Energy Materials, 9 (2019) 1802906.

简介：本文提出了摩擦纳米发电机的发展技术路线图和四大研究领域：微纳能源、自驱动传感、蓝色能源、高压电源。

ABSTRACT As the world is marching into the era of the internet of things (IoTs) and artificial intelligence, the most vital development for hardware is a multifunctional array of sensing systems, which forms the foundation of the fourth industrial revolution toward an intelligent world. Given the need for mobility of these multitudes of sensors, the success of the IoTs calls for distributed energy sources, which can be provided by solar, thermal, wind, and mechanical triggering/vibrations. The triboelectric nanogenerator (TENG) for mechanical energy harvesting developed by Z.L. Wang's group is one of the best choices for this energy for the new era, since triboelectrification is a universal and ubiquitous effect with an abundant choice of materials. The development of self-powered active sensors enabled by TENGs is revolutionary compared to externally powered passive sensors, similar to the advance from wired to wireless communication. In this paper, the fundamental theory, experiments, and applications of TENGs are reviewed as a foundation of the energy for the new era with four major application fields: micro/nano power sources, self-powered sensors, large-scale blue energy, and direct high-voltage power sources. A roadmap is proposed for the research and commercialization of TENG in the next 10 years.

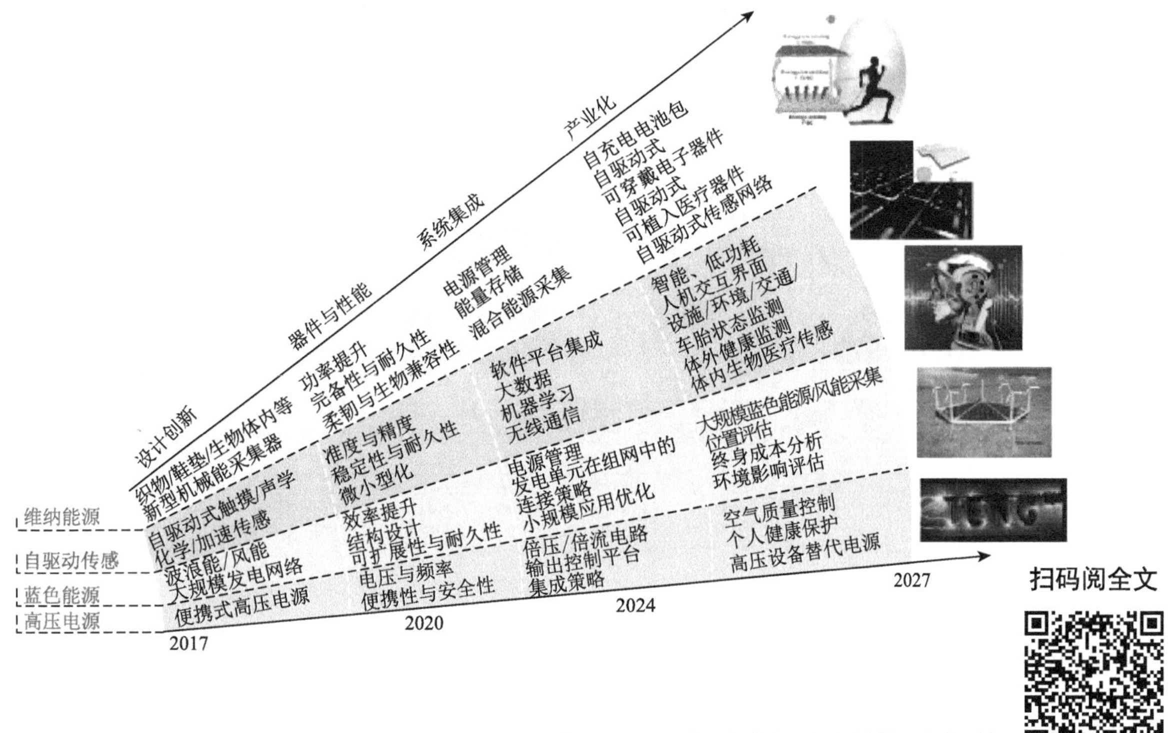

扫码阅全文

http://www.binn.cas.cn/book/202106/W020210610505823257678.pdf

6. Self-Powered Cardiovascular Electronic Devices and Systems

Qiang Zheng, Qizhu Tang, Zhong Lin Wang* ***and*** **Zhou Li***

Nature Reviews Cardiology, 18 (2021) 7-21.

简介：本文首次总结了应用于植入式生物医疗器件的纳米发电机及其相关技术，为自驱动的医学传感打下基础。

ABSTRACT Cardiovascular electronic devices have enormous benefits for health and quality of life but the long-term operation of these implantable and wearable devices remains a huge challenge owing to the limited life of batteries, which increases the risk of device failure and causes uncertainty among patients. A possible approach to overcoming the challenge of limited battery life is to harvest energy from the body and its ambient environment, including biomechanical, solar, thermal and biochemical energy, so that the devices can be self-powered. This strategy could allow the development of advanced features for cardiovascular electronic devices, such as extended life, miniaturization to improve comfort and conformability, and functions that integrate with real-time data transmission, mobile data processing and smart power utilization. In this Review, we present an update on self-powered cardiovascular implantable electronic devices and wearable active sensors. We summarize the existing self-powered technologies and their fundamental features. We then review the current applications of self-powered electronic devices in the cardiovascular field, which have two main goals. The first is to harvest energy from the body as a sustainable power source for cardiovascular electronic devices, such as cardiac pacemakers. The second is to use self-powered devices with low power consumption and high performance as active sensors to monitor physiological signals (for example, for active endocardial monitoring). Finally, we present the current challenges and future perspectives for the field.

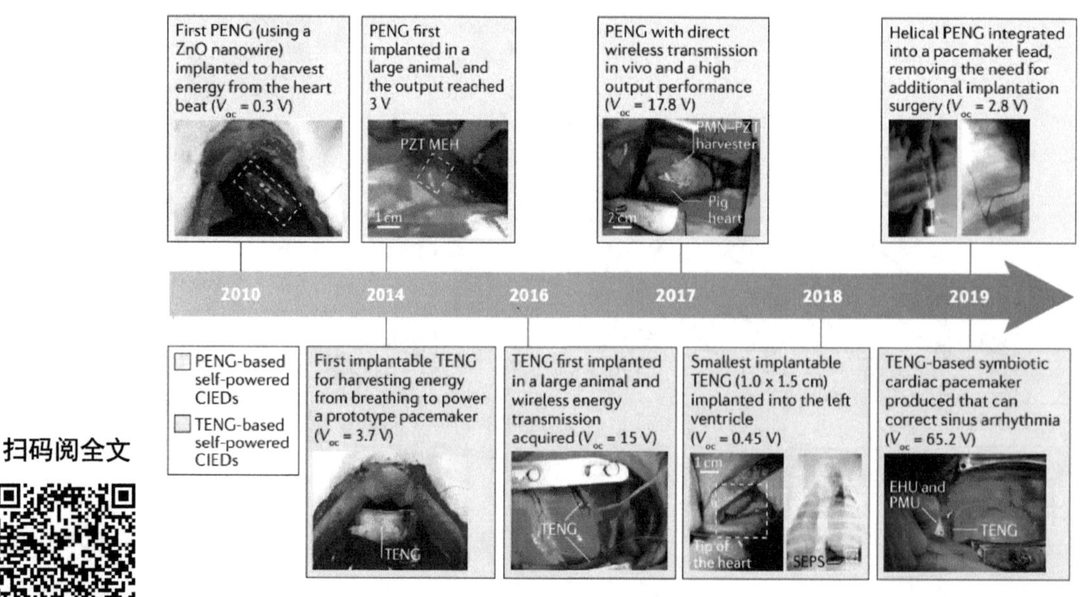

扫码阅全文

http://www.binn.cas.cn/book/202106/W020210610505824536067.pdf

7. Triboelectric Nanogenerator (TENG)—Sparking an Energy and Sensor Revolution

Zhong Lin Wang

Advanced Energy Materials, (2020) 2000137.

简介：本文总结了摩擦纳米发电机作为新时代能源与传感技术的最新进展。

ABSTRACT The study presents the fundamental scientific understanding of electron transfer in contact electrification in solid-solid and liquid-solid cases and a newly revised model for the formation of electric double layer. The potential revolutionary impacts of triboelectric nanogenerators as energy sources and sensors are presented in the fields of health care, environmental science, wearable electronics, internet of things, human-machine interfacing, robotics, and artificial intelligence.

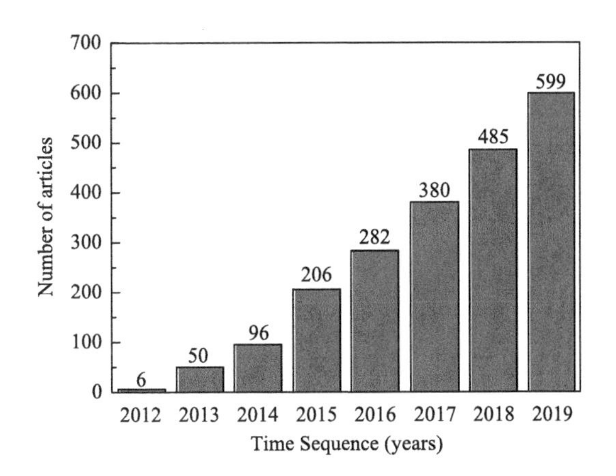

扫码阅全文

http://www.binn.cas.cn/book/202106/W020220106517965927555.pdf

8. The Triboelectric Nanogenerator as an Innovative Technology toward Intelligent Sports

Jianjun Luo, Wenchao Gao *and* Zhong Lin Wang*

Advanced Materials, (2021) 2004178.

简介：本文总结了摩擦纳米发电机在智能体育方面的应用。

ABSTRACT In the new era of the Internet-of-Things, athletic big data collection and analysis based on widely distributed sensing networks are particularly important in the development of intelligent sports. Conventional sensors usually require an external power supply, with limitations such as limited lifetime and high maintenance cost. As a newly developed mechanical energy harvesting and self-powered sensing technology, the triboelectric nanogenerator (TENG) shows great potential to overcome these limitations. Most importantly, TENGs can be fabricated using wood, paper, fibers, and polymers, which are the most frequently used materials for sports. Recent progress on the development of TENGs for the field of intelligent sports is summarized. First, the working mechanism of TENG and its association with athletic big data are introduced. Subsequently, the development of TENG-based sports sensing systems, including smart sports facilities and wearable equipment is highlighted. At last, the remaining challenges and open opportunities are also discussed.

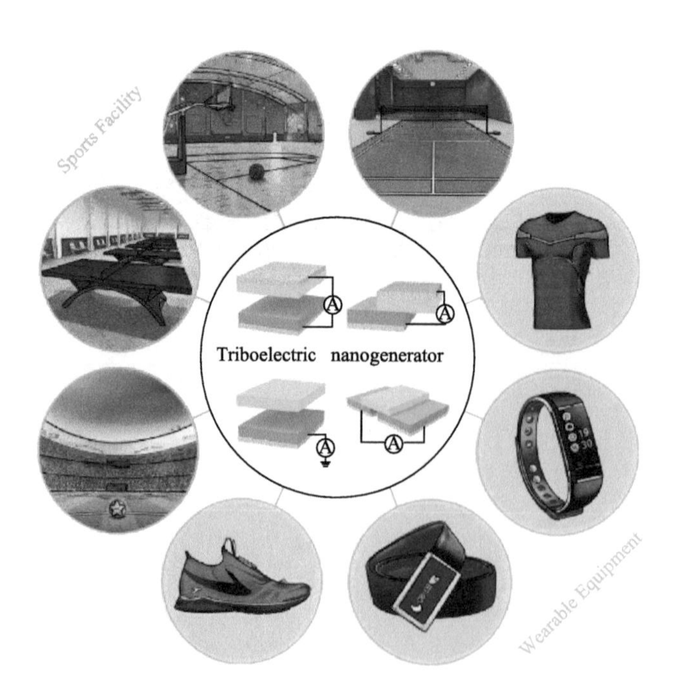

扫码阅全文

http://www.binn.cas.cn/book/202106/W020210610505825532930.pdf

9. Paper-Based Triboelectric Nanogenerators and Their Applications: A review

Jing Han, Nuo Xu, Yuchen Liang, Mei Ding, Junyi Zhai, Qijun Sun* *and* **Zhong Lin Wang***

Beilstein Journal of Nanotechnology, 12 (2021) 151-171.

简介：本文总结了纸质的摩擦纳米发电机及其应用。

ABSTRACT The development of industry and of the Internet of Things (IoTs) have brought energy issues and huge challenges to the environment. The emergence of triboelectric nanogenerators (TENGs) has attracted wide attention due to their advantages, such as self-powering, lightweight, and facile fabrication. Similarly to paper and other fiber-based materials, which are biocompatible, biodegradable, environmentally friendly, and are everywhere in daily life, paper-based TENGs (P-TENGs) have shown great potential for various energy harvesting and interactive applications. Here, a detailed summary of P-TENGs with two-dimensional patterns and three-dimensional structures is reported. P-TENGs have the potential to be used in many practical applications, including self-powered sensing devices, human-machine interaction, electrochemistry, and highly efficient energy harvesting devices. This leads to a simple yet effective way for the next generation of energy devices and paper electronics.

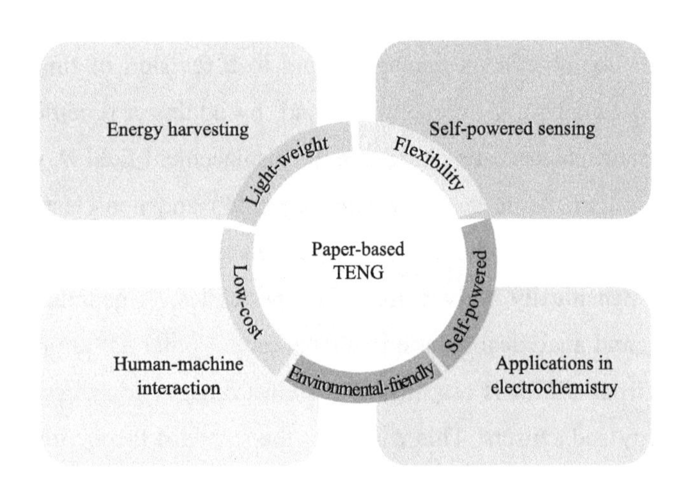

扫码阅全文

http://www.binn.cas.cn/book/202106/W020210610505826328996.pdf

10. From Contact-Electrication to Triboelectric Nanogenerators

Zhong Lin Wang

Reports on Progress in Physics，84（2021）096502

简介：本文系统地总结了接触起电的机理、摩擦纳米发电机的系统理论体系、评判标准，以及在五大领域的重大应用和智能体育方面的应用。这是迄今为止最全面的一篇综述，可以做摩擦纳米发电机基础理论的教本。

ABSTRACT Although the contact electrification (CE) (or usually called "triboelectrification (TE)") effect has been known for over 2600 years, its scientific mechanism still remains debated after decades. Interest in studying CE has been recently revisited due to the invention of the triboelectric nanogenerators (TENGs), which are the most effective approach for converting random, low-frequency mechanical energy into electric power for distributed energy applications. This review is composed of three parts that are coherently linked ranging from basic physics, through classical electrodynamics, and to technological advances and engineering applications. First, the mechanisms of CE are studied for general cases involving solids, liquids and gas phases. Various physics models are presented to explain the fundamentals of CE by illustrating that electron transfer is the dominant mechanism for CE for solid-solid interfaces. Electron transfer also occurs in the CE at liquid-solid and liquid-liquid interfaces. An electron-cloud overlap model is proposed to explain CE in general. This electron transfer model is extended to liquid-solid interfaces, leading to a revision of the formation mechanism of the electric-double-layer (EDL) at liquid-solid interfaces. Second, by adding a time-dependent polarization term $\boldsymbol{P}_s$ created by the CE induced surface electrostatic charges in the displacement field $\boldsymbol{D}$, we expand Maxwell equations to include both the medium polarizations due to electric field ($\boldsymbol{P}$) and non-electric field (such as strain) ($\boldsymbol{P}_s$) induced polarization terms. From these, the output power, electromagnetic behaviour and current transport equation for a TENG are systematically derived from first principles. A general solution is presented for the modified Maxwell equations, and analytical solutions for the output potential are provided for a few cases. The displacement current arising from $\varepsilon\partial \boldsymbol{E}/\partial t$ is responsible for electromagnetic waves, while the newly added term $\partial \boldsymbol{P}_s/\partial t$ is responsible for energy and sensors. This work sets the standard theory for quantifying the performance and electromagnetic behaviour of TENGs in general. Finally, we review the applications of the TENGs for harvesting all kinds of available mechanical energy that is wasted in our daily life, such as human motion, walking, vibration, mechanical triggering, rotating tires, wind, flowing water and more. A summary is provided about the applications of TENG in energy science, environmental protection, wearable electronics, self-powered sensors, medical science, robotics and artificial intelligence.

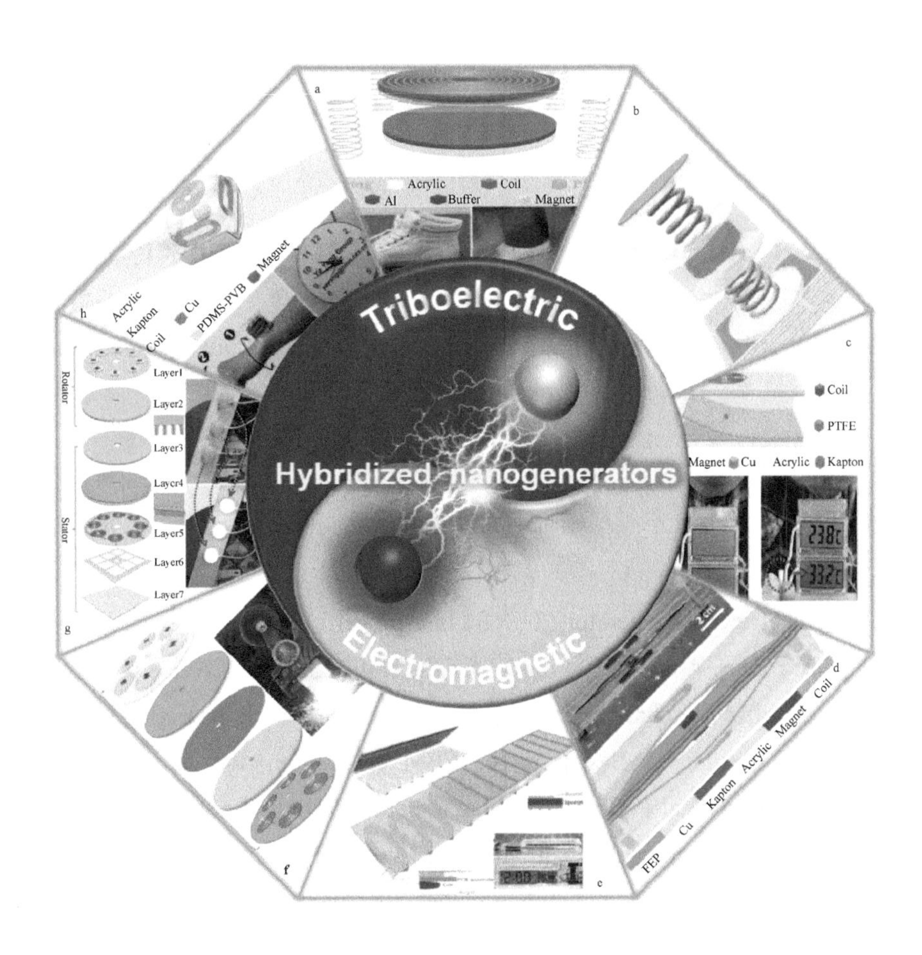

扫码阅全文

http://www.binn.cas.cn/book/202106/W020210617409566484665.pdf

Ⅱ.7 摩擦纳米发电机：等效电路理论
Triboelectric Nanogenerators—Equivalent Circuit Theory

1. Theory of Sliding-Mode Triboelectric Nanogenerators

Simiao Niu, Ying Liu, Sihong Wang, Long Lin, Yu Sheng Zhou, Youfan Hu *and* Zhong Lin Wang*

Advanced Materials, 25 (2013) 6184-6193.

简介：本文首次发展了基于等效电路模型的滑动式摩擦纳米发电机的基本模拟计算方法。

ABSTRACT The triboelectric nanogenerator (TENG) is a powerful approach toward new energy technology, especially for portable electronics. A theoretical model for the sliding-mode TENG is presented in this work. The finite element method was utilized to characterize the distributions of electric potential, electric field, and charges on the metal electrodes of the TENG. Based on the FEM calculation, the semi-analytical results from the interpolation method and the analytical *V-Q-x* relationship are built to study the sliding-mode TENG. The analytical *V-Q-x* equation is validated through comparison with the semi-analytical results. Furthermore, based on the analytical *V-Q-x* equation, dynamic output performance of sliding-mode TENG is calculated with arbitrary load resistance, and good agreement with experimental data is achieved. The theory presented here is a milestone work for in-depth understanding of the working mechanism of the sliding-mode TENG, and provides a theoretical basis for further enhancement of the sliding-mode TENG for both energy scavenging and self-powered sensor applications.

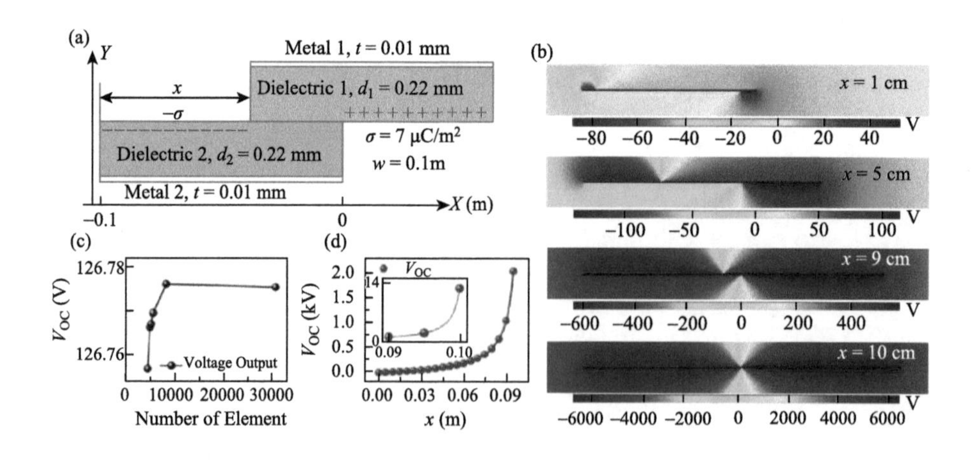

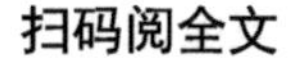

http://www.binn.cas.cn/book/202106/W020210610508075278551.pdf

2. Theoretical Study of Contact-Mode Triboelectric Nanogenerators as an Effective Power Source

Simiao Niu, Sihong Wang, Long Lin, Ying Liu, Yu Sheng Zhou, Youfan Hu *and* **Zhong Lin Wang***

Energy & Environmental Science, 6 (2013) 3576-3583.

简介：本文首次发展了基于等效电路模型的接触分离式摩擦纳米发电机的基本模拟计算方法。

ABSTRACT A theoretical model for contact-mode TENGs was constructed in this paper. Based on the theoretical model, its real-time output characteristics and the relationship between the optimum resistance and TENG parameters were derived. The theory presented here is the first in-depth interpretation of the contact-mode TENG, which can serve as important guidance for rational design of the TENG structure in specific applications.

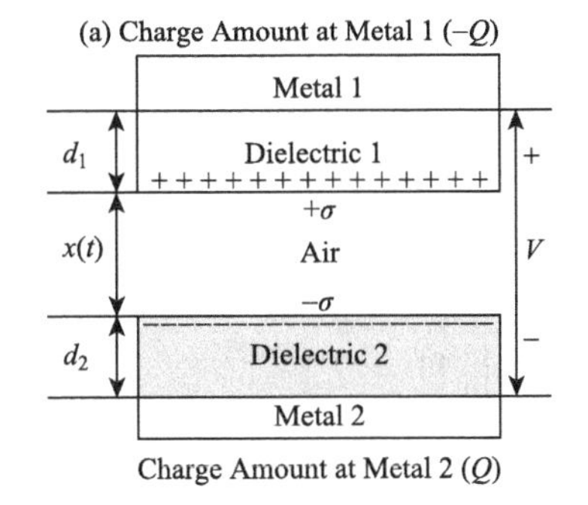

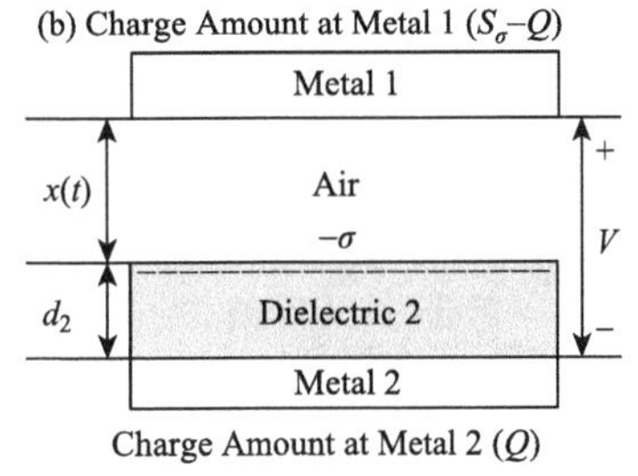

扫码阅全文

http://www.binn.cas.cn/book/202106/W020210610508075813412.pdf

3. A Theoretical Study of Grating Structured Triboelectric Nanogenerators

Simiao Niu, Sihong Wang, Ying Liu, Yu Sheng Zhou, Long Lin, Youfan Hu, Ken C. Pradel *and* **Zhong Lin Wang***

Energy & Environmental Science, 7 (2014) 2339-2349.

简介：本文首次发展了基于等效电路模型具有光栅结构的滑动式摩擦纳米发电机的基本模拟计算方法。

ABSTRACT Triboelectric nanogenerator (TENG) technology is an emerging new mechanical energy harvesting technology with numerous advantages. Amongst the multitude of TENG designs, the grating structure is not only the most promising for ultra-high output power but also the most complicated. In this manuscript, the first theoretical model of the grating structured TENG is presented with in-depth interpretation and analysis of its working principle. Two different categories of grating structured TENGs—grating structured TENGs with equal plate-length and with unequal plate-length are discussed in detail to illustrate their difference in output characteristics. Then for each of these two categories, a study of the basic output profiles and an in-depth discussion on the influence of the electrode structure, number of grating units, and thickness of the dielectric layers were made to obtain a strategy for optimization.

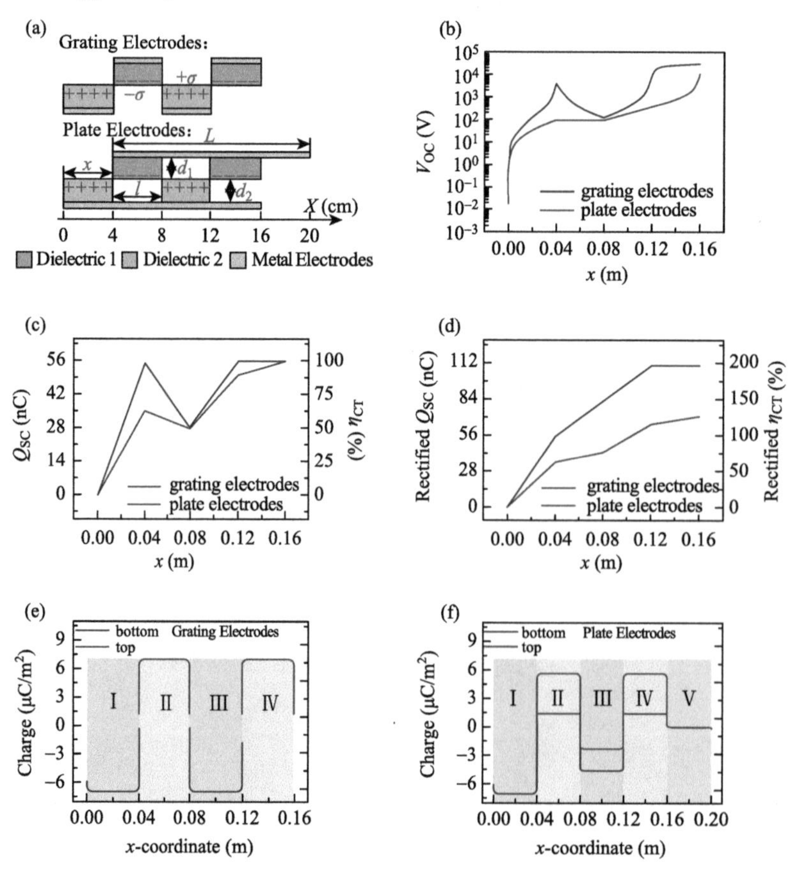

扫码阅全文

http://www.binn.cas.cn/book/202106/W020210610508076073548.pdf

4. Theoretical Investigation and Structural Optimization of Single-Electrode Triboelectric Nanogenerators

Simiao Niu, Ying Liu, Sihong Wang, Long Lin, Yu Sheng Zhou, Youfan Hu *and* **Zhong Lin Wang***

Advanced Functional Materials, 24 (2014) 3332-3340.

简介：本文首次发展了基于等效电路模型的单电极式摩擦纳米发电机的基本模拟计算方法。

ABSTRACT Single-electrode triboelectric nanogenerators (SETENGs) significantly expand the application of triboelectric nanogenerators in various circumstances, such as touch-pad technologies. In this work, a theoretical model of SETENGs is presented with in-depth interpretation and analysis of their working principle. Electrostatic shield effect from the primary electrode is the main consideration in the design of such SETENGs. On the basis of this analysis, the impacts of two important structural parameters, that is, the electrode gap distance and the area size, on the output performance are theoretically investigated. An optimized electrode gap distance and an optimized area size are observed to provide a maximum transit output power. Parallel connection of multiple SETENGs with micro-scale size and relatively larger spacing should be utilized as the scaling-up strategy. The discussion of the basic working principle and the influence of structural parameters on the whole performance of the device can serve as an important guidance for rational design of the device structure towards the optimum output in specific applications.

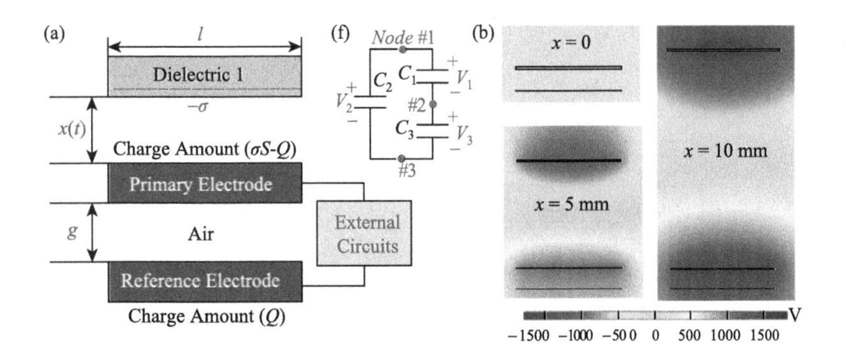

扫码阅全文

http://www.binn.cas.cn/book/202106/W020210610508076372720.pdf

5. Theory of Freestanding Triboelectric-Layer-Based Nanogenerators

Simiao Niu, Ying Liu, Xiangyu Chen, Sihong Wang, Yu Sheng Zhou, Long Lin, Yannan Xie, Zhong Lin Wang*

Nano Energy, 12 (2015) 760-774.

简介：本文首次发展了基于等效电路模型的自由悬浮式摩擦纳米发电机的基本模拟计算方法。

ABSTRACT Triboelectric nanogenerator technology is emerging as a promising candidate for mechanical energy harvesting from ambient environment. Freestanding triboelectric-layer-based nanogenerators (FTENGs) are one of the fundamental operation modes with many advantages. In this paper, the first theoretical model of FTENGs is proposed with thorough analysis of their operation principle. Both contact-mode and sliding-mode FTENGs are discussed to fully uncover their unique characteristics. Contact-mode FTENGs have superior linearity, which is highly beneficial for both energy-harvesting and self-powered sensing applications. Sliding-mode FTENGs have two subcategories based on the material of their freestanding layer. For both of the two subcategories, the coupling effect of the height of the freestanding layer and the electrode gap on their output characteristics are discussed in detail to obtain the strategy for their structural optimization.

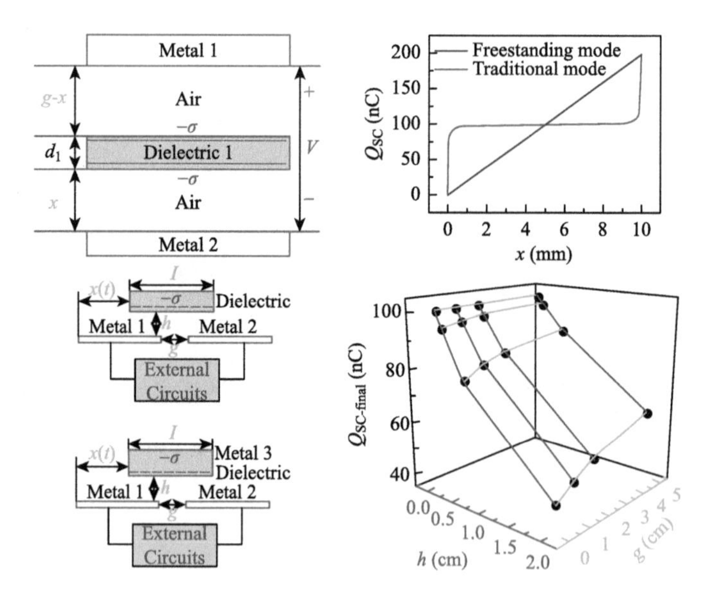

扫码阅全文

http://www.binn.cas.cn/book/202106/W020210610508076870556.pdf

6. Simulation Method for Optimizing the Performance of an Integrated Triboelectric Nanogenerator Energy Harvesting System

Simiao Niu, Yu Sheng Zhou, Sihong Wang, Ying Liu, Long Lin, Yoshio Bando, Zhong Lin Wang*

Nano Energy, 8 (2014)150-156.

简介：本文首次发展了基于等效电路模型的优化摩擦纳米发电机输出功率的基本模拟计算方法。

ABSTRACT Demonstrating integrated triboelectric nanogenerator energy harvesting systems that contain triboelectric nanogenerators, power management circuits, signal processing circuits, energy storage elements, and/or load circuits are core steps for practical applications of triboelectric nanogenerators. Through the design flow of such systems, theoretical simulation plays a critical role. In this manuscript, we provided a new theoretical simulation method for integrated triboelectric nanogenerator systems through integrating the equivalent circuit model of triboelectric nanogenerators into SPICE software. This new simulation method was validated by comparing its results with analytical solutions in some specific triboelectric nanogenerator systems. Finally, we employed this new simulator to analyze the performance of an integrated triboelectric nanogenerator system with a power management circuit. From the study of the influence of different circuit parameters, we outline the design strategy for such kind of triboelectric nanogenerator systems.

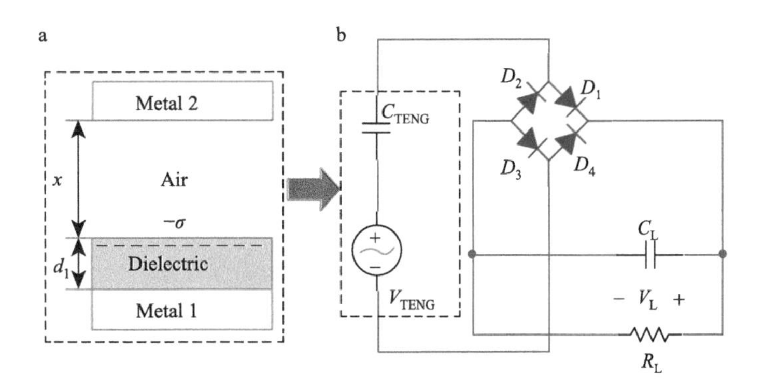

扫码阅全文

http://www.binn.cas.cn/book/202106/W020210610508077919965.pdf

7. Theoretical Study of Rotary Freestanding Triboelectric Nanogenerators

Tao Jiang, Xiangyu Chen, Chang Bao Han, Wei Tang *and* Zhong Lin Wang*

Advanced Functional Materials, 25 (2015) 2928-2938.

简介：本文首次发展了基于等效电路模型的旋转滑动式摩擦纳米发电机的基本模拟计算方法。

ABSTRACT The use of triboelectric nanogenerators (TENG) is a highly effective technology for harvesting ambient mechanical energy to produce electricity. In this work, a theoretical model of rotary freestanding TENG with grating structure is constructed, including the conductor-to-dielectric and dielectric-to-dielectric categories. The finite element simulations are performed to capture the fundamental physics of rotary freestanding TENG. Based on the simulations and derivations of approximate analytical equations, the real-time output characteristics of TENG with arbitrary load resistance are calculated. Furthermore, the influences of structural parameters of TENG and rotation rate on the output performance are investigated. The theory presented here facilitates a deep understanding of the working mechanism of rotary freestanding TENG and provides useful guidance for designing high performance TENG for energy harvesting applications.

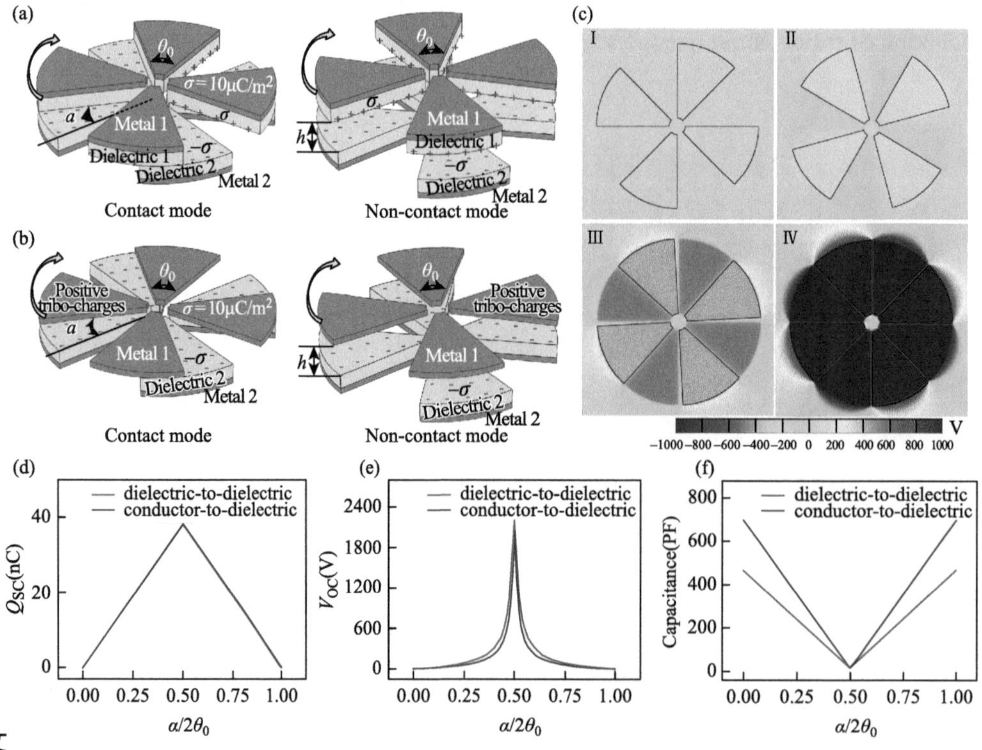

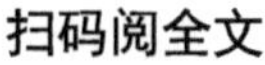
扫码阅全文

http://www.binn.cas.cn/book/202106/W020210610508078609616.pdf

8. Theoretical Systems of Triboelectric Nanogenerators

Simiao Niu, Zhong Lin Wang*

Nano Energy, 14 (2015) 161-192.

简介：首篇总结基于等效电路模型的摩擦纳米发电机的基本模拟计算方法的综述文章。

ABSTRACT Triboelectric nanogenerator (TENG) technology based on contact electrification and electrostatic induction is an emerging new mechanical energy harvesting technology with numerous advantages. The current area power density of TENGs has reached 313W/m^2 and their volume energy density has reached 490 kW/m^3. In this review, we systematically analyzed the theoretical system of triboelectric nanogenerators. Starting from the physics of TENGs, we thoroughly discussed their fundamental working principle and simulation method. Then the intrinsic output characteristics, load characteristics, and optimization strategy is in-depth discussed. TENGs have inherent capacitive behavior and their governing equation is their *V-Q-x* relationship. There are two capacitance formed between the tribo-charged dielectric surface and the two metal electrodes, respectively. The ratio of these two capacitances changes with the position of this dielectric surface, inducing electrons to transfer between the metal electrodes under short circuit conditions. This is the core working mechanism of triboelectric generators and different TENG fundamental modes can be classified based on the changing behavior of these two capacitances. Their first-order lumped-parameter equivalent circuit model is a voltage source in series with a capacitor. Their resistive load characteristics have a "three-working-region" behavior because of the impedance match mechanism. Besides, when TENGs are utilized to charge a capacitor with a bridge rectifier in multiple motion cycles, it is equivalent to utilizing a constant DC voltage source with an internal resistance to charge. The optimization techniques for all TENG fundamental modes are also discussed in detail. The theoretical system reviewed in this work provides a theoretical basis of TENGs and can be utilized as a guideline for TENG designers to continue improving TENG output performance.

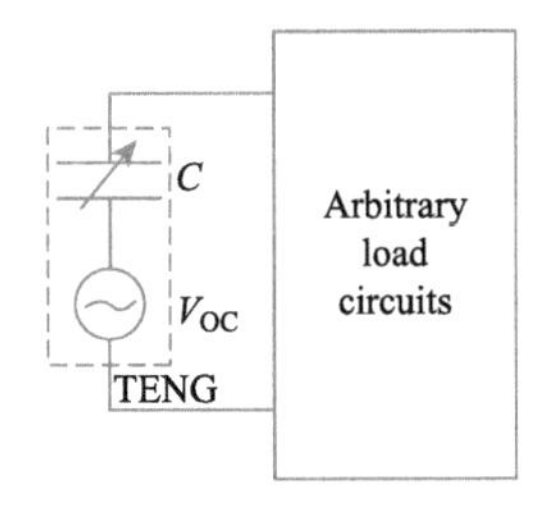

扫码阅全文

http://www.binn.cas.cn/book/202106/W020210610508079109832.pdf

II.8 摩擦纳米发电机：麦克斯韦位移电流理论
Triboelectric Nanogenerators—Maxwell Displacement Current Theory

1. On Maxwell's Displacement Current for Energy and Sensors: The Origin of Nanogenerators

Zhong Lin Wang

Materials Today, 20 (2017) 74-82.

简介：本文首次提出了驱动纳米发电机的是麦克斯韦位移电流，并提出了基本的理论构架，建立了纳米发电机的理论基础，是纳米发电机理论的开山之作。

ABSTRACT Self-powered system is a system that can sustainably operate without an external power supply for sensing, detection, data processing and data transmission. Nanogenerators were first developed for self-powered systems based on piezoelectric effect and triboelectrification effect for converting tiny mechanical energy into electricity, which have applications in internet of things, environmental/infrastructural monitoring, medical science and security. In this paper, we present the fundamental theory of the nanogenerators starting from the Maxwell equations. In the Maxwell's displacement current, the first term $\varepsilon_0 \frac{\partial \boldsymbol{E}}{\partial t}$ gives the birth of electromagnetic wave, which is the foundation of wireless communication, radar and later the information technology. Our study indicates that the second term $\frac{\partial \boldsymbol{P}}{\partial t}$ in the Maxwell's displacement current is directly related to the output electric current of the nanogenerator, meaning that our nanogenerators are the applications of Maxwell's displacement current in energy and sensors. By contrast, electromagnetic generators are built based on Lorentz force driven flow of free electrons in a conductor. This study presents the similarity and differences between piezoelectric nanogenerator and triboelectric nanogenerator, as well as the classical electromagnetic generator, so that the impact and uniqueness of the nanogenerators can be clearly understood. We also present the three major applications of nanogenerators as micro/nano-power source, self-powered sensors and blue energy.

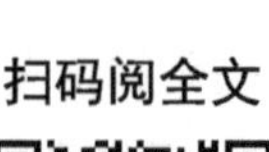

http://www.binn.cas.cn/book/202106/W020210610509677903386.pdf

2. On the First Principle Theory of Nanogenerators from Maxwell's Equations

Zhong Lin Wang

Nano Energy, 68 (2020) 104272

简介：本文首次全面和系统地从麦克斯韦方程组出发，在位移矢量中引入了由于表面电荷的存在，介质形状在外力作用下发生变化而产生的极化项 $\boldsymbol{P}_s$，拓展了麦克斯韦方程组在能源与传感方面的应用，并从理论上给出了计算纳米发电机的输出电流、电压和功率的解析表达式。该文是纳米发电机的理论基石和该领域的开山之作。

ABSTRACT Nanogenerators (NGs) are a field that uses Maxwell's displacement current as the driving force for effectively converting mechanical energy into electric power/signal, which have broad applications in energy science, environmental protection, wearable electronics, self-powered sensors, medical science, robotics and artificial intelligence. NGs are usually based on three effects: piezoelectricity, triboelectricity (contact electrification), and pyroelectricity. In this paper, a formal theory for NGs is presented starting from Maxwell's equations. Besides the general expression for displacement vector $\boldsymbol{D}=\varepsilon\boldsymbol{E}$ used for deriving classical electromagnetic dynamics, we added an additional term $\boldsymbol{P}_s$ in $\boldsymbol{D}$, which represents the polarization created by the electrostatic surface charges owing to piezoelectricity and/or triboelectricity as a result of mechanical triggering in NG. In contrast to $\boldsymbol{P}$ that is resulted from the electric field induced medium polarization and vanishes if $\boldsymbol{E}=0$, $\boldsymbol{P}_s$ remains even when there is no external electric field. We reformulated the Maxwell equations that include both the medium polarizations due to electric field ($\boldsymbol{P}$) and non-electric field (such as strain) ($\boldsymbol{P}_s$) induced polarization terms, from which, the output power, electromagnetic behavior and current transport equation for a NG are systematically derived. A general solution is presented for the modified Maxwell equations, and analytical solutions about the output potential are provided for a few cases. The displacement current arising from $\varepsilon\partial\boldsymbol{E}/\partial t$ is responsible for the electromagnetic waves, while the newly added term $\partial\boldsymbol{P}_s/\partial t$ is the application of Maxwell's equations in energy and sensors. This work sets the first principle theory for quantifying the performance and electromagnetic behavior of a nanogenerator in general.

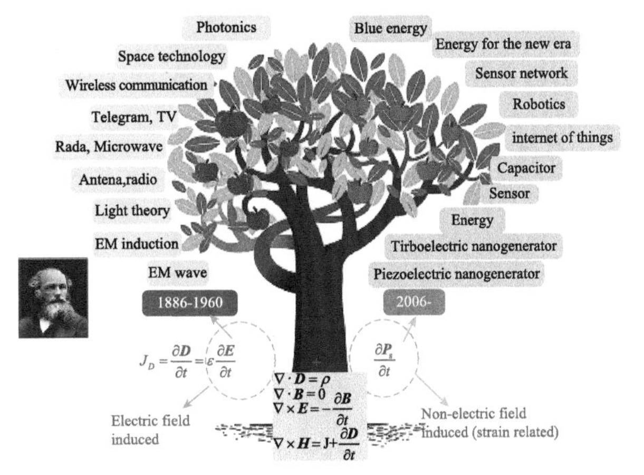

扫码阅全文

http://www.binn.cas.cn/book/202106/W020210610509678921003.pdf

3. Quantifying the Power Output and Structural Figure-of-Merits of Triboelectric Nanogenerators in a Charging System Starting from the Maxwell's Displacement Current

Jiajia Shao, Morten Willatzen, Tao Jiang, Wei Tang, Xiangyu Chen, Jie Wang, Zhong Lin Wang*

Nano Energy, 59 (2019) 380-389.

简介：本文从拓展的麦克斯韦方程组出发，模拟和计算了摩擦纳米发电机的输出特性。

ABSTRACT Conversion of mechanical energy into electricity using triboelectric nanogenerators (TENGs) is a rapidly expanding research area. Although the theoretical origin of TENGs has been proven using the Maxwell's displacement current (ID), a profound quantitative understanding of its generation is not available. Moreover, a comprehensive analysis of the fundamental charging behavior of TENGs and building a standard to evaluate each TENG's unique charging characteristic are critical to ensure efficient use of them in practice. We present a thorough analysis of TENG's charging behavior through which a more complete evaluation of TENG charging is proposed by introducing the structural figure of merit (FOMCs) in a charging system (powering capacitors). The analysis is based on Maxwell's displacement current and results are verified experimentally. To achieve this, according to the distance-dependent electric field model, we provide a systematic discussion on the generation of ID in TENGs, along with the derived analytical formula and numerical calculations. This work suggests a new way to deeply understand the nature of the ID generated within the TENGs; and the modified FOMCS can be used to predict the charging characteristics of TENGs in an energy storage system, allowing us to utilize the TENGs more efficiently towards different applications.

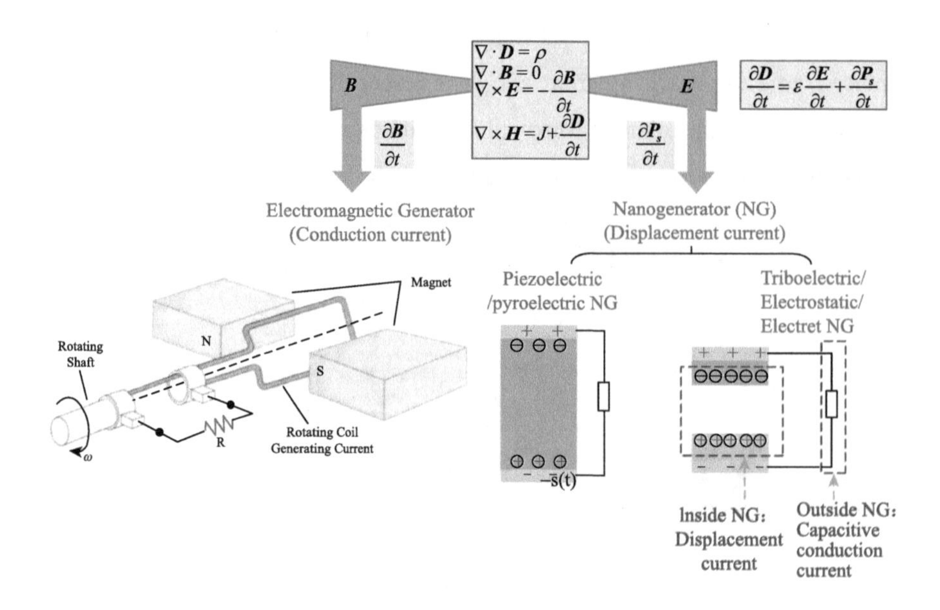

扫码阅全文

http://www.binn.cas.cn/book/202106/W020210610509679354505.pdf

4. Three-Dimensional Modeling of Alternating Current Triboelectric Nanogenerator in the Linear Sliding Mode

Jiajia Shao, Di Liu, Morten Willatzen* *and* **Zhong Lin Wang***

Applied Physics Reviews, 7 (2020) 011405.

简介：本文从麦克斯韦位移电流出发，模拟和计算了摩擦纳米发电机在滑动模式的输出特性。

ABSTRACT Energy harvesting from mechanical motions has immense applications such as self-powered sensors and renewable energy sources powered by ocean waves. In this context, the triboelectric nanogenerator is the cutting-edge technology which can effectively convert ambient mechanical energy into electricity through the Maxwell's displacement current. While further improvements of the energy conversion efficiency of triboelectric nanogenerators critically depend on theoretical modeling of the energy conversion process, to date only models based on single-relative-motion processes have been explored. Here, we analyze energy harvesting of triboelectric nanogenerators using a three-dimensional model in a linear-sliding mode and demonstrate a design of triboelectric nanogenerators that have a 77.5% enhancement in the average power in comparison with previous approaches. Moreover, our model shows the existence of a DC-like bias voltage contained in the basic AC output from the energy conversion, which makes the triboelectric nanogenerators an energy source more pliable than the traditional AC power generation systems. The present work provides a framework for systematic modeling of triboelectric nanogenerators and reveals the importance of obtaining direct analytical insight in understanding the current output characteristics of the triboelectric nanogenerators. Incorporating our model analysis in future designs of triboelectric nanogenerators is beneficial for increasing the energy conversion power and may provide insights that can be used in engineering the profile of the output current of the nanogenerators.

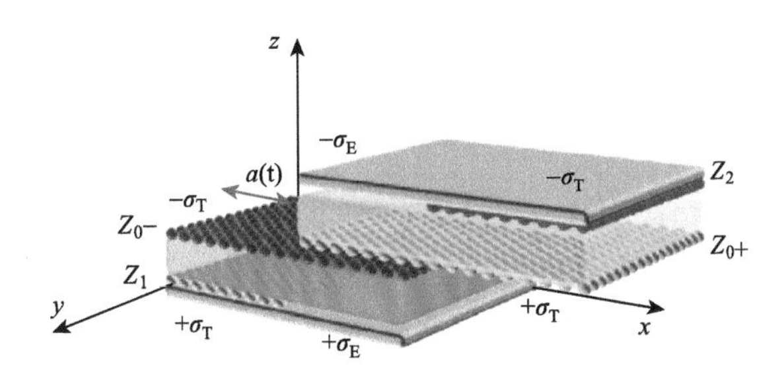

扫码阅全文

http://www.binn.cas.cn/book/202106/W020210610509680774827.pdf

5. Theoretical Modelling of Triboelectric Nanogenerators (TENGs)

Jiajia Shao, Morten Willatzen* *and* Zhong Lin Wang*

Journal of Applied Physics, 128 (2020) 111101.

简介：本文是首篇从麦克斯韦位移电流出发，综述了计算摩擦纳米发电机输出特性的文献，是目前最标准和全面的一个理论总结。

ABSTRACT Triboelectric nanogenerators (TENGs), using Maxwell's displacement current as the driving force, can effectively convert mechanical energy into electricity. In this work, an extensive review of theoretical models of TENGs is presented. Based on Maxwell's equations, a formal physical model is established referred to as the quasi-electrostatic model of a TENG. Since a TENG is electrically neutral at any time owing to the low operation frequency, it is conveniently regarded as a lumped circuit element. Then, using the lumped parameter equivalent circuit theory, the conventional capacitive model and Norton's equivalent circuit model are derived. Optimal conditions for power, voltage, and total energy conversion efficiency can be calculated. The presented TENG models provide an effective theoretical foundation for understanding and predicting the performance of TENGs for practical applications.

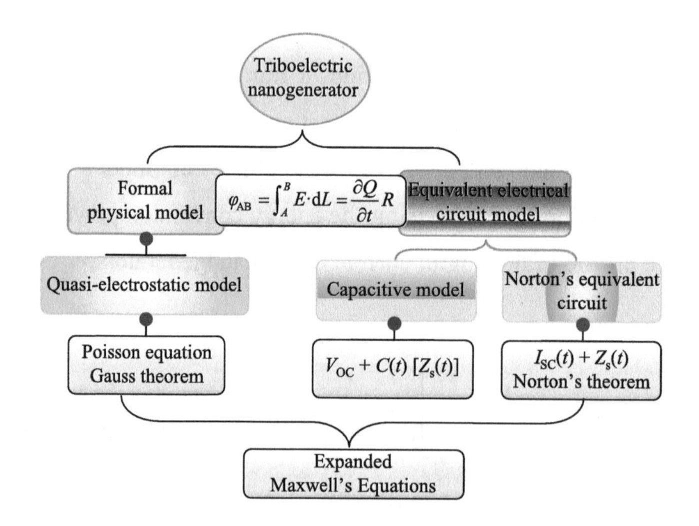

扫码阅全文

http://www.binn.cas.cn/book/202106/W020210610509681826706.pdf

6. Designing Rules and Optimization of Triboelectric Nanogenerator Arrays

Jiajia Shao, Yi Yang, Ou Yang, Jie Wang, Morten Willatzen* *and* **Zhong Lin Wang***

Advanced Energy Materials, 11 (2021) 2100065.

简介：对于接触分离式摩擦纳米发电机，考虑到内场的屏蔽作用和边缘效应，TENG 的输出功率不是和板面积成正比，相反，对于一个无穷大板的 TENG，其输出应该为零。然而对于一个极板太小的 TENG，由于漏电场的作用以及板面所带电荷有限，其输出功率也必然很小。因此，极板尺寸的选择很重要。本文首次探索了阵列型 TENG 的几何参数，给出了什么样的长宽比、TENG 间隔比将给出最大有效单位面积输出功率。这于实际应用有重要指导价值。

ABSTRACT A triboelectric nanogenerator (TENG) is an effective means for the conversion of mechanical energy into electricity. Although Maxwell's displacement current is the underline mechanism of TENGs, the spatial and temporal variations of the electric field and electric displacement remain elusive, which prohibits an effective optimization of the energy conversion process. Here, the electric field distribution and energy dynamics of TENGs is determined using 3D mathematical modeling. The electrical energies stored in TENGs and extracted into the external circuit are calculated quantitatively whereby the ratio of the two, defined as the output efficiency, is obtained. Then, the power density and energy density of TENGs are defined. Utilizing the principle of virtual work, the minimum required external force-time relationship is evaluated. The influence of device parameters, geometry, and optimum conditions are discussed systematically so as to determine general optimization guidelines for TENGs. In addition, although the fringing electric field is practically inevitable, adjustments of the gap distance between neighboring TENG devices to assure that an optimized fringing electric field is "leaked", is demonstrated to lead to improvement in a TENG array. Then, for the first time, this work presents universal design rules and holistic optimization strategies for the network structure of TENGs.

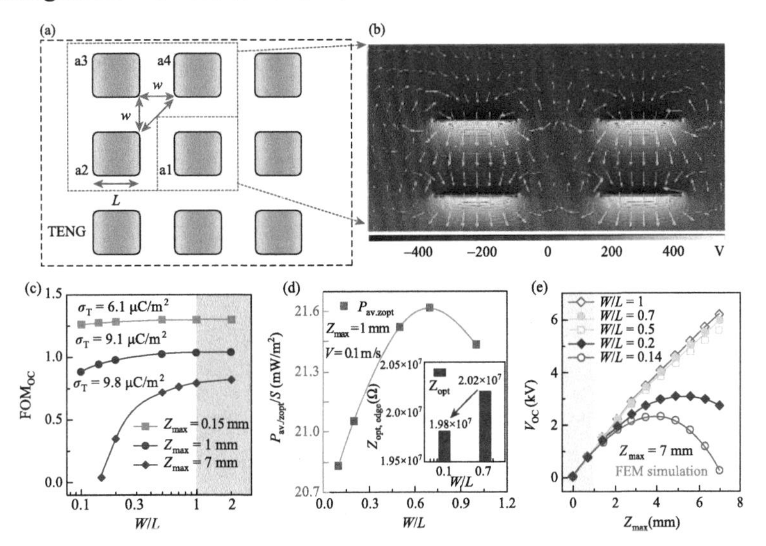

扫码阅全文

http://www.binn.cas.cn/book/202106/W020210610509683126177.pdf

7. On the Expanded Maxwell's Equations for Moving Charged Media System—General Theory, Mathematical Solutions and Applications in TENG

Zhong Lin Wang*

Materials Today
(in press)

简介：我们熟悉的微分型麦克斯韦方程组是在假设介质的体积和表面形状及其分布不变化的情况下而推导出的。本文从积分式的麦克斯韦方程组出发，推导出了有运动介质情况下的扩展型微分式麦克斯韦方程组，并首次给出了几种求解这些方程组的数学方法和理论，同时也把这些方程组应用到摩擦纳米发电机的基础理论中。扩展型的麦克斯韦方程组对计算飞行物体（例如导弹、战机、天体等）的电磁波反射与物质相互作用提供了完整的理论构架，有可能对实际测量提出相关的修正，具有广阔的应用领域。

ABSTRACT The conventional Maxwell's equations are for media whose boundaries and volumes are fixed. But for cases that involve moving media and time-dependent configuration, the equations have to be expanded. Here, starting from the integral form of the Maxwell's equations for general cases, we first derived the expanded Maxwell's equations in differential form by assuming that the medium is moving as a rigid translation object. Secondly, the expanded Maxwell's equations are further developed with including the polarization density term $\boldsymbol{P}_s$ in displacement vector owing to electrostatic charges on medium surfaces as produced by effect such as triboelectrification, based on which the first principle theory for the triboelectric nanogenerators (TENGs) is developed. The expanded equations are the most comprehensive governing equations including both electromagnetic interaction and power generation as well as their coupling. Thirdly, general approaches are presented for solving the expanded Maxwell's equations using vector and scalar potentials as well as perturbation theory, so that the scheme for numerical calculations is set. Finally, we investigated the conservation of energy as governed by the expanded Maxwell's equations, and derived the general approach for calculating the displacement current $\frac{\partial}{\partial t}\boldsymbol{P}_s$ for the output power of TENGs. The current theory is general and it may impact the electromagnetic scattering and reflection behavior of flight jets, missiles, comet, and even galaxy stars observed from earth.

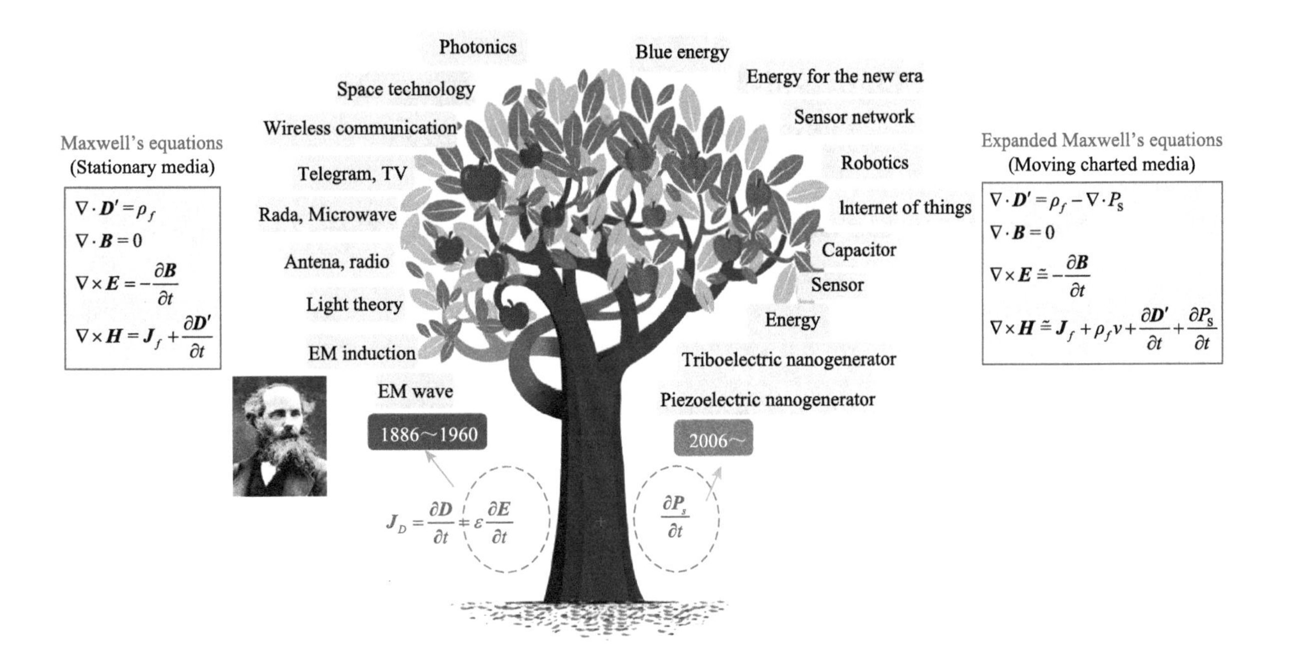

扫码阅全文

http://www.binn.cas.cn/book/202106/W020211209356566239311.pdf

Ⅱ.9　摩擦纳米发电机：技术标准
Triboelectric Nanogenerators—Technical Standards

1. Standards and Figure-of-Merits for Quantifying the Performance of Triboelectric Nanogenerators

Yunlong Zi, Simiao Niu, Jie Wang, Zhen Wen, Wei Tang *and* Zhong Lin Wang*

Nature Communications, 6 (2015) 8376.

简介：本文首次提出了摩擦纳米发电机输出功率的评判标准：figure of merits，包括表面起电电荷密度和TENG结构因子等，奠定了定标摩擦纳米发电机的国际标准。

ABSTRACT　Triboelectric nanogenerators have been invented as a highly efficient, cost-effective and easy scalable energy-harvesting technology for converting ambient mechanical energy into electricity. Four basic working modes have been demonstrated, each of which has different designs to accommodate the corresponding mechanical triggering conditions. A common standard is thus required to quantify the performance of the triboelectric nanogenerators so that their outputs can be compared and evaluated. Here we report figure-of-merits for defining the performance of a triboelectric nanogenerator, which is composed of a structural figure-of-merit related to the structure and a material figure of merit that is the square of the surface charge density. The structural figure-of-merit is derived and simulated to compare the triboelectric nanogenerators with different configurations. A standard method is introduced to quantify the material figure-of-merit for a general surface. This study is likely to establish the standards for developing TENGs towards practical applications and industrialization.

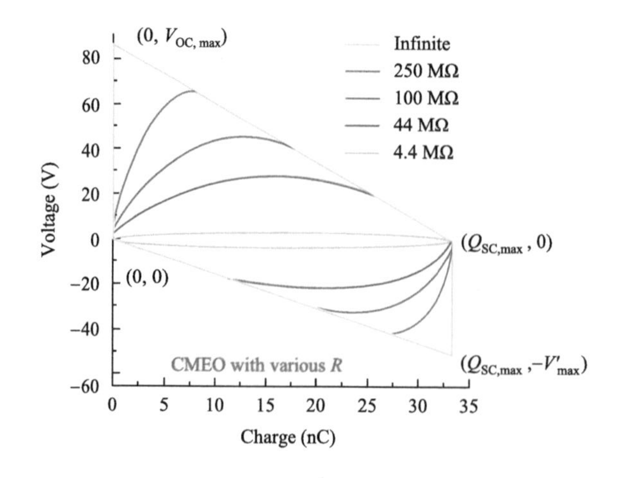

扫码阅全文

http://www.binn.cas.cn/book/202106/W020210610511267085980.pdf

2. Effective Energy Storage from a Triboelectric Nanogenerator

Yunlong Zi, Jie Wang, Sihong Wang, Shengming Li, Zhen Wen, Hengyu Guo, Zhong Lin Wang*

Nature Communications, 7 (2016) 10987.

简介：本文首次探索了如何高效存储摩擦纳米发电机的输出能量。

ABSTRACT To sustainably power electronics by harvesting mechanical energy using nanogenerators, energy storage is essential to supply a regulated and stable electric output, which is traditionally realized by a direct connection between the two components through a rectifier. However, this may lead to low energy-storage efficiency. Here, we rationally design a charging cycle to maximize energy-storage efficiency by modulating the charge flow in the system, which is demonstrated on a triboelectric nanogenerator by adding a motion-triggered switch. Both theoretical and experimental comparisons show that the designed charging cycle can enhance the charging rate, improve the maximum energy-storage efficiency by up to 50% and promote the saturation voltage by at least a factor of two. This represents a progress to effectively store the energy harvested by nanogenerators with the aim to utilize ambient mechanical energy to drive portable/wearable/implantable electronics.

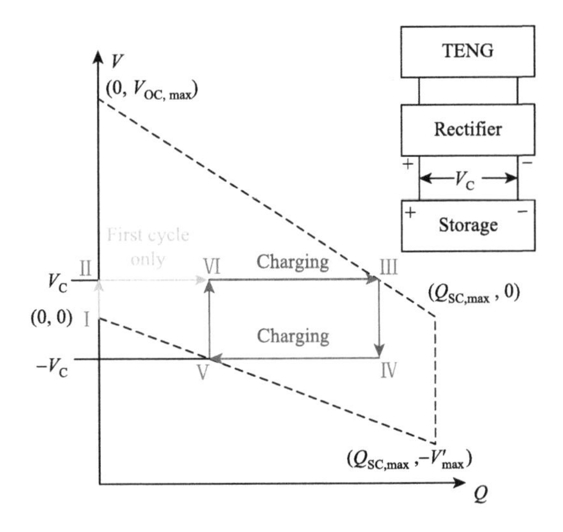

扫码阅全文

http://www.binn.cas.cn/book/202106/W020210610511267304378.pdf

3. Quantifying the Triboelectric Series

Haiyang Zou, Ying Zhang, Litong Guo, Peihong Wang, Xu He, Guozhang Dai, Haiwu Zheng, Chaoyu Chen, Aurelia Chi Wang, Cheng Xu *and* Zhong Lin Wang*

Nature Communications, 10 (2019) 1427.

简介：本文首次提出了如何测试薄膜材料表面接触起电的电荷密度的方法与设备标准，并首次给出了55种高分子材料的表面电荷密度的测试结果。从此，摩擦起电可以定标化、比较化。

ABSTRACT Triboelectrification is a well-known phenomenon that commonly occurs in nature and in our lives at any time and any place. Although each and every material exhibits triboelectrification, its quantification has not been standardized. A triboelectric series has been qualitatively ranked with regards to triboelectric polarization. Here, we introduce a universal standard method to quantify the triboelectric series for a wide range of polymers, establishing quantitative triboelectrification as a fundamental materials property. By measuring the tested materials with a liquid metal in an environment under well-defined conditions, the proposed method standardizes the experimental set up for uniformly quantifying the surface triboelectrification of general materials. The normalized triboelectric charge density is derived to reveal the intrinsic character of polymers for gaining or losing electrons. This quantitative triboelectric series may serve as a textbook standard for implementing the application of triboelectrification for energy harvesting and self-powered sensing.

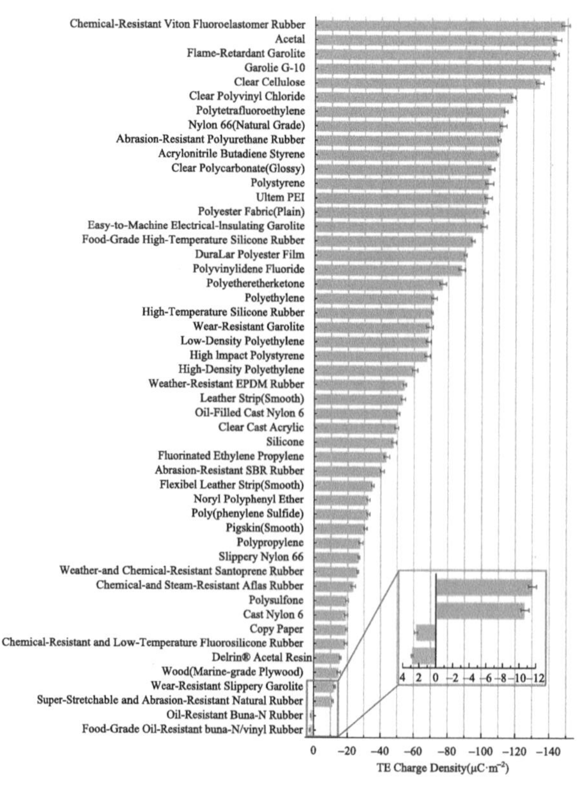

扫码阅全文

http://www.binn.cas.cn/book/202106/W020210610511267575746.pdf

4. Quantifying and Understanding the Triboelectric Series of Inorganic Non-Metallic Materials

Haiyang Zou, Litong Guo, Hao Xue, Ying Zhang, Xiaofang Shen, Xiaoting Liu, Peihong Wang, Xu He, Guozhang Dai, Peng Jiang, Haiwu Zheng, Binbin Zhang, Cheng Xu *and* Zhong Lin Wang*

Nature Communications,11 (2020) 2093.

简介：本文首次给出了35种功能氧化物薄膜材料的表面电荷密度的测试结果。

ABSTRACT Contact-electrification is a universal effect for all existing materials, but it still lacks a quantitative materials database to systematically understand its scientific mechanisms. Using an established measurement method, this study quantifies the triboelectric charge densities of nearly 30 inorganic nonmetallic materials. From the matrix of their triboelectric charge densities and band structures, it is found that the triboelectric output is strongly related to the work functions of the materials. Our study verifies that contact-electrification is an electronic quantum transition effect under ambient conditions. The basic driving force for contact-electrification is that electrons seek to fill the lowest available states once two materials are forced to reach atomically close distance so that electron transitions are possible through strongly overlapping electron wave functions. We hope that the quantified series could serve as a textbook standard and a fundamental database for scientific research, practical manufacturing, and engineering.

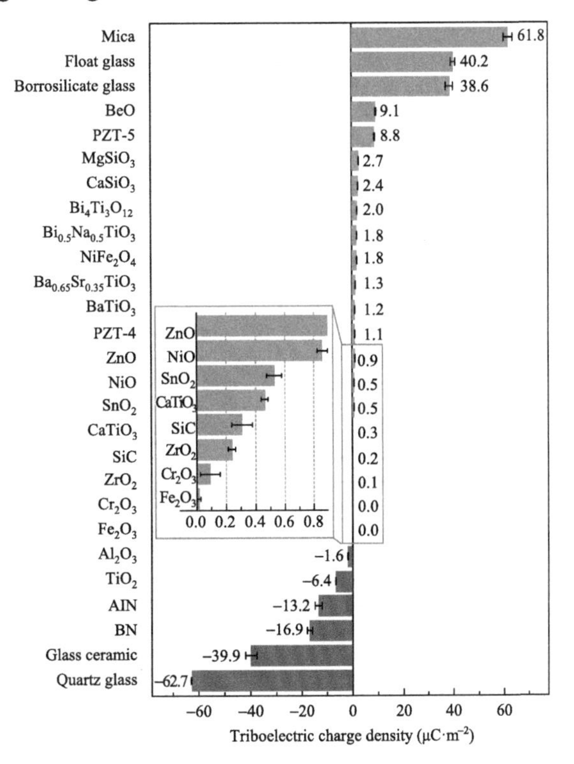

扫码阅全文

http://www.binn.cas.cn/book/202106/W020210610511268209558.pdf

II.10 摩擦纳米发电机：功率优化与管理
Triboelectric Nanogenerators—Power Maximization and Management

1. Pulsed Nanogenerator with Huge Instantaneous Output Power Density

Gang Cheng, Zong-Hong Lin, Long Lin, Zu-liang Du *and* Zhong Lin Wang*

ACS Nano, 7 (2013) 7383-7391.

简介：由于摩擦纳米发电机具有输出电压高、内阻抗大和电流小的特性，进行输出功率的管理是有效利用所产生的能量的重要方法。本文首次探索如何解决这个问题。

ABSTRACT A nanogenerator (NG) usually gives a high output voltage but low output current, so that the output power is low. In this paper, we developed a general approach that gives a hugely improved instantaneous output power of the NG, while the entire output energy stays the same. Our design is based on an off-on-off contact based switching during mechanical triggering that largely reduces the duration of the charging/discharge process, so that the instantaneous output current pulse is hugely improved without sacrificing the output voltage. For a vertical contact-separation mode triboelectric NG (TENG), the instantaneous output current and power peak can reach as high as 0.53 A and 142 W at a load of 500 Ω, respectively. The corresponding instantaneous output current and power density peak even approach 1325 A/m^2 and 3.6×10^5 W/m^2, which are more than 2500 and 1100 times higher than the previous records of TENG, respectively. For the rotation disk based TENG in the lateral sliding mode, the instantaneous output current and power density of 104 A/m^2 and 1.4×10^4 W/m^2 have been demonstrated at a frequency of 106.7 Hz. The approach presented here applies to both a piezoelectric NG and a triboelectric NG, and it is a major advance toward practical applications of a NG as a high pulsed power source.

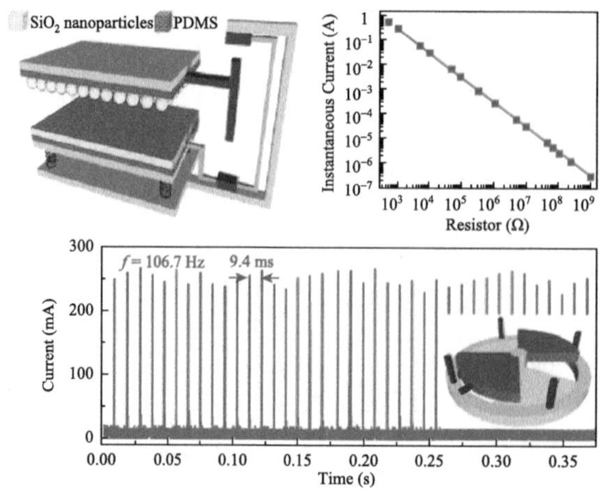

扫码阅全文

http://www.binn.cas.cn/book/202106/W020210610512733987169.pdf

2. Universal Power Management Strategy for Triboelectric Nanogenerator

Fengben Xi, Yaokun Pang, Wei Li, Tao Jiang, Limin Zhang, Tong Guo, Guoxu Liu, Chi Zhang*, Zhong Lin Wang*

Nano Energy, 37 (2017) 168-176.

简介：本文提出了管理摩擦纳米发电机输出功率的普适方法。

ABSTRACT Effective power management has always been the difficulty and bottleneck for practicability of triboelectric nanogenerator (TENG). Here we propose a universal power management strategy for TENG by maximizing energy transfer, direct current (DC) buck conversion, and self-management mechanism. With the implemented power management module (PMM), about 85% energy can be autonomously released from the TENG and output as a steady and continuous DC voltage on the load resistance. The DC component and ripple have been systematically investigated with different circuit parameters. At a low frequency of 1 Hz with the PMM, the matched impedance of the TENG has been converted from 35 MΩ to 1 MΩ at 80% efficiency, and the stored energy has been dramatically improved in charging a capacitor. The universality of this strategy has been greatly demonstrated by various TENGs with the PMM for harvesting human kinetic and environmental mechanical energy. The universal power management strategy for TENG is promising for a complete micro-energy solution in powering wearable electronics and industrial wireless networks. With the new coupling mode of triboelectricity and semiconductor in the PMM, the tribotronics has been extended and a new branch of power-tribotronics is proposed for manageable triboelectric power by electronics.

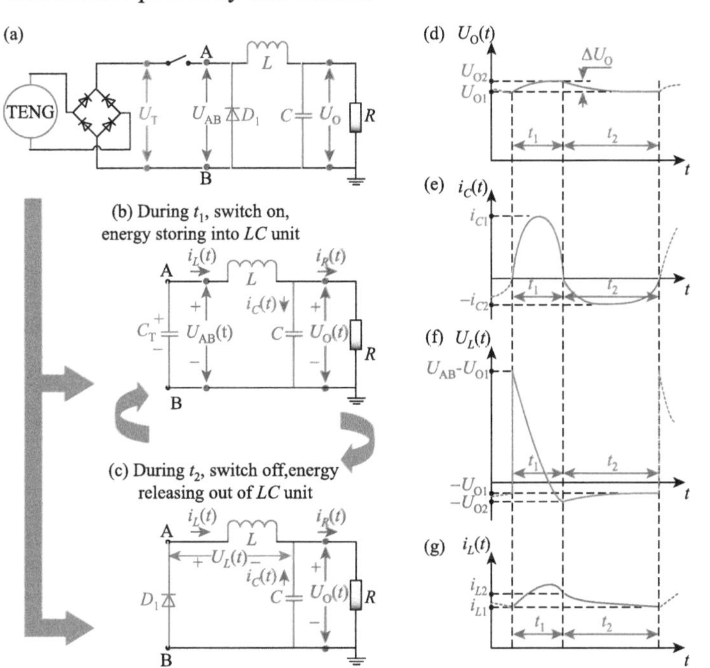

扫码阅全文

http://www.binn.cas.cn/book/202106/W020210610512734261066.pdf

3. Ultrahigh Charge Density Realized by Charge Pumping at Ambient Conditions for Triboelectric Nanogenerators

Liang Xu, Tian Zhao Bu, Xiao Dan Yang, Chi Zhang*, Zhong Lin Wang*

Nano Energy, 49 (2018) 625-633.

简介：摩擦纳米发电机的输出功率正比于表面电荷密度的平方。本文首先提出了如何利用两个摩擦纳米发电机泵电荷的方法来实现表面电荷密度的增加，其中一个充当了泵电荷的作用，另外一个是真正的发电输出，进而提升输出功率密度。

ABSTRACT As a promising technology to harvest mechanical energy from environment, triboelectric nanogenerators (TENGs) impose great importance to further enhance the power density, which is closely related to the charge density on the dielectric surface. A few approaches have been proposed to meet the challenge to improve the charge density, while certain preconditions restrict their applications. Here, a facile and universal method using floating layer structure and charge pump is proposed, based on which an integrated self-charge-pumping TENG device is fabricated. The device adopts a floating layer to accumulate and bind charges for electrostatic induction, while a charge pump is devised to pump charges into the floating layer simultaneously. With elaborately designed structures, this device can achieve ultrahigh effective surface charge density of 1020 μC·m^{-2} in ambient conditions, which is 4 times of that of the density corresponding to air breakdown, presenting a simple and robust strategy to greatly enhance the output of TENGs which should be crucial for developing high performance energy harvesting devices and self-powered systems.

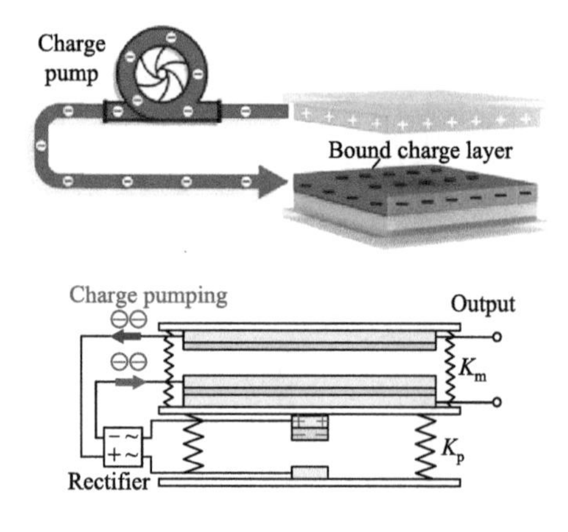

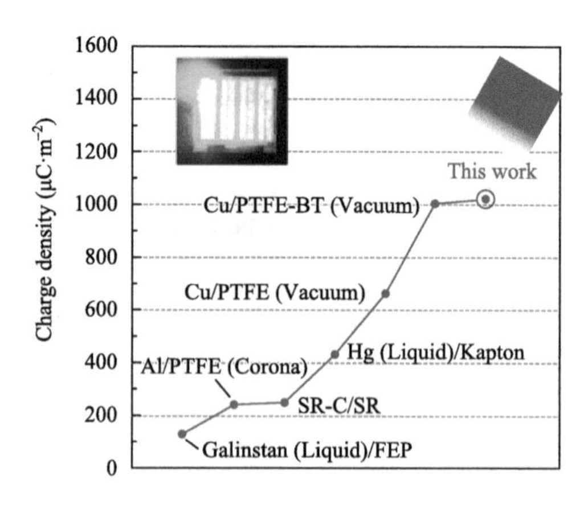

扫码阅全文

http://www.binn.cas.cn/book/202106/W020210610512734749411.pdf

4. Pumping Up the Charge Density of a Triboelectric Nanogenerator by Charge-Shuttling

Huamei Wang, Liang Xu, Yu Bai *and* **Zhong Lin Wang***

Nature Communications, 11 (2020) 4203.

简介：本文首次探索了如何利用两个摩擦纳米发电机相互电荷运载的方法来实现表面电荷密度的提升，大大增强输出功率。

ABSTRACT As an emerging technology for harvesting mechanical energy, low surface charge density greatly hinders the practical applications of triboelectric nanogenerators (TENGs). Here, a high-performance TENG based on charge shuttling is demonstrated. Unlike conventional TENGs with static charges fully constrained on the dielectric surface, the device works based on the shuttling of charges corralled in conduction domains. Driven by the interaction of two quasi-symmetrical domains, shuttling of two mirror charge carriers can be achieved to double the charge output. Based on the mechanism, an ultrahigh projected charge density of 1.85 $mC \cdot m^{-2}$ is obtained in ambient conditions. An integrated device for water wave energy harvesting is also presented, confirming its feasibility for practical applications. The device provides insights into new modes of TENGs using unfixed charges in domains, shedding a new light on high-performance mechanical energy harvesting technology.

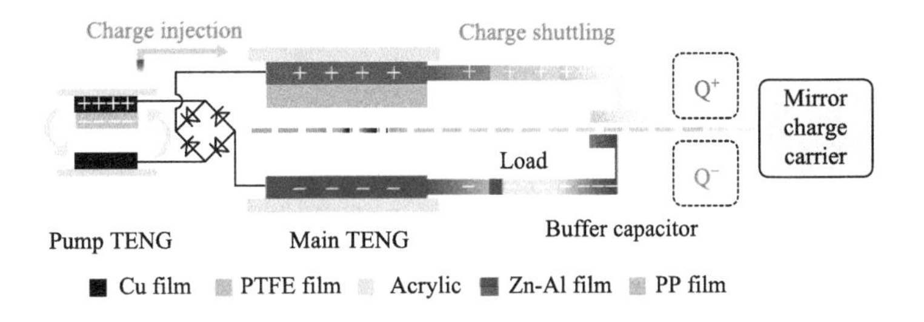

扫码阅全文

http://www.binn.cas.cn/book/202106/W020210610512735417056.pdf

5. Integrated Charge Excitation Triboelectric Nanogenerator

Wenlin Liu, Zhao Wang, Gao Wang, Guanlin Liu, Jie Chen, Xianjie Pu, Yi Xi, Xue Wang, Hengyu Guo*, Chenguo Hu* *and* Zhong Lin Wang*

Natrue Communications, 10 (2019) 1426.

简介：本文首次探索了如何利用一个摩擦纳米发电机进行自泵电荷的方法来实现表面电荷密度的提升，大大增强输出功率。

ABSTRACT Performance of triboelectric nanogenerators is limited by low and unstable charge density on tribo-layers. An external-charge pumping method was recently developed and presents a promising and efficient strategy towards high-output triboelectric nanogenerators. However, integratibility and charge accumulation efficiency of the system is rather low. Inspired by the historical development of electromagnetic generators, here, we propose and realize a self-charge excitation triboelectric nanogenerator system towards high and stable output in analogy to the principle of traditional magnetic excitation generators. By rational design of the voltage-multiplying circuits, the completed external and self-charge excitation modes with stable and tailorable output over 1.25 $mC\cdot m^{-2}$ in contact-separation mode have been realized in ambient condition. The realization of the charge excitation system in this work may provide a promising strategy for achieving high-output triboelectric nanogenerators towards practical applications.

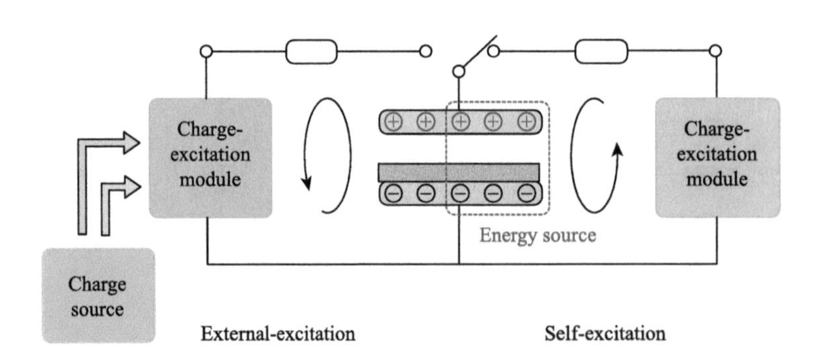

扫码阅全文

http://www.binn.cas.cn/book/202106/W020210610512735984512.pdf

6. Switched-Capacitor-Convertors Based on Fractal Design for Output Power Management of Triboelectric Nanogenerator

Wenlin Liu, Zhao Wang, Gao Wang, Qixuan Zeng, Wencong He, Liyu Liu, Xue Wang , Yi Xi, Hengyu Guo*, Chenguo Hu* and Zhong Lin Wang*

Nature Communications, 11 (2020) 1883.

简介：本文首次探索了如何利用一个摩擦纳米发电机通过开关电容转换器的方法进行自泵电荷，进而实现输出功率的很大增强。脉冲型输出瞬时功率高达 964 W/m^2，能量管理效率高达 94%。

ABSTRACT Owing to the advantages of integration and being magnet-free and light-weight, the switched-capacitor-convertor plays an increasing role compared to traditional transformer in some specific power supply systems. However, the high output impedance and switching loss largely reduces its power efficiency, due to imperfect topology and transistors. Herein, we propose a fractal-design based switched-capacitor-convertors with characteristics including high conversion efficiency, minimum output impedance, and electrostatic voltage applicability. As a double-function output power management system for triboelectric nanogenerators, it delivers over 67 times charge boosting and 954 W·m^{-2} power density in pulse mode, and achieves over 94% total energy transfer efficiency in constant mode. The establishment of the fractal-design switched-capacitor-convertors provides significant guidance for the development of power management toward multi-functional output for numerous applications. The successful demonstration in triboelectric nanogenerators also declares its great potential in electric vehicles, DC micro-grids etc.

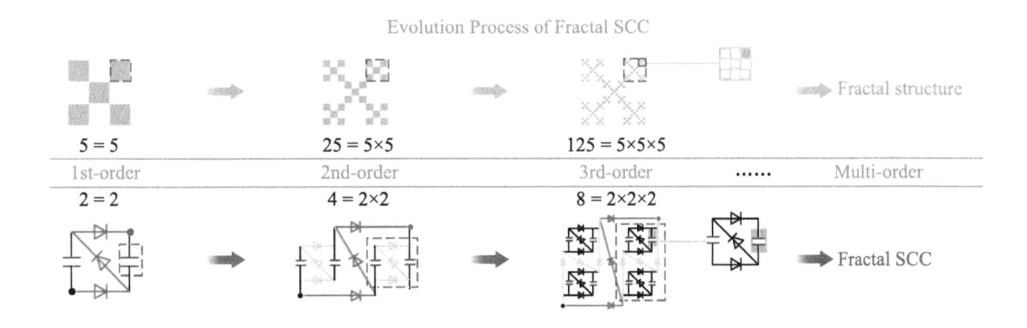

扫码阅全文

http://www.binn.cas.cn/book/202106/W020210610512737159714.pdf

II.11 固体-固体接触起电的机理
The Origins of Solid-Solid Contact-Electrification

1. Manipulating Nanoscale Contact Electrification by an Applied Electric Field

Yu Sheng Zhou, Sihong Wang, Ya Yang, Guang Zhu, Simiao Niu, Zong-Hong Lin, Ying Liu *and* Zhong Lin Wang*

Nano Letters, 14 (2014) 1567-1572.

简介：原子力显微镜是研究接触起电的最重要方法之一。本文首次利用施加到开尔文探针上电压的正负变化而导致金属-绝缘体界面电荷转移符合的反转，证明了金属-绝缘体界面电荷转移是电子转移。

ABSTRACT Contact electrification is about the charge transfer between the surfaces of two materials in a contact-separation process. This effect has been widely utilized in particle separation and energy harvesting, where the charge transfer is preferred to be maximized. However, this effect is always undesirable in some areas such as electronic circuit systems due to the damage from the accumulated electrostatic charges. Herein, we introduced an approach to purposely manipulate the contact electrification process both in polarity and magnitude of the charge transfer through an applied electric field between two materials. Theoretical modeling and the corresponding experiments for controlling the charge transfer between a Pt coated atomic force microscopy tip and Parylene film have been demonstrated. The modulation effect of the electric field on contact electrification is enhanced for a thinner dielectric layer. This work can potentially be utilized to enhance the output performance of energy harvesting devices or nullify contact electric charge transfer in applications where this effect is undesirable.

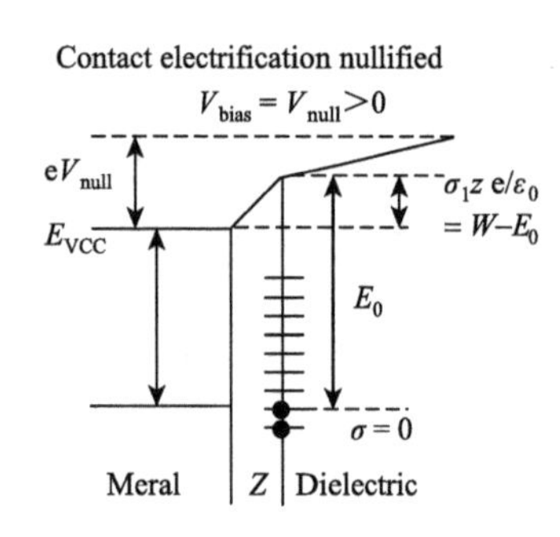

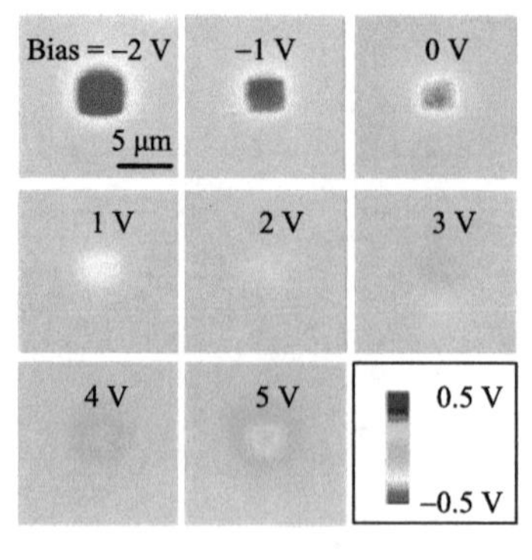

扫码阅全文

http://www.binn.cas.cn/book/202106/W020210610515181766390.pdf

2. Effect of Contact- and Sliding-Mode Electrification on Nanoscale Charge Transfer for Energy Harvesting

Yu Sheng Zhou, Shengming Li, Simiao Niu *and* **Zhong Lin Wang***

Nano Research, 9 (2016) 3705-3713.

简介：本文首次探索了在接触与滑动两种模式下表面起电的差异，证明了在两种实验条件下最后达到的表面稳态电荷和过程没有关系。

ABSTRACT The process of charge transfer based on triboelectrification (TE) and contact electrification (CE) has been recently utilized as the basis for a new and promising energy harvesting technology, i.e., triboelectric nanogenerators, as well as self-powered sensors and systems. The electrostatic charge transfer between two surfaces can occur in both the TE and the CE modes depending on the involvement of relative sliding friction. Does the sliding behavior in TE induce any fundamental difference in the charge transfer from the CE? Few studies are available on this comparison because of the challenges in ruling out the effect of the contact area using traditional macro-scale characterization methods. This paper provides the first study on the fundamental differences in CE and TE at the nanoscale based on scanning probe microscopic methods. A quantitative comparison of the two processes at equivalent contact time and force is provided, and the results suggest that the charge transfer from TE is much faster than that from CE, but the saturation value of the transferred charge density is the same. The measured frictional energy dissipation of ～11 eV when the tip scans over distance of 1 Å sheds light on a potential mechanism: The friction may facilitate the charge transfer process via electronic excitation. These results provide fundamental guidance for the selection of materials and device structures to enable the TE or the CE in different applications; the CE mode is favorable for frequent moderate contact such as vibration energy harvesting and the TE mode is favorable for instant movement such as harvesting of energy from human walking.

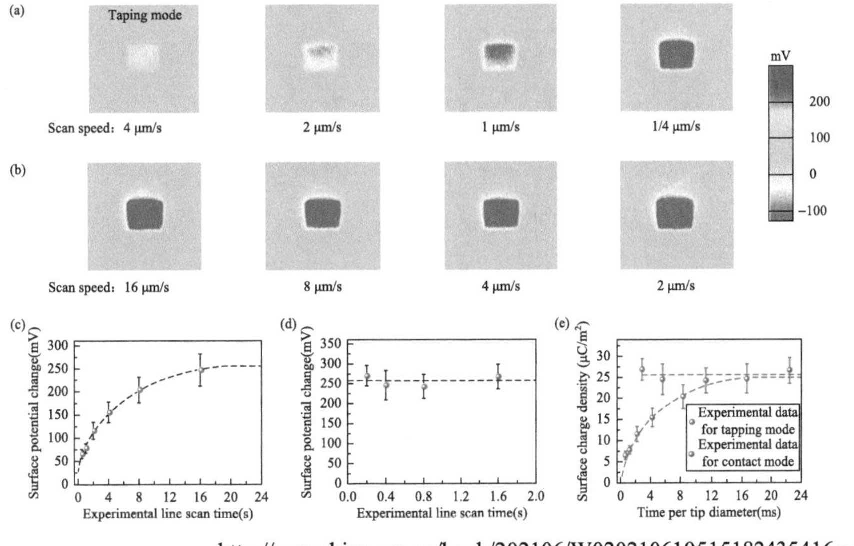

扫码阅全文

http://www.binn.cas.cn/book/202106/W020210610515182435416.pdf

3. On the Electron-Transfer Mechanism in the Contact-Electrification Effect

Cheng Xu, Yunlong Zi, Aurelia Chi Wang, Haiyang Zou, Yejing Dai, Xu He, Peihong Wang, Yi-Cheng Wang, Peizhong Feng, Dawei Li *and* Zhong Lin Wang*

Advanced Materials, 30 (2018) 1706790.

简介：本文首次探索摩擦纳米发电机的最高工作温度可达 300℃，其原因是在高温下传输的电子在表面被热离子发射，保留不下来。本文也首次提出了接触起电的最重要途径是通过两个物质原子之间在外压力作用下实现电子云的高度重叠，进而实现电子的跨原子跃迁，这个跃迁被称为“王氏跃迁”。这个模型是一个解释所有相之间接触起电的基本模式，其核心就是“电子转移是接触起电的根本机理”。

ABSTRACT A long debate on the charge identity and the associated mechanisms occurring in contact-electrification (CE) (or triboelectrification) has persisted for many decades, while a conclusive model has not yet been reached for explaining this phenomenon known for more than 2600 years! Here, a new method is reported to quantitatively investigate real-time charge transfer in CE via triboelectric nanogenerator as a function of temperature, which reveals that electron transfer is the dominant process for CE between two inorganic solids. A study on the surface charge density evolution with time at various high temperatures is consistent with the electron thermionic emission theory for triboelectric pairs composed of Ti-SiO_2 and Ti-Al_2O_3. Moreover, it is found that a potential barrier exists at the surface that prevents the charges generated by CE from flowing back to the solid where they are escaping from the surface after the contacting. This pinpoints the main reason why the charges generated in CE are readily retained by the material as electrostatic charges for hours at room temperature. Furthermore, an electron-cloud-potential-well model is proposed based on the electron-emission-dominated charge-transfer mechanism, which can be generally applied to explain all types of CE in conventional materials.

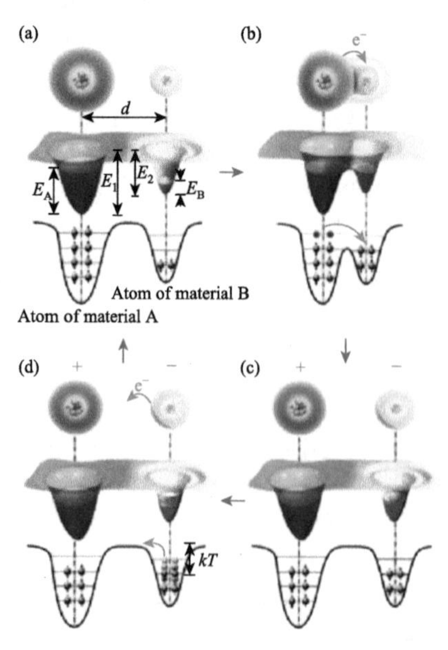

扫码阅全文

http://www.binn.cas.cn/book/202106/W020210610515182932546.pdf

4. Raising the Working Temperature of a Triboelectric Nanogenerator by Quenching Down Electron Thermionic Emission in Contact-Electrification

Cheng Xu, Aurelia Chi Wang, Haiyang Zou, Binbin Zhang, Chunli Zhang, Yunlong Zi, Lun Pan, Peihong Wang, Peizhong Feng, Zhiqun Lin *and* Zhong Lin Wang*

Advanced Materials, 30 (2018) 1803968.

简介：本文把摩擦纳米发电机的最高工作温度提高到 400℃。本文也首次提出了在外力作用下的接触起电与在温度作用下的原子振动所引起的电子云的高度重叠都是实现化学反应的重要途径，这就是机械化学的概念。

ABSTRACT As previously demonstrated, contact-electrification (CE) is strongly dependent on temperature, however the highest temperature in which a triboelectric nanogenerator (TENG) can still function is unknown. Here, by designing and preparing a rotating free-standing mode Ti/SiO_2 TENG, the relationship between CE and temperature is revealed. It is found that the dominant deterring factor of CE at high temperatures is the electron thermionic emission. Although it is normally difficult for CE to occur at temperatures higher than 583 K, the working temperature of the rotating TENG can be raised to 673 K when thermionic emission is prevented by direct physical contact of the two materials via preannealing. The surface states model is proposed for explaining the experimental phenomenon. Moreover, the developed electron cloud-potential well model accounts for the CE mechanism with temperature effects for all types of materials. The model indicates that besides thermionic emission of electrons, the atomic thermal vibration also influences CE. This study is fundamentally important for understanding triboelectrification, which will impact the design and improve the TENG for practical applications in a high temperature environment.

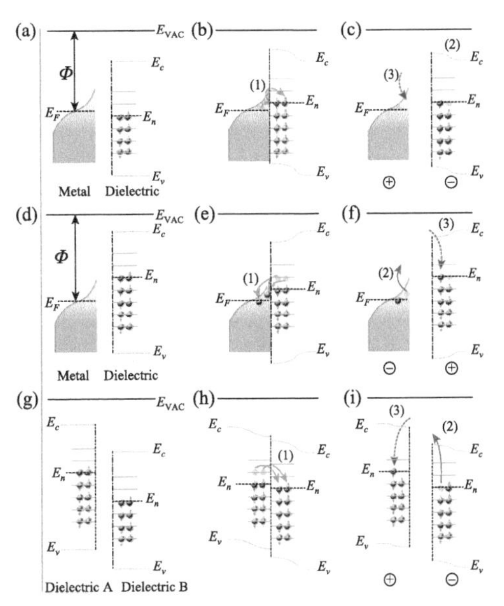

扫码阅全文

http://www.binn.cas.cn/book/202106/W020210610515183315815.pdf

5. Contact-Electrification between Two Identical Materials: Curvature Effect

Cheng Xu, Binbin Zhang, Aurelia Chi Wang, Haiyang Zou, Guanlin Liu, Wenbo Ding, Changsheng Wu, Ming Ma, Peizhong Feng, Zhiqun Lin *and* Zhong Lin Wang*

ACS Nano, 13 (2019) 2034-2041.

简介：接触起电一般是相对于两种不同的材料而言，但实验发现同种材料的接触也会有微弱的电荷转移。本文首次提出了同种材料的电荷转移是由于表面曲率变化所引起的表面态的能级位移而产生。

ABSTRACT It is known that contact-electrification (or triboelectrification) usually occurs between two different materials, which could be explained by several models for different materials systems (*Adv. Mater*, 2018, 30, 1706790; *Adv. Mater*, 2018, 30, 1803968). But contact between two pieces of the chemically same material could also result in electrostatic charges, although the charge density is rather low, which is hard to understand from a physics point of view. In this paper, by preparing a contact-separation mode triboelectric nanogenerator using two pieces of an identical material, the direction of charge transfer during contact-electrification is studied regarding its dependence on curvatures of the sample surfaces. For materials such as polytetrafluoroethylene, fluorinated ethylene propylene, Kapton, polyester, and nylon, the positive curvature surfaces are net negatively charged, while the negative curvature surfaces tend to be net positively charged. Further verification of the above-mentioned trends was obtained under vacuum (~1 Pa) and higher temperature (≤358 K) conditions. Based on the received data acquired for gentle contacting cases, we propose a curvature-dependent charge transfer model by introducing curvature-induced energy shifts of the surface states. However, this model is subject to be revised if the mutual contact mode turns into a sliding mode or more complicated hard-pressed contact mode, in which a rigorous contact between the two pieces of the same material could result in nanoscale damage/fracture and possible species transfer. Our study provides a primitive step toward understanding the basics of contact-electrification.

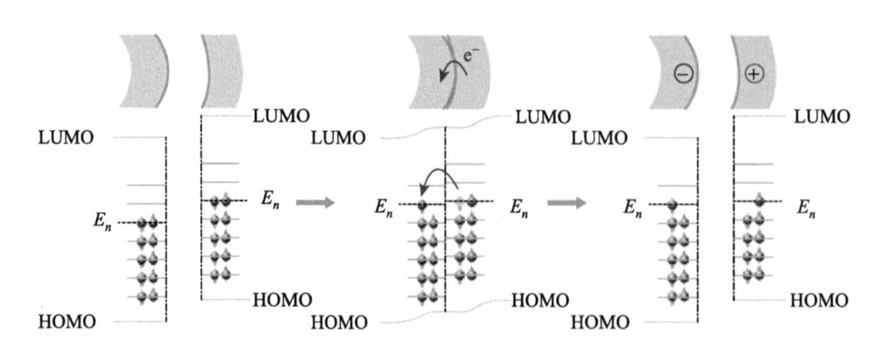

扫码阅全文

http://www.binn.cas.cn/book/202106/W020210610515183891455.pdf

6. Electron Transfer in Nanoscale Contact Electrification: Photon Excitation Effect

Shiquan Lin, Liang Xu, Laipan Zhu, Xiangyu Chen *and* Zhong Lin Wang*

Advanced Materials, (2019) 1901418.

简介：本文发现利用紫外光照射是除掉表面因接触起电所带电荷的途径之一。

ABSTRACT Contact electrification (CE) (or triboelectrification) is a well-known phenomenon, and the identity of the charge carriers and their transfer mechanism have been discussed for decades. Recently, the species of transferred charges in the CE between a metal and a ceramic was revealed as electron transfer and its subsequent release is dominated by the thermionic emission process. Here, the release of CE-induced electrostatic charges on a dielectric surface under photon excitation is studied by varying the light intensity and wavelength, but under no significant raise in temperature. The results suggest that there exists a threshold photon energy for releasing the triboelectric charges from the surface, which is 4.1 eV (light wavelength at 300 nm) for SiO_2 and 3.4 eV (light wavelength at 360 nm) for PVC; photons with energy smaller than this cannot effectively excite the surface electrostatic charges. This process is attributed to the photoelectron emission of the charges trapped in the surface states of the dielectric material. Further, a photoelectron emission model is proposed to describe light-induced charge decay on a dielectric surface. The findings provide an additional strong evidence about the electron transfer process in the CE between metals and dielectrics as well as polymers.

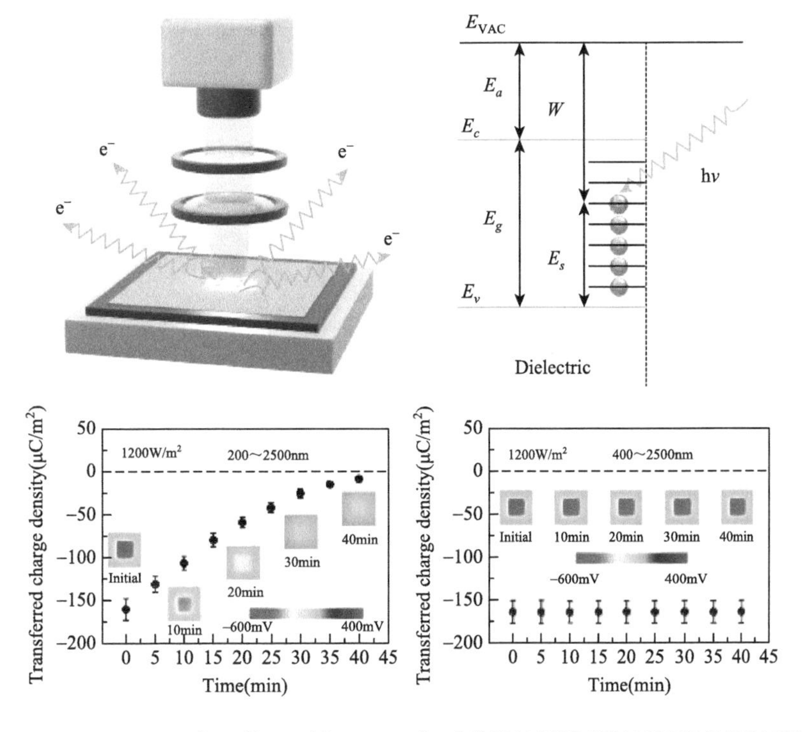

扫码阅全文

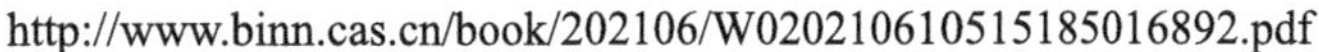
http://www.binn.cas.cn/book/202106/W020210610515185016892.pdf

7. Electron Transfer in Nanoscale Contact Electrification: Effect of Temperature in the Metal-Dielectric Case

Shiquan Lin, Liang Xu, Cheng Xu, Xiangyu Chen, Aurelia C. Wang, Binbin Zhang, Pei Lin, Ya Yang, Huabo Zhao *and* Zhong Lin Wang*

Advanced Materials, 31 (2019) 1808197.

简介：通过调制施加在针尖上和样品上的温度差，本文进一步证明了金属-介电质之间的电荷转移是电子转移。

ABSTRACT The phenomenon of contact electrification (CE) has been known for thousands of years, but the nature of the charge carriers and their transfer mechanisms are still under debate. Here, the CE and triboelectric charging process are studied for a metal-dielectric case at different thermal conditions by using atomic force microscopy and Kelvin probe force microscopy. The charge transfer process at the nanoscale is found to follow the modified thermionic-emission model. In particular, the focus here is on the effect of a temperature difference between two contacting materials on the CE. It is revealed that hotter solids tend to receive positive triboelectric charges, while cooler solids tend to be negatively charged, which suggests that the temperature-difference-induced charge transfer can be attributed to the thermionic-emission effect, in which the electrons are thermally excited and transfer from a hotter surface to a cooler one. Further, a thermionic-emission band-structure model is proposed to describe the electron transfer between two solids at different temperatures. The findings also suggest that CE can occur between two identical materials owing to the existence of a local temperature difference arising from the nanoscale rubbing of surfaces with different curvatures/roughness.

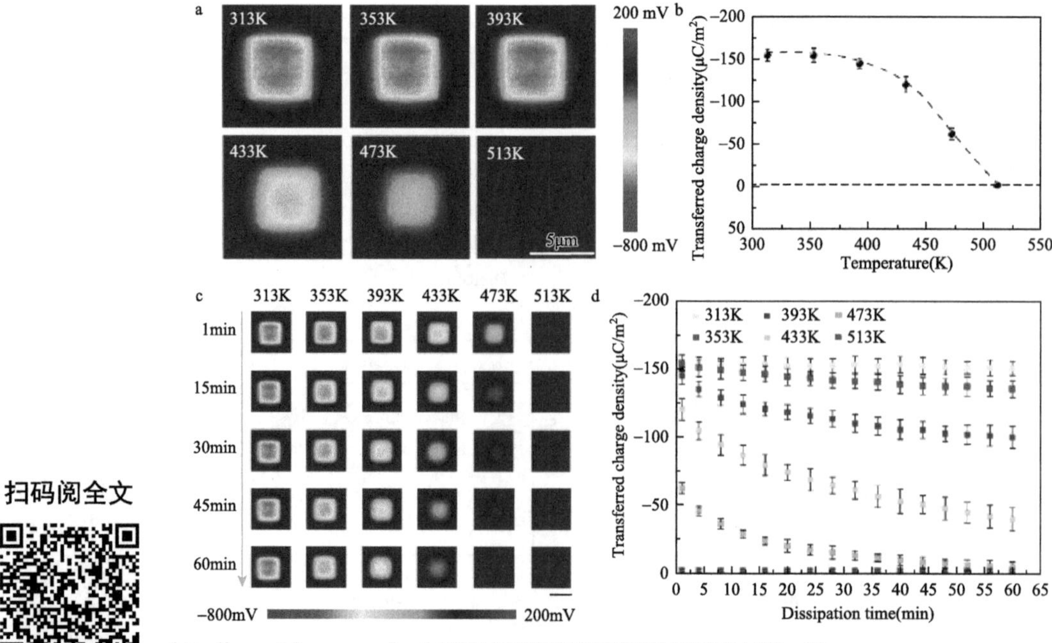

扫码阅全文

http://www.binn.cas.cn/book/202106/W020210610515185634679.pdf

8. The Overlapped Electron-Cloud Model for Electron Transfer in Contact Electrification

Shiquan Lin, Cheng Xu, Liang Xu *and* **Zhong Lin Wang***

Advanced Functional Materials, 30 (2020) 1909724.

简介：通过调制施加在针尖上的偏压，本文进一步深化了电子云高度重叠模型是接触起电的基础模型。

ABSTRACT Contact electrification (CE) is one of the oldest topics in physics, which has been discussed for more than 2600 years. Recently, an overlapped electron-cloud (OEC) model was proposed by Wang to explain all types of CE phenomena for general materials in which a deep overlapping of electron clouds belonging to two atoms results in a lowered potential barrier for electron transfer from one to the other by applying a compressive force, which is simply referred to as Wang transition for CE. Here, the degree of electron-cloud overlap between two atoms is controlled by using tapping mode atomic force microscopy. A temperature difference and electric field are applied between atoms of two surfaces. It is found that electron transfer only occurs when the contact tip and sample interact in the repulsive-force region, which corresponds to a strong overlap of the electron clouds. Such a fact even preserves if the tip is at a higher temperature than the sample for 120 K. Alternatively, by applying a bias, electron tunneling would occur when the tip is in the attractive-force region within which normal electron transfer would not occur. These studies solidify the overlapped electron-cloud model first proposed by Wang. Further, the temperature and bias effects on the CE are explained based on a modified OEC model.

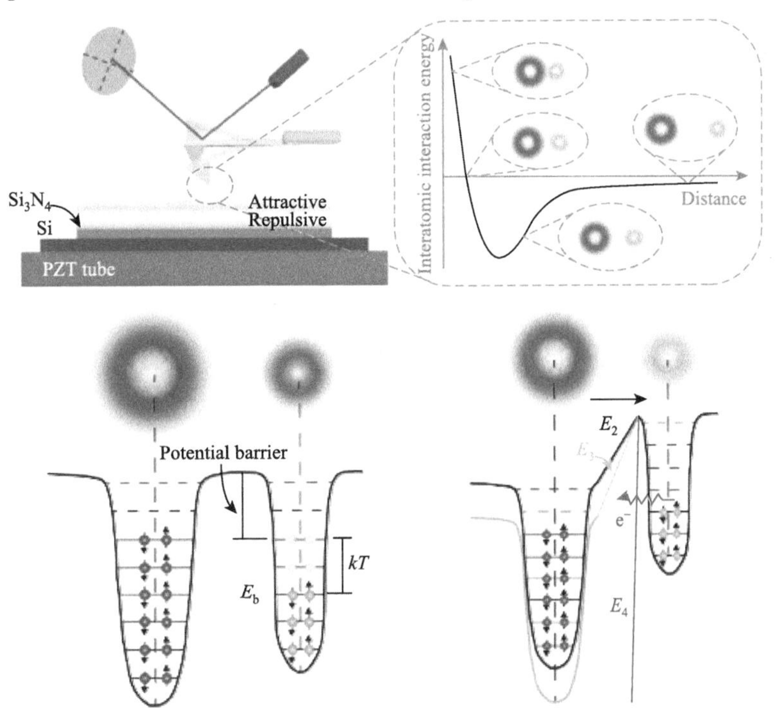

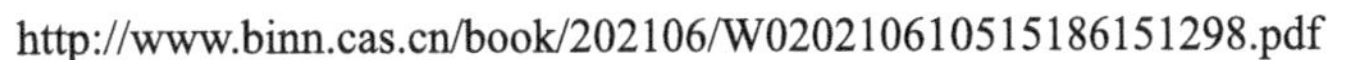
http://www.binn.cas.cn/book/202106/W020210610515186151298.pdf

扫码阅全文

9. Direct Current Triboelectric Cell by Sliding an n-Type Semiconductor on a p-Type Semiconductor

Ran Xu, Qing Zhang*, Jing Yuan Wang, Di Liu, Jie Wang, Zhong Lin Wang*

Nano Energy, 66 (2019) 104185.

简介：本文报道了在 p 型半导体上滑动一个 n 型半导体片可以在外电路观测到直流电信号，这和以前报道的利用绝缘体-绝缘体所做的摩擦纳米发电机是不一样的，引出了研究接触起电的一个有趣的现象：摩擦伏特效应。

ABSTRACT Triboelectric generators are a type of devices for harvesting mechanical energy by utilizing triboelectric and electrostatic induction effects. When two materials of different electron affinities are brought into contact, electrons transfer across the contacted surfaces owing to triboelectrification. When the two materials are slid against each other with a variation of the contacting surface area, an alternating current is electrostatically induced in the external circuit. In this paper, a novel electric generator, called triboelectric cell, based on sliding friction between n- and p-type doped semiconductors without changing the contacting area is reported. A direct current is generated in the direction of the built-in electric field in the dynamic p-n junction across the contacted surfaces, flowing from the p-semiconductor through the external circuit to the n-semiconductor. The generated currents and voltages are studied through sliding speeds, accelerations, contacting forces, operation temperatures and the geometries of the top electrode. The direct current generation can be attributed to electron-hole pairs generated at the two sliding surfaces, which are then swept out the dynamic junction by the built-in electric field.

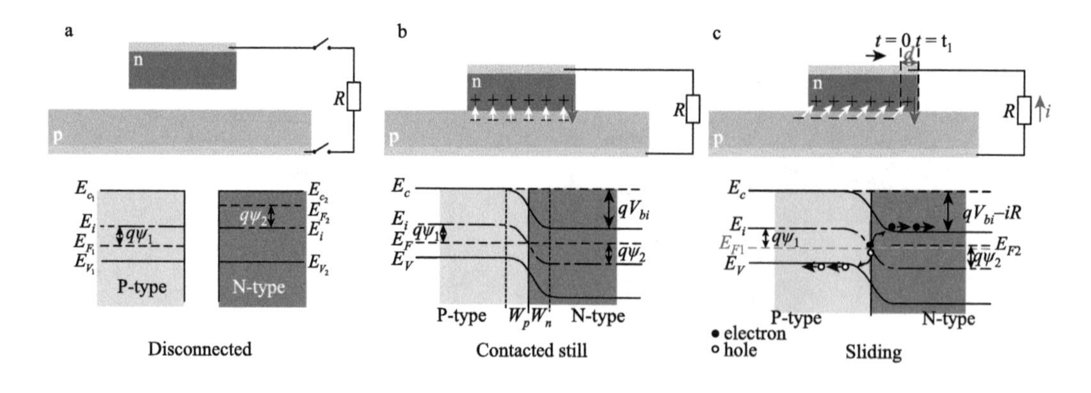

扫码阅全文

http://www.binn.cas.cn/book/202106/W020210610515187007397.pdf

10. On the Origin of Contact-Electrification

Zhong Lin Wang[*], Aurelia Chi Wang[*]

Materials Today, 30 (2019) 34-51.

简介：本文是综述固体-固体界面接触起电机理的经典文献。从现有的数据出发，明确地提出了电子输运是接触起电的主要机理，并把该机理延伸到了包括液体、气体、固体多项接触的体系。这一结论终结了百年来关于接触起电不断争论而无定论的局面。本文预言了两个结果：关于固液界面双电层形成的两步走的模型；关于半导体界面的摩擦伏特效应。这两个预言后来都得到了实验验证。

ABSTRACT Although contact electrification (triboelectrification) (CE) has been documented since 2600 years ago, its scientific understanding remains inconclusive, unclear, and un-unified. This paper reviews the updated progress for studying the fundamental mechanism of CE using Kelvin probe force microscopy for solid-solid cases. Our conclusion is that electron transfer is the dominant mechanism for CE between solid-solid pairs. Electron transfer occurs only when the interatomic distance between the two materials is shorter than the normal bonding length (typically ~0.2 nm) in the region of repulsive forces. A strong electron cloud overlap (or wave function overlap) between the two atoms/molecules in the repulsive region leads to electron transition between the atoms/molecules, owing to the reduced interatomic potential barrier. The role played by contact/friction force is to induce strong overlap between the electron clouds (or wave function in physics, bonding in chemistry). The electrostatic charges on the surfaces can be released from the surface by electron thermionic emission and/or photon excitation, so these electrostatic charges may not remain on the surface if sample temperature is higher than ~300-400℃. The electron transfer model could be extended to liquid-solid, liquid-gas and even liquid-liquid cases. As for the liquid-solid case, molecules in the liquid would have electron cloud overlap with the atoms on the solid surface at the very first contact with a virginal solid surface, and electron transfer is required in order to create the first layer of electrostatic charges on the solid surface. This step only occurs for the very first contact of the liquid with the solid. Then, ion transfer is the second step and is the dominant process thereafter, which is a redistribution of the ions in solution considering electrostatic interactions with the charged solid surface. This is proposed as a two-step formation process of the electric double layer (EDL) at the liquid-solid interface. Charge transfer in the liquid-gas case is believed to be due to electron transfer once a gas molecule strikes the liquid surface to induce the overlapping electron cloud under pressure. In general, electron transfer due to the overlapping electron cloud under mechanical force/pressure is proposed as the dominant mechanism for initiating CE between solids, liquids and gases. This study provides not only the first systematic understanding about the physics of CE, but also demonstrates that the triboelectric nanogenerator (TENG) is an effective method for studying the nature of CE between any materials.

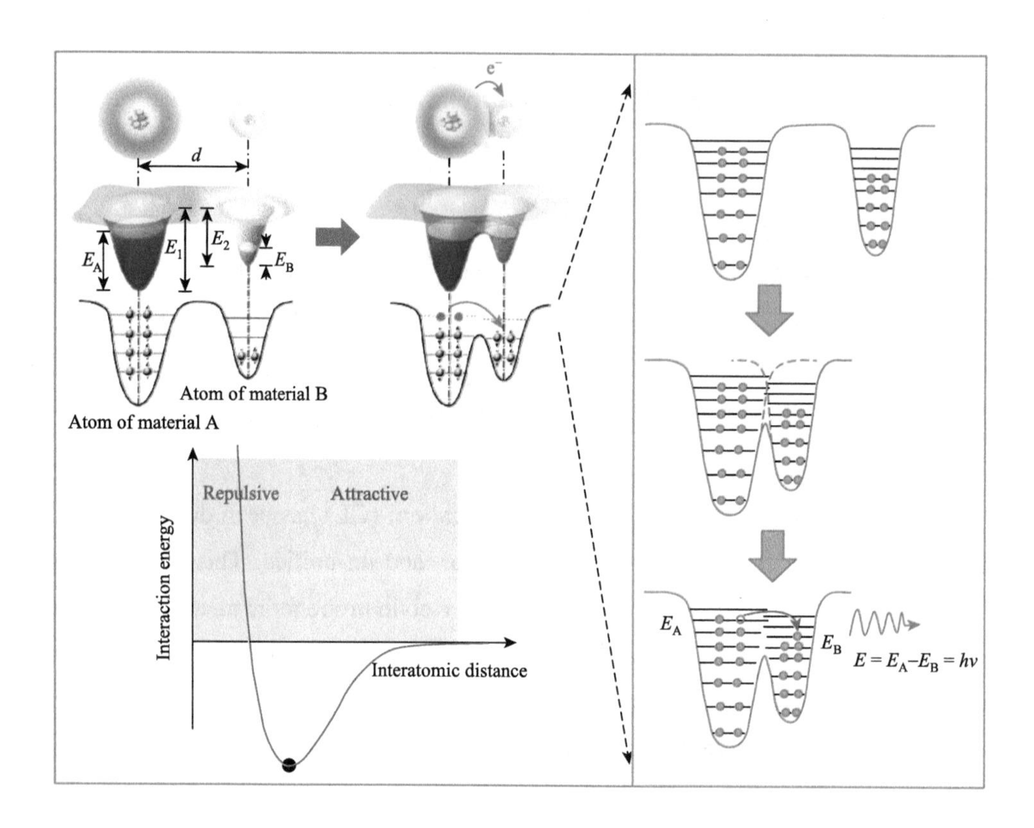

扫码阅全文

http://www.binn.cas.cn/book/202106/W020210610515187414924.pdf

II.12　液体-固体接触起电的机理
The Origins of Liquid-Solid Contact-Electrification

1. Quantifying Electron-Transfer in Liquid-Solid Contact Electrification and the Formation of Electric Double-Layer

Shiquan Lin, Liang Xu, Aurelia Chi Wang *and* **Zhong Lin Wang***

Nature Communications, 11 (2020) 399.

简介：本文开创了量化液体-固体界面电子与离子转移的新方法，并证实了王中林提出的双电层形成的两步走模型：先是电子转移，然后才是离子吸附。该模型修正了传统文献对双电层形成的解释机理，是一篇研究液体-固体界面的经典文献。

ABSTRACT　Contact electrification (CE) has been known for more than 2600 years but the nature of charge carriers and their transfer mechanisms still remain poorly understood, especially for the cases of liquid-solid CE. Here, we study the CE between liquids and solids and investigate the decay of CE charges on the solid surfaces after liquid-solid CE at different thermal conditions. The contribution of electron transfer is distinguished from that of ion transfer on the charged surfaces by using the theory of electron thermionic emission. Our study shows that there are both electron transfer and ion transfer in the liquid-solid CE. We reveal that solutes in the solution, pH value of the solution and the hydrophilicity of the solid affect the ratio of electron transfers to ion transfers. Further, we propose a two-step model of electron or/and ion transfer and demonstrate the formation of electric double-layer in liquid-solid CE.

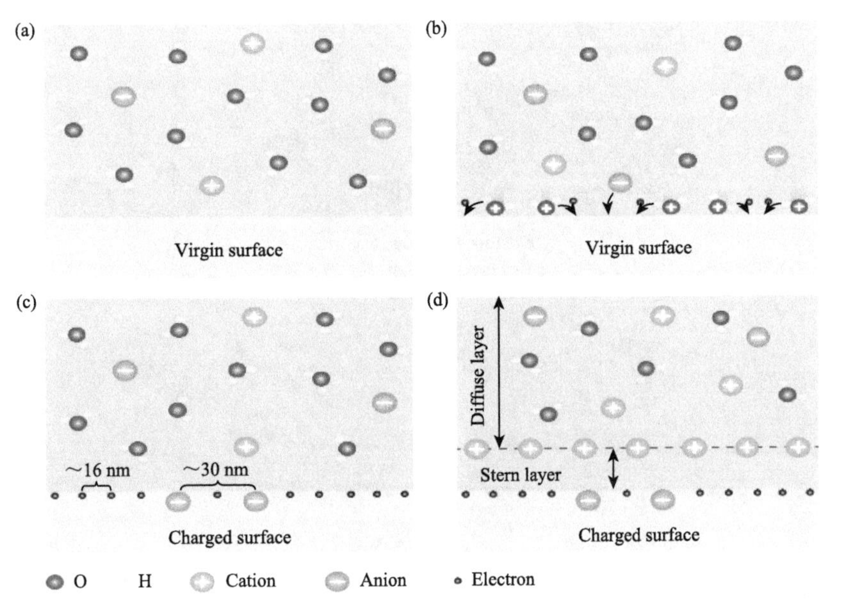

扫码阅全文

http://www.binn.cas.cn/book/202106/W020210610518631611119.pdf

2. Electron Transfer as a Liquid Droplet Contacting a Polymer Surface

Fei Zhan, Aurelia C. Wang, Liang Xu, Shiquan Lin, Jiajia Shao, Xiangyu Chen* *and* Zhong Lin Wang*

ACS Nano, 14 (2020) 17565-17573.

简介：本文首次把单电极式摩擦纳米发电机作为一个探针来研究液体-固体接触起电。根据水滴与高分子膜表面接触逐步增加表面电荷密度的分析，确定了双电层形成的两步走模型。该文章也提出了摩擦纳米发电机的第五大应用领域：研究液体-固体界面起电的探针。

ABSTRACT It has been demonstrated that substantial electric power can be produced by a liquid-based triboelectric nanogenerator (TENG). However, the mechanisms regarding the electrification between a liquid and a solid surface remain to be extensively investigated. Here, the working mechanism of a droplet-TENG was proposed based on the study of its dynamic saturation process. Moreover, the charge-transfer mechanism at the liquid-solid interface was verified as the hybrid effects of electron transfer and ion adsorption by a simple but valid method. Thus, we proposed a model for the charge distribution at the liquid-solid interface, named Wang's hybrid layer, which involves the electron transfer, the ionization reaction, and the van der Waals force. Our work not only proves that TENG is a probe for investigating charge transfer at interface of all phases, such as solid-solid and liquid-solid, but also may have great significance to water energy harvesting and may revolutionize the traditional understanding of the liquid-solid interface used in many fields such as electrochemistry, catalysis, colloidal science, and even cell biology.

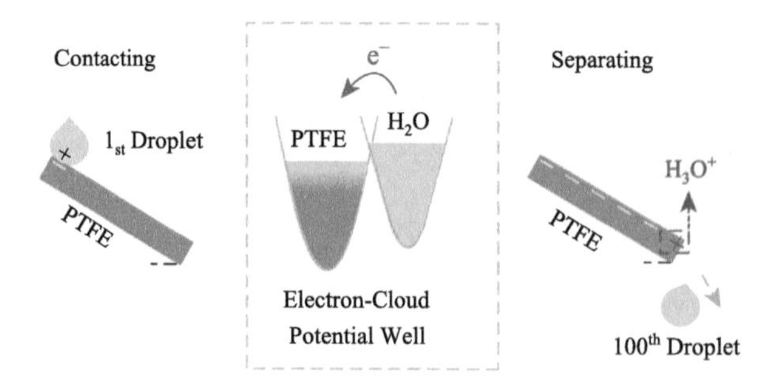

扫码阅全文

http://www.binn.cas.cn/book/202106/W020210610518631954006.pdf

3. Probing Contact-Electrification-Induced Electron and Ion Transfers at a Liquid-Solid Interface

Jinhui Nie, Zewei Ren, Liang Xu, Shiquan Lin, Fei Zhan, Xiangyu Chen* *and* **Zhong Lin Wang***

Advanced Materials, 32 (2020) 1905696.

简介：本文首次通过改变液体-固体接触面积来研究界面的起电，确定了双电层形成的两步走模型，修正了传统文献对双电层形成的解释机理。

ABSTRACT As a well-known phenomenon, contact electrification (CE) has been studied for decades. Although recent studies have proven that CE between two solids is primarily due to electron transfer, the mechanism for CE between liquid and solid remains controversial. The CE process between different liquids and polytetrafluoroethylene (PTFE) film is systematically studied to clarify the electrification mechanism of the solid-liquid interface. The CE between deionized water and PTFE can produce a surface charges density in the scale of 1 $nC \cdot cm^{-2}$, which is ten times higher than the calculation based on the pure ion-transfer model. Hence, electron transfer is likely the dominating effect for this liquid-solid electrification process. Meanwhile, as ion concentration increases, the ion adsorption on the PTFE hinders electron transfer and results in the suppression of the transferred charge amount. Furthermore, there is an obvious charge transfer between oil and PTFE, which further confirms the presence of electron transfer between liquid and solid, simply because there are no ions in oil droplets. It is demonstrated that electron transfer plays the dominant role during CE between liquids and solids, which directly impacts the traditional understanding of the formation of an electric double layer (EDL) at a liquid-solid interface in physical chemistry.

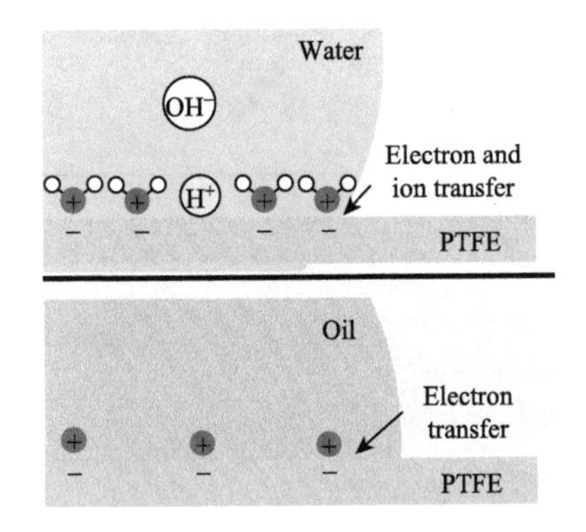

扫码阅全文

http://www.binn.cas.cn/book/202106/W020210610518632698005.pdf

4. Harvesting Water Drop Energy by a Sequential Contact-Electrification and Electrostatic-Induction Process

Zong-Hong Lin, Gang Cheng, Sangmin Lee, Ken C. Pradel *and* Zhong Lin Wang*

Advanced Materials, 26 (2014) 4690-4696.

简介：本文是首篇应用液体-固体接触起电机理来利用雨水发电的开山之作，具有一定的实用价值。

ABSTRACT A new prototype triboelectric nanogenerator with superhydrophobic and self-cleaning features is invented to harvest water drop energy based on a sequential contact electrification and electrostatic induction process. Because of the easy-fabrication, cost-effectiveness, and robust properties, the developed triboelectric nanogenerator expands the potential applications to harvesting energy from household wastewater and raindrops.

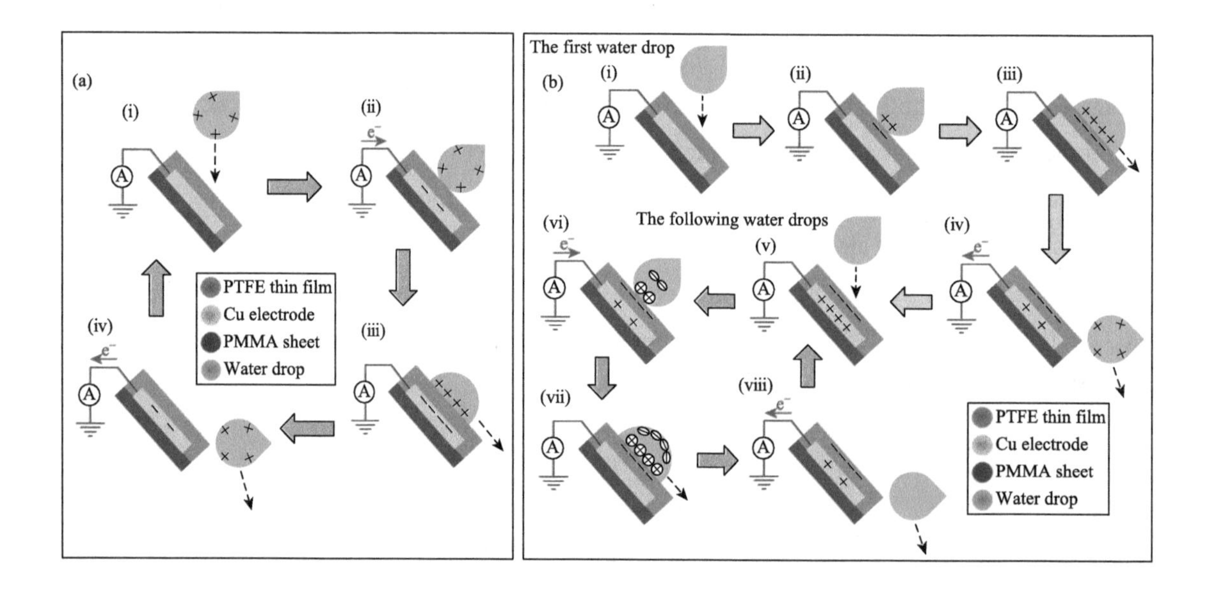

扫码阅全文

http://www.binn.cas.cn/book/202106/W020210610518633112937.pdf

5. Harvesting Water Wave Energy by Asymmetric Screening of Electrostatic Charges on a Nanostructured Hydrophobic Thin-Film Surface

Guang Zhu, Yuanjie Su, Peng Bai, Jun Chen, Qingshen Jing, Weiqing Yang *and* Zhong Lin Wang*

ACS Nano, 8 (2014) 6031-6037.

简介：本文是首篇应用液体-固体接触起电和不对称的电极设计来利用雨水发电的开山之作，具有很高的实用价值。

ABSTRACT Energy harvesting from ambient water motions is a desirable but underexplored solution to on-site energy demand for self-powered electronics. Here we report a liquid-solid electrification-enabled generator based on a fluorinated ethylene propylene thin film, below which an array of electrodes are fabricated. The surface of the thin film is charged first due to the water-solid contact electrification. Aligned nanowires created on the thin film make it hydrophobic and also increase the surface area. Then the asymmetric screening to the surface charges by the waving water during emerging and submerging processes causes the free electrons on the electrodes to flow through an external load, resulting in power generation. The generator produces sufficient output power for driving an array of small electronics during direct interaction with water bodies, including surface waves and falling drops. Polymer-nanowire-based surface modification increases the contact area at the liquid-solid interface, leading to enhanced surface charging density and thus electric output at an efficiency of 7.7%. Our planar-structured generator features an all-in-one design without separate and movable components for capturing and transmitting mechanical energy. It has extremely lightweight and small volume, making it a portable, flexible, and convenient power solution that can be applied on the ocean/river surface, at coastal/offshore areas, and even in rainy places. Considering the demonstrated scalability, it can also be possibly used in large-scale energy generation if layers of planar sheets are connected into a network.

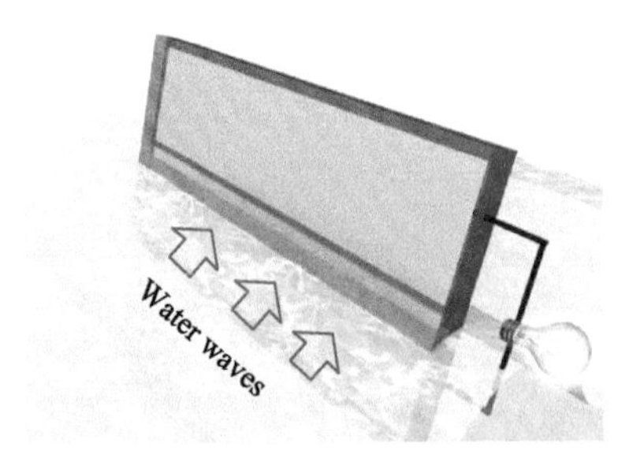

扫码阅全文

http://www.binn.cas.cn/book/202106/W020210610518633667795.pdf

6. Self-Powered Triboelectric Micro Liquid/Gas Flow Sensor for Microfluidics

Jie Chen, Hengyu Guo, Jiangeng Zheng, Yingzhou Huang, Guanlin Liu, Chenguo Hu* *and* **Zhong Lin Wang***

ACS Nano, 10 (2016) 8104-8112.

简介：本文是首篇利用液体-固体接触起电来做监测液滴运动轨迹的文章。

ABSTRACT Liquid and gas flow sensors are important components of the micro total analysis systems (μTAS) for modern analytical sciences. In this paper, we proposed a self-powered triboelectric microfluidic sensor (TMS) by utilizing the signals produced from the droplet/bubble via the capillary and the triboelectrification effects on the liquid/solid interface for real-time liquid and gas flow detection. By alternating capillary with different diameters, the sensor's detecting range and sensitivity can be adjusted. Both the relationship between the droplet/bubble and capillary size, and the output signal of the sensor are systematically studied. By demonstrating the monitoring of the transfusion process for a patient and the gas flow produced from an injector, it shows that TMS has a great potential in building a self-powered micro total analysis system.

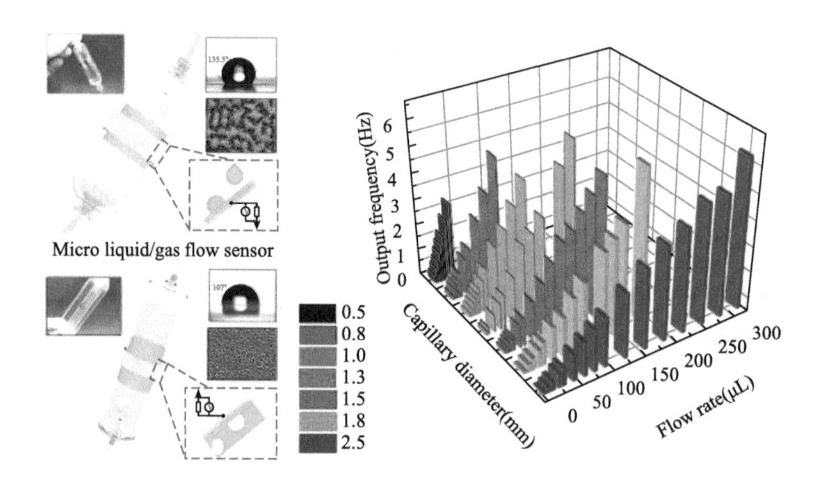

扫码阅全文

http://www.binn.cas.cn/book/202106/W020210610518634347500.pdf

7. Self-Powered Triboelectric Nanosensor for Microfluidics and Cavity-Confined Solution Chemistry

Xiuhan Li, Min-Hsin Yeh, Zong-Hong Lin, Hengyu Guo, Po-Kang Yang, Jie Wang, Sihong Wang, Ruomeng Yu, Tiejun Zhang *and* Zhong Lin Wang*

ACS Nano, 9 (2015) 11056-11063.

简介：本文是首篇把摩擦纳米发电机应用于微流体监测的文章。

ABSTRACT Micro total analysis system (μTAS) is one of the important tools for modern analytical sciences. In this paper, we not only propose the concept of integrating the self-powered triboelectric microfluidic nanosensor (TMN) with μTAS, but also demonstrate that the developed system can be used as an *in situ* tool to quantify the flowing liquid for microfluidics and solution chemistry. The TMN automatically generates electric outputs when the fluid passing through it and the outputs are affected by the solution temperature, polarity, ionic concentration, and fluid flow velocity. The self-powered TMN can detect the flowing water velocity, position, reaction temperature, ethanol, and salt concentrations. We also integrate the TMNs in a μTAS platform to directly characterize the synthesis of Au nanoparticles by a chemical reduction method.

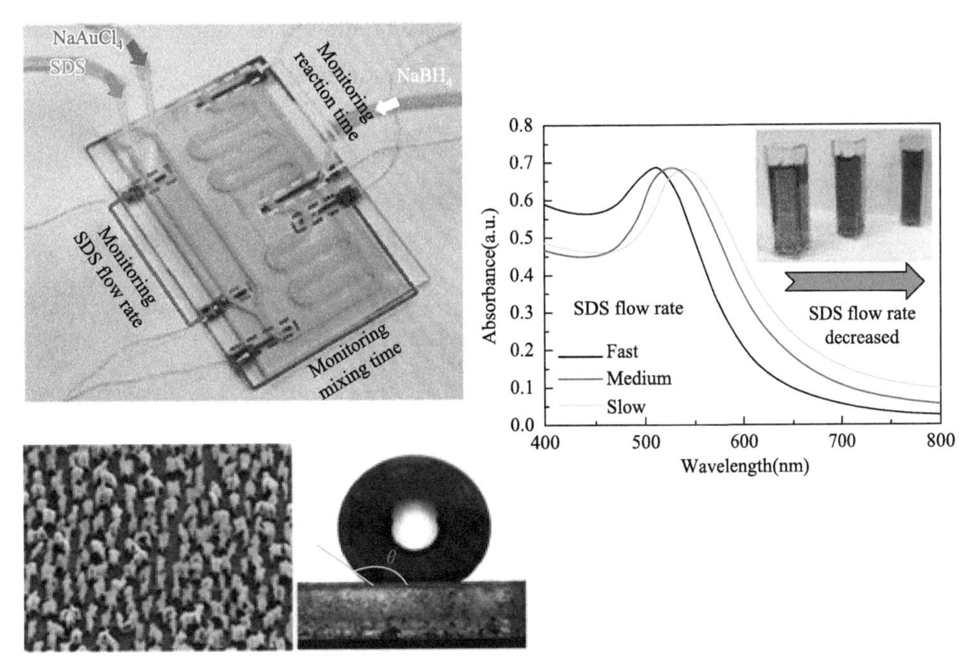

扫码阅全文

http://www.binn.cas.cn/book/202106/W020210610518635661305.pdf

8. Signal Output of Triboelectric Nanogenerator at Oil-Water-Solid Multiphase Interfaces and its Application for Dual-Signal Chemical Sensing

Peng Jiang, Lei Zhang, Hengyu Guo, Chaoyu Chen, Changsheng Wu, Steven Zhang *and* Zhong Lin Wang*

Advanced Materials, 31 (2019) 1902793.

简介：本文是首篇利用摩擦纳米发电机做探针来研究油水界面带电的第一篇文章，并发现液体-液体界面确实有电荷转移。

ABSTRACT A liquid-solid contact triboelectric nanogenerator (TENG) based on poly(tetrafluoroethylene) (PTFE) film, a copper electrode, and a glass substrate for harvesting energy in oil/water multiphases is reported. There are two distinctive signals being generated, one is from the contact electrification and electrostatic induction between the liquid (water/oil) and the PTFE film (V_{TENG} and I_{TENG}); and the other is from the electrostatic induction in the copper electrode by the oil/water interfacial charges ($\Delta V_{interface}$ and $I_{interface}$), which is generated only when the liquid-solid contact TENG is inserted across the oil/water interface. The two signals show interesting opposite changing trends that the V_{TENG} and I_{TENG} decrease while the oil/water interfacial signals of $\Delta V_{interface}$ and $I_{interface}$ increase after coating a layer of polydopamine on the surfaces of PTFE and glass via self-polymerization. As an application of the observed phenomena, both the values of I_{TENG} and $I_{interface}$ have a good linear relationship versus the natural logarithm of the concentration of the dopamine. Based on this, the first self-powered dual-signal detection of dopamine using TENG is demonstrated.

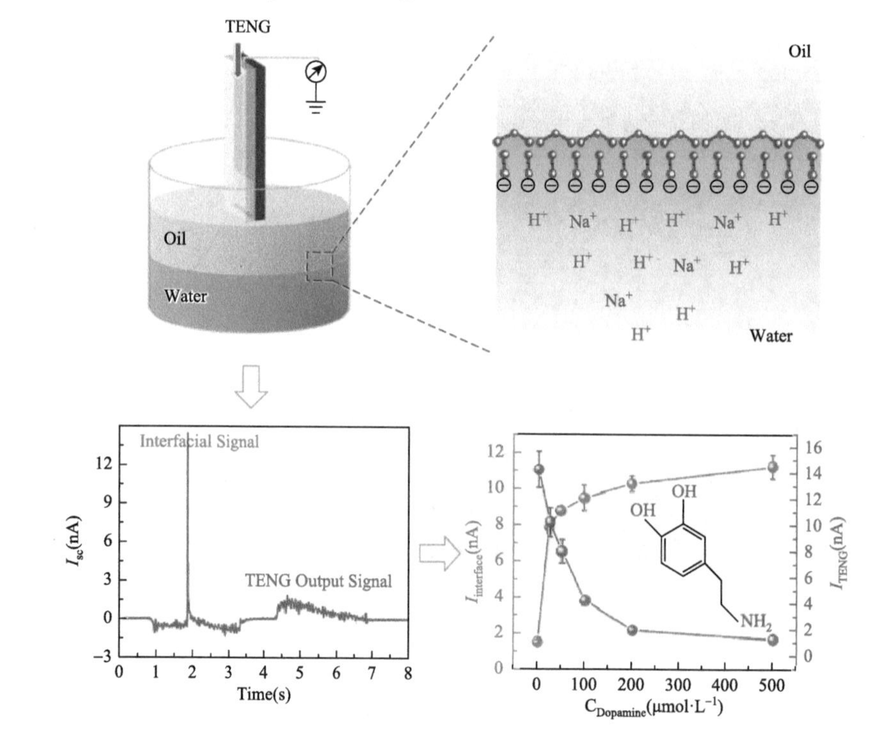

扫码阅全文

http://www.binn.cas.cn/book/202106/W020210610518636333358.pdf

9. A Streaming Potential/Current-Based Microfluidic Direct Current Generator for Self-Powered Nanosystems

Rui Zhang, Sihong Wang, Min-Hsin Yeh, Caofeng Pan, Long Lin, Ruomeng Yu, Yan Zhang, Li Zheng, Zongxia Jiao *and* Zhong Lin Wang*

Advanced Materials, 27 (2015) 6482-6487.

简介：本文是利用液体-固体界面存在的双电层结构随着压力造成的离子浓度变化而产生的束流来发电，而压力的变化是利用多孔材料来实现的。

ABSTRACT A simple but practical method to convert the hydroenergy of microfluidics into continuous electrical output is reported. Based on the principle of streaming potential/current, a microfluidic generator (MFG) is demonstrated using patterned micropillar arrays as a quasi-porous flow channel. The continuous electrical output makes this MFG particularly suitable as a power source in self-powered systems. Using the proposed MFG to power a single nanowire-based pH sensor, a self-powered fluid sensor system is demonstrated.

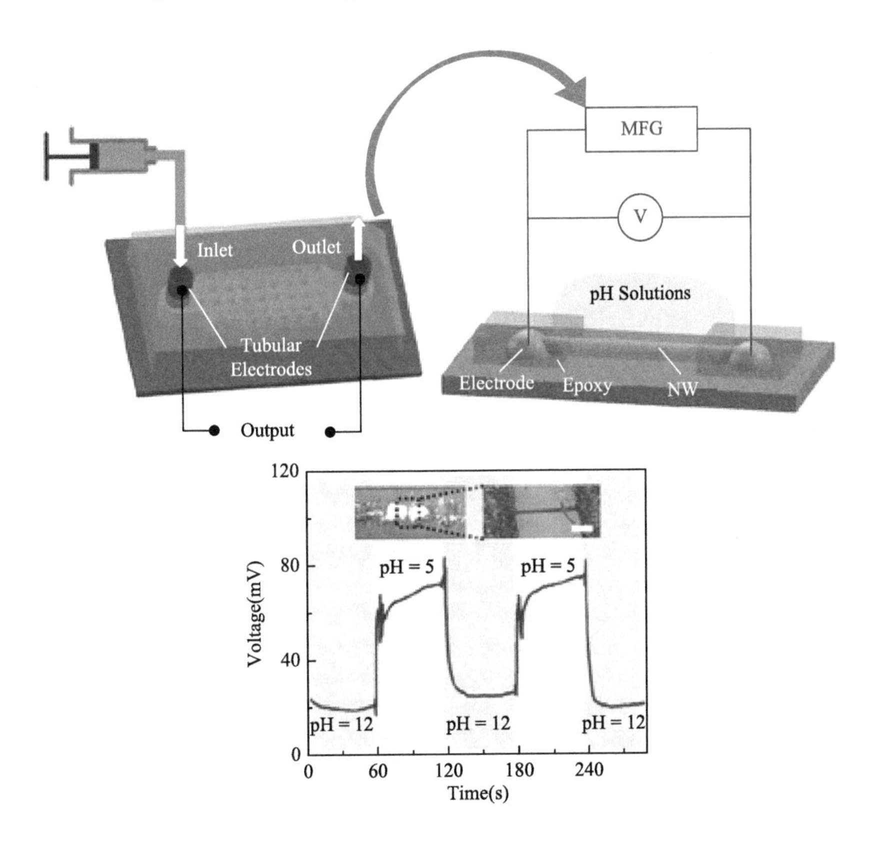

扫码阅全文

http://www.binn.cas.cn/book/202106/W020210610518636782680.pdf

10. Contact Electrification at Liquid-Solid Interface

Shiquan Lin, Xiangyu Chen *and* **Zhong Lin Wang***

Chemical Reviews, (2021)
(in press)

简介：本文是关于液体-固体界面接触起电的经典综述文章，主要包括两方面的内容：界面处电子的转移过程与形成双电层的两步走模型。

ABSTRACT Interfaces between liquid-solid (L-S) are the most important surface science in chemistry, catalysis, energy and even biology. The formation of electric double layer (EDL) at the L-S interface has been attributed due to the adsorption of a layer of ions at the solid surface, which causes the ions in the liquid to redistribute. Although the existence of a layer of charges on a solid surface is always assumed, but the origin of the charges is not extensively explored. Recent studies of contact-electrification (CE) between liquid and solid suggest that electron transfer plays a dominant role at the initial stage for forming the charge layer at the L-S interface. Here, we review the recent works about electron transfer in liquid-solid CE, including sceneries such as liquid-insulator, liquid-semiconductor and liquid-metal. The formation about the EDL is revisited with considering the existence of electron transfer at the L-S interface. Furthermore, a technique of triboelectric nanogenerator (TENG) based on the liquid-solid CE is introduced, which can be used not only for harvesting mechanical energy from liquid, but also as a probe for probing the charge transfer at liquid-solid interfaces.

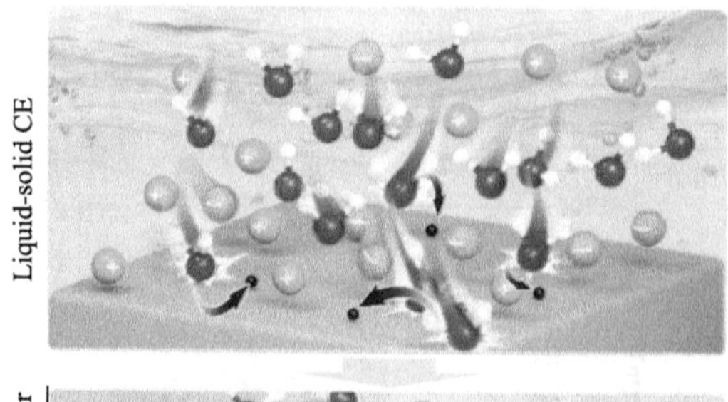

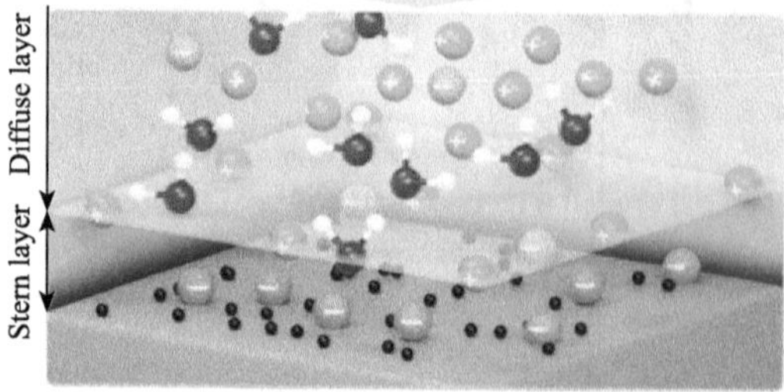

扫码阅全文

http://www.binn.cas.cn/book/202106/W020210702349251266710.pdf

Ⅱ.13　液体-液体接触起电的机理

The Origins of Liquid-Liquid Contact-Electrification

1. Studying of Contact Electrification and Electron Transfer at Liquid-Liquid Interface

Xiuzhong Zhao, Xiao Lu, Qiwei Zheng, Lin Fang, Li Zheng*, Xiangyu Chen*, Zhong Lin Wang*

Nano Energy, 87 (2021) 106191

简介：本文利用单电极摩擦纳米发电机作为一个探针，首次研究了液体-液体界面接触起电的现象，并证实了电子传输也在液体-液体界面发生，并再次确认了王氏跃迁可以解释其电荷转移过程。

ABSTRACT　Triboelectric nanogenerator (TENG) provides an effective approach for studying transferred charges at solid-solid or liquid-solid interfaces. Here, by dripping a liquid droplet through an immiscible organic solution (transformer oil), we used single-electrode mode TENG to measure the transferred charges on the liquid droplet once it goes down in the solution. The falling droplets (with volume of 25 μL) can generate negative charges on its surface and the output charges amount can reach −5.3 pC by passing through the transformer oil. In addition, we found that both the pH value and the electrostatic shielding effect of free ions greatly hinder the electron transfer efficiency at L-L contact electrification. Combining the experimental results and previous research, we infer that there are both ion transfer and electron transfer happening in the L-L electrification.

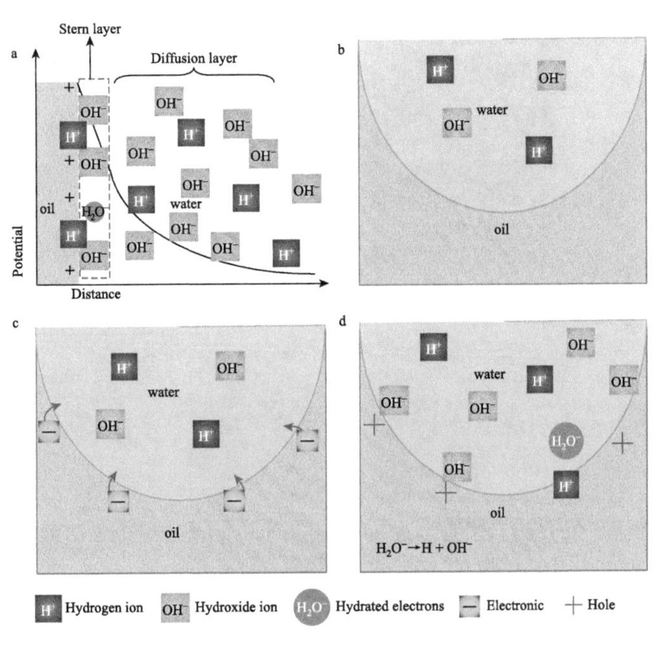

http://www.binn.cas.cn/book/202106/W020210617400970260459.pdf

2. Quantifying Contact-Electrification Induced Charge Transfer on a Liquid Droplet after Contacting with a Liquid or Solid

Zhen Tang, Shiquan Lin *and* Zhong Lin Wang*

Advanced Materials, 33 (2021) 2102886.

简介：本文利用超声悬浮液滴的方法，通过电场导致的液滴偏转，定量测试了液体-固体与液体-液体界面处的电荷转移，证实了电子传输也在液体-液体界面发生，并再次确认了王氏跃迁可以解释其电荷转移过程。

ABSTRACT Contact electrification (CE) is a common physical phenomenon, and its mechanisms for solid-solid and liquid-solid cases have been widely discussed. However, the studies about liquid-liquid CE are hindered by the lack of proper techniques. Here, a contactless method is proposed for quantifying the charges on a liquid droplet based on the combination electric field and acoustic field, in which the liquid droplet is suspended in an acoustic field, an electric field force is created on the droplet to balance the acoustic trap force, and the amount of charges on the droplet is thus calculated based on the equilibrium of forces. Further, the liquid-solid and liquid-liquid CE are both studied by using the method, and we focus on the latter. The behavior of negatively pre-charged liquid droplet in the liquid-liquid CE is found to be different from that of the positively pre-charged one. The results show that the silicone oil droplet prefers to receive negative charges from a negatively charged aqueous droplet rather than positive charges from a positively charged aqueous droplet, which provides a strong evidence about the dominant role played by electron transfer in the liquid-liquid CE.

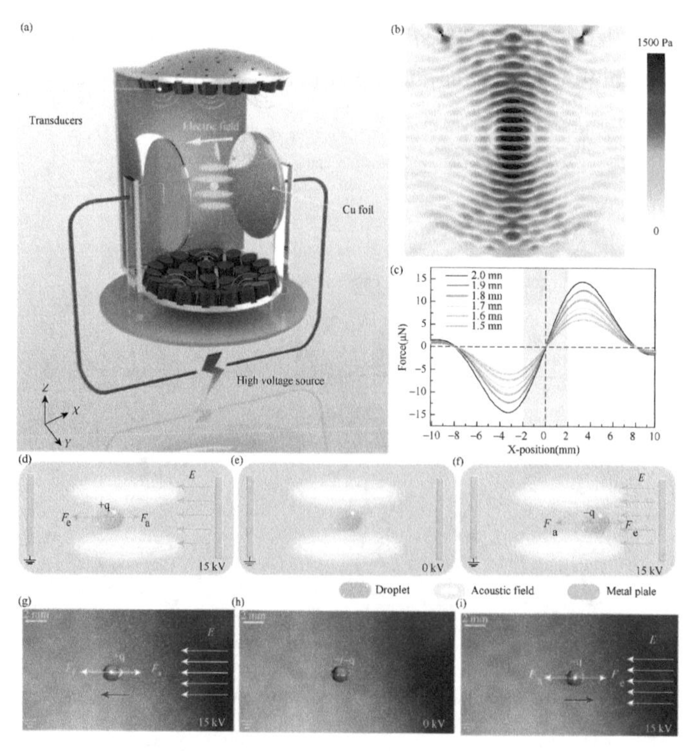

扫码阅全文

http://www.binn.cas.cn/book/202106/W020211026465249189906.pdf

II.14 气体-固体接触起电的机理

The Origins of Gas-Solid Contact-Electrification

1. Probing Contact Electrification between Gas and Solid Surface

Lin Lin Sun, Ziming Wang, Chengyu Li, Wei Tang *and* Zhong Lin Wang

(in press)

简介：本文利用单电极摩擦纳米发电机作为一个探针，首次研究了气体-固体界面接触起电的现象，并证实了电子传输也在气体-固体界面发生，并再次确认了王氏跃迁可以解释其电荷转移过程及其他现象的普适性。

ABSTRACT Contact electrification exists everywhere and between any phase of matters, while its mechanism still remains to be studied. Recent triboelectric nanogenerator serves as a probe and discloses some new clues about the mechanism present in solid-solid, solid-liquid, and liquid-liquid contact electrification. The gas-solid model still remains to be exploited. Here, we investigated the contact electrification between gas and solid based on the single-electrode triboelectric nanogenerator. Our work shows that the amount of transferred charges between gas and solid particles increases as the surface area, the movement distance, and/or the initial charges of particle increase. And we find that the initial charges on the particle surface can enhance the gas adsorption and attract more polar molecules. Since there are bare ions existing in the gas/solid contact, we speculate that the gas/solid contact electrification is mainly based on electrons transferring. Further, we propose a theoretical model about the gas-solid contact electrification and the gas adsorption model involving the initial charges of the particle. Our study may have great significance to the gas-solid interface chemistry.

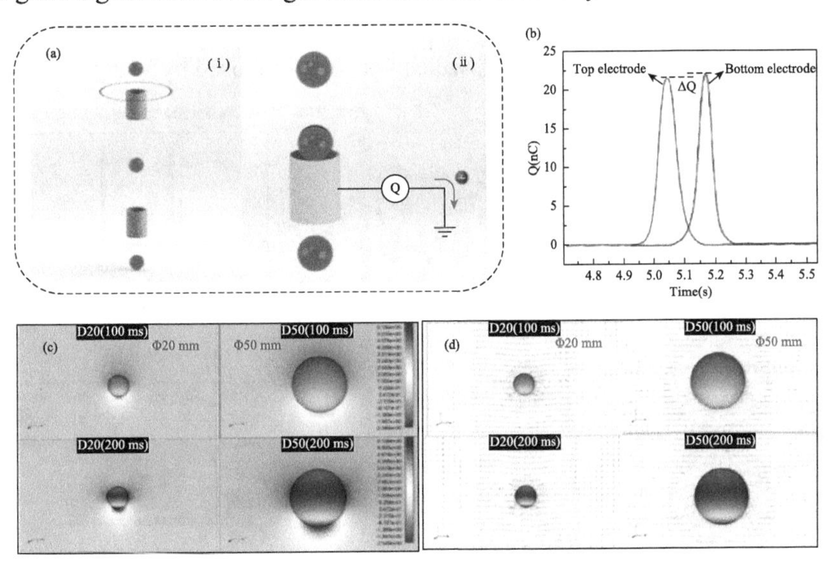

扫码阅全文

http://www.binn.cas.cn/book/202106/t20210610_6082006.html

II.15 摩擦伏特效应
Tribovoltaic Effect

1. Photovoltaic Effect and Tribovoltaic Effect at Liquid-Semiconductor Interface

Mingli Zheng, Shiquan Lin, Zhen Tang, Yawei Feng, Zhong Lin Wang*

Nano Energy, 83 (2021) 105810.

简介：本文通过比较光伏效应与摩擦伏特效应的输出类似性，首次明确提出了摩擦伏特效应的基本物理机制：当 n 型半导体与 p 型半导体相互滑动时，在新形成的界面处由于原子与原子形成新的键合，必然释放出一个能量子，即结合能量子，该能量子激发 pn 结区域的电子-空穴对，内建电场把电子与空穴分开，在外电路就观测到直流电。

ABSTRACT Contact electrification involving semiconductors has attracted attention for that it generates direct current. But its mechanism is still under debate, especially for the liquid-semiconductor cases. Here, the tribo-current is generated by sliding a DI water droplet on a semiconductor wafer, such as Si and TiO_2, under the light irradiation. It is revealed that the photoexcited electron-hole pairs at the interface will contribute to the tribo-current, and the enhanced tribo-current increases with the increased light intensity or the decreased light wavelength. The results suggest that the tribo-current at the DI water-semiconductor interfaces is induced by the tribovoltaic effect, in which electron-hole pairs are excited during contact owing to the energy released by the newly formed bonds, which can be named as "bindington". The electron-hole pairs are further driven by the built-in electric field to move from one side to the other side at the interfaces, generating a direct current. The findings imply that the electron transfer exist at the liquid-solid interface in the CE, and support the "two-step" model for the formation of the electric-double layer, which was first proposed by Wang.

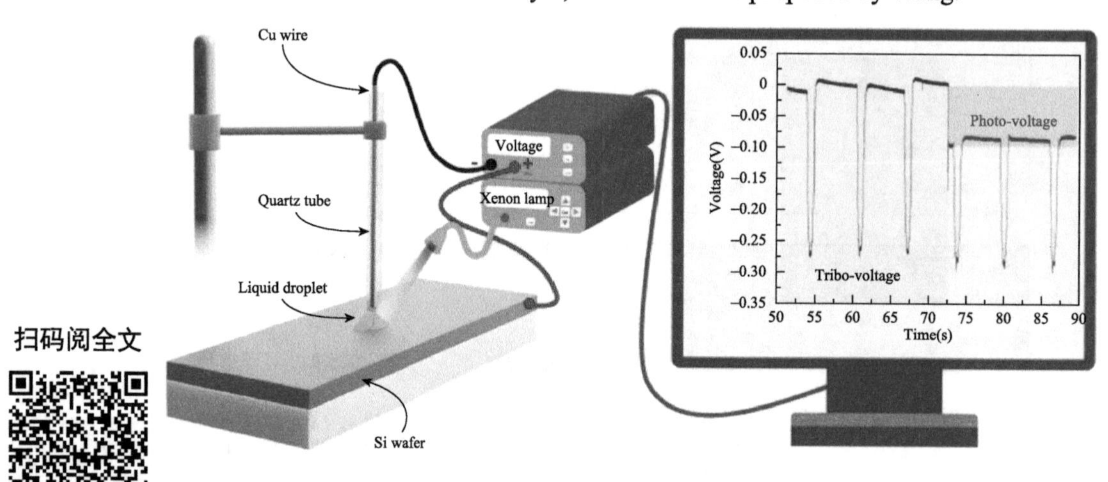

扫码阅全文

http://www.binn.cas.cn/book/202106/W020210610543041881289.pdf

2. The Tribovoltaic Effect and Electron Transfer at a Liquid-Semiconductor Interface

Shiquan Lin, Xiangyu Chen, Zhong Lin Wang*

Nano Energy, 76 (2020) 105070.

简介：本文通过在半导体表面滑动液滴而得到直流电输出，首次证实了摩擦伏特效应在液体-半导体界面也存在。

ABSTRACT Contact electrification (CE) is a ubiquitous phenomenon, which occurs at almost any interfaces between any phases, but its mechanism still remains ambiguous, especially for the liquid-solid case. Here, the tribo-voltage and the DC tribo-current are generated by sliding a water droplet over a silicon surface. By varying the doping types and concentrations, the direction of the tribo-voltage and the tribo-current is found highly depending on the built-in electric field at the liquid-solid junction. It is suggested that the electrical output is induced by the triboelectric effect at the water-silicon interface, in which electron-hole pairs are excited at the silicon surface by the energy released at interface during sliding, and further separated by the built-in electric field, which is analogous to that occurs in the photovoltaic effect. The only difference is that the energy in tribovoltaic effect is provided by electron transfer across the sliding interface owing to CE and bonding interaction between liquid molecules and semiconductor surface, rather than by photon. It is demonstrated that the electron transfer exists at the liquid-solid interface, and the "two-step" model for the formation of the electric double-layer is further verified, in which the electron transfer is considered to occur in the first step when the liquid contacts a solid in the very first time, which is then followed by ion chemical/physical adsorption as the second step. The tribovoltaic effect provides a new approach for studying of the mechanism at liquid-solid CE.

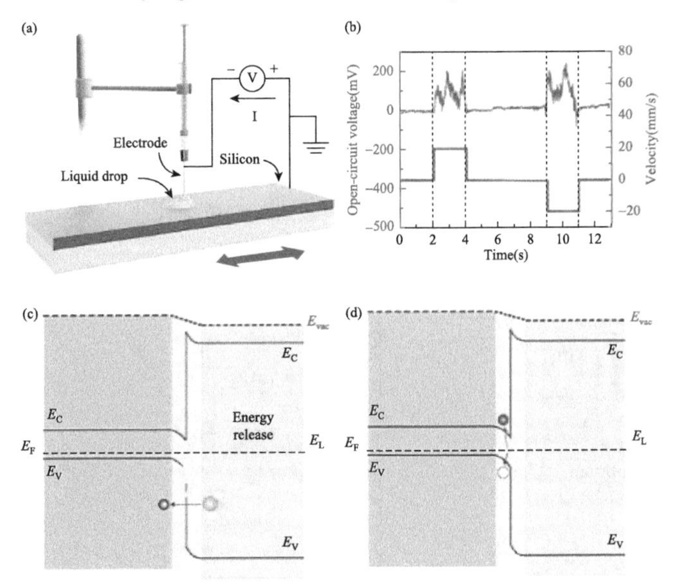

扫码阅全文

http://www.binn.cas.cn/book/202106/W020210610543043085497.pdf

3. Tribovoltaic Effect on Metal-Semiconductor Interface for Direct-Current Low-Impedance Triboelectric Nanogenerators

Zhi Zhang, Dongdong Jiang, Junqing Zhao, Guoxu Liu, Tianzhao Bu, Chi Zhang* *and* **Zhong Lin Wang***

Advanced Energy Materials, 10 (2020) 1903713.

简介：本文首次证实摩擦伏特效应在金属-固体界面的存在。

ABSTRACT The triboelectric nanogenerator (TENG) is a new energy technology that is enabled by coupled contact electrification and electrostatic induction. The conventional TENGs are usually based on organic polymer insulator materials, which have the limitations and disadvantages of high impedance and alternating output current. Here, a tribovoltaic effect based metal-semiconductor direct-current triboelectric nanogenerator (MSDC-TENG) is reported. The tribovoltaic effect is facilitated by direct voltage and current by rubbing a metal/semiconductor on another semiconductor. The frictional energy released by the forming atomic bonds excites nonequilibrium carriers, which are directionally separated to form a current under the built-in electric field. The continuous average open-circuit voltage (10-20 mV), short-circuit direct-current output (10-20 μA), and low impedance characteristic (0.55-5 kΩ) of the MSDC-TENG can be observed during relative sliding of the metal and silicon. The working parameters are systematically studied for electric output and impedance characteristics. The results reveal that faster velocity, larger pressure, and smaller area can improve the maximum power density. The internal resistance is mainly determined by the velocity and the electrical resistance of semiconductor. This work not only expands the material candidates of TENGs from organic polymers to semiconductors, but also demonstrates a tribovoltaic effect based electric energy conversion mechanism.

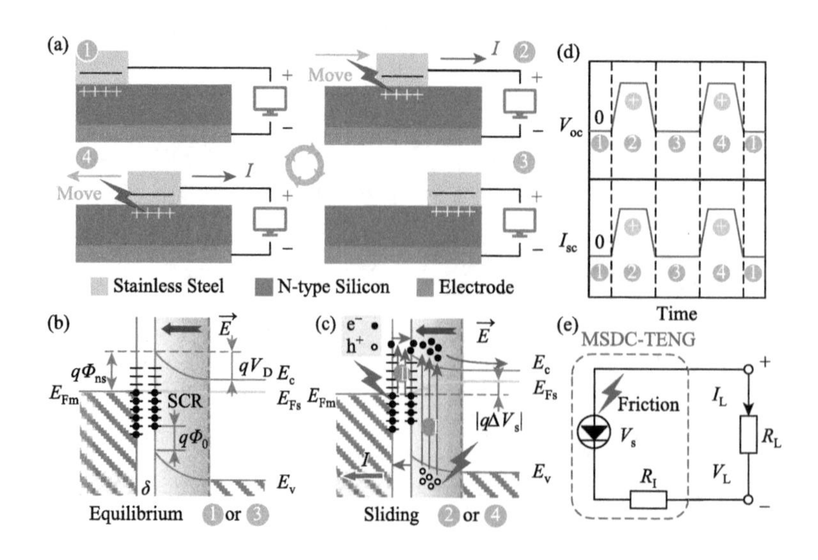

扫码阅全文

http://www.binn.cas.cn/book/202106/W020210610543043935052.pdf

4. Scanning Probing of the Tribovoltaic Effect at the Sliding Interface of Two Semiconductors

Mingli Zheng, Shiquan Lin, Liang Xu, Laipan Zhu *and* **Zhong Lin Wang***

Advanced Materials, (2020) 2000928.

简介：利用半导体扫描探针在半导体表面的滑动，从微观上证明了摩擦伏特效应的存在。

ABSTRACT Contact electrification (CE or triboelectrification) is a common phenomenon, which can occur for almost all types of materials. In previous studies, the CE between insulators and metals has been widely discussed, while CE involving semiconductors is only recently. Here, a tribo-current is generated by sliding an N-type diamond coated tip on a P-type or N-type Si wafers. The density of surface states of the Si wafer is changed by introducing different densities of doping. It is found that the tribo-current between two sliding semiconductors increases with increasing density of surface states of the semiconductor and the sliding load. The results suggest that the tribo-current is induced by the tribovoltaic effect, in which the electron-hole pairs at the sliding interface are excited by the energy release during friction, which may be due to the transition of electrons between the surface states during contact, or bond formation across the sliding interface. The electron-hole pairs at the sliding interface are subsequently separated by the built-in electric field at the PN or NN heterojunctions, which results in a tribo-current, in analogy to that which occurs in the photovoltaic effect.

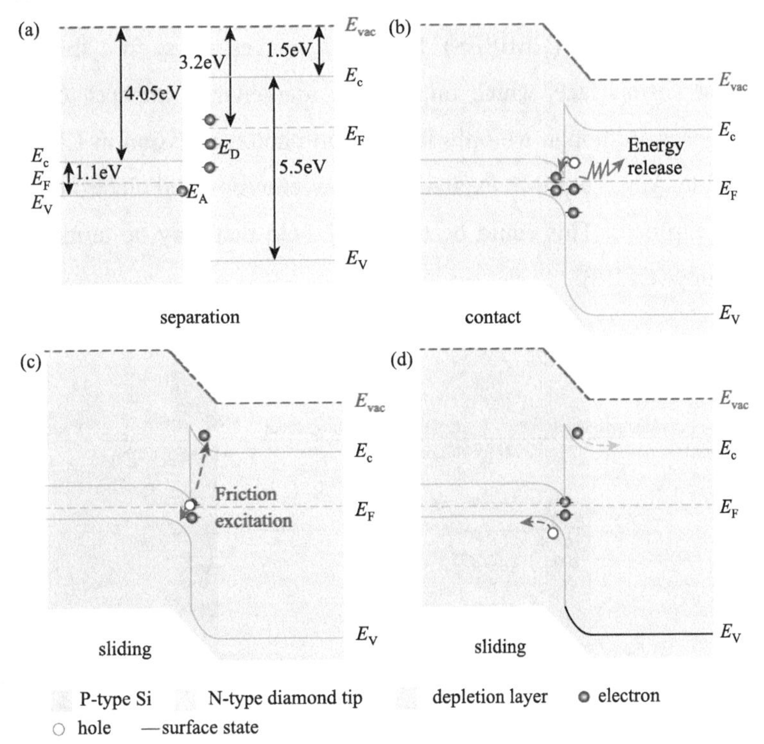

扫码阅全文

http://www.binn.cas.cn/book/202106/W020210610543044251866.pdf

Ⅱ.16 接触起电界面谱仪学

Contact Electrification Induced Interface Spectroscopy (CEIIS)

1. Interface Inter-atomic Electron-transition Induced Photon Emission in Contact-electrification

Ding Li, Cheng Xu, Yanjun Liao, Wenzhe Cai, Yongqiao Zhu, Zhong Lin Wang*

Science Advances, 7 (2021) eabj0349.

简介：本文首次报道了由界面处两种材料接触时，因跨原子电子跃迁所导致的光子辐射现象，有可能发展出分析界面电子跃迁结构的新谱仪学。这也是王氏跃迁的必然结果。

ABSTRACT Contact-electrification (CE), which is the scientific term for the well-known triboelectrification (TE), means that the charges are produced due to physical contact. CE is a universal phenomenon in both our daily life and nature world. And CE was recorded first as early as 2600 years ago during ancient Greek civilization. Here we report the first atomic featured photon emission spectra during CE between two solid materials. The photon emissions are the evidences that electrons transfer takes place at the interface from one atom in one material to another atom in the other material during CE. This process is the contact-electrification induced interface photon emission spectroscopy (CEIIPES). It naturally paves a way to a spectroscopy corresponding to the contact-electrification at an interface, which might have fundamental impacts to understand the interaction among solids, liquids and gases. Although we only focused on photon emission in CE for solid-solid case in this study, it could be expanded to Auger electron excitation, X-ray emission and electron emission in CE for general cases, which remain to be explored. This could be a general field that may be termed as contact electrification induced interface spectroscopy (CEIIS).

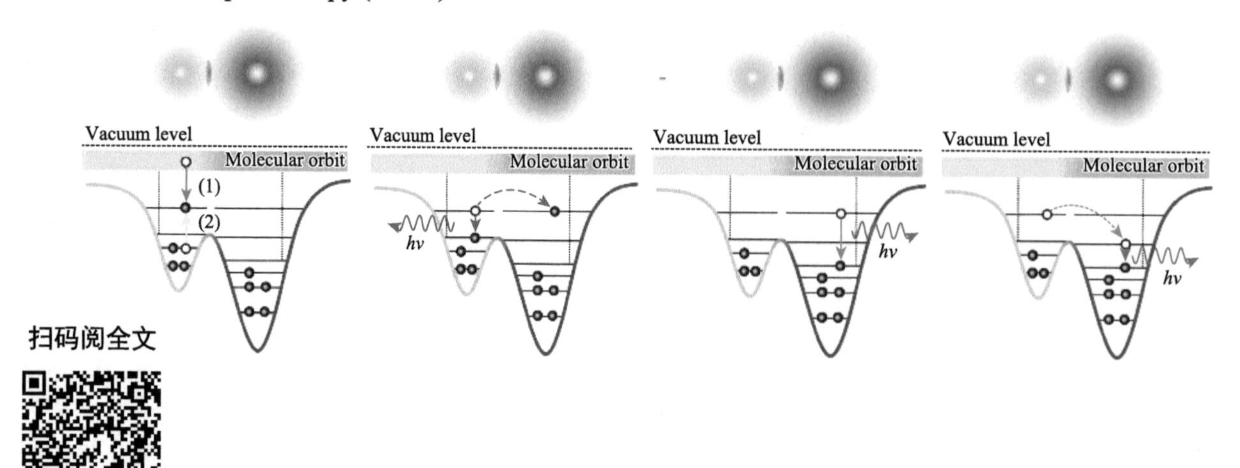

扫码阅全文

http://www.binn.cas.cn/book/202106/W020210929447442478438.pdf

II.17 摩擦纳米发电机：一种扫描探针技术

Triboelectric Nanogenerator—A New Scanning Tip Technology

1. Scanning Triboelectric Nanogenerator as a Nanoscale Probe for Measuring Local Surface Charge Density on a Dielectric Film

Shiquan Lin *and* **Zhong Lin Wang***

Applied Physics Letters, 118 (2021) 193901.

简介：本文首次把单电极式摩擦纳米发电机原理设计到探针的结构中，发展成为一种通过测试输出电流来测试表面电荷密度的扫描探针方法。

ABSTRACT Inspired by the contact-separation mode triboelectric nanogenerator (TENG), we propose a new technique for local surface charge density measurement based on atomic force microscopy. It is named scanning TENG, in which a conductive tip tapping above a charged dielectric surface produces an AC current between the tip and the dielectric bottom electrode due to electrostatic induction. The Fourier analysis shows that the amplitude of the first harmonic of the AC current is linearly related to surface charge density. The results demonstrate that the scanning TENG is a powerful tool for probing nanoscale charge transfer in contact-electrification.

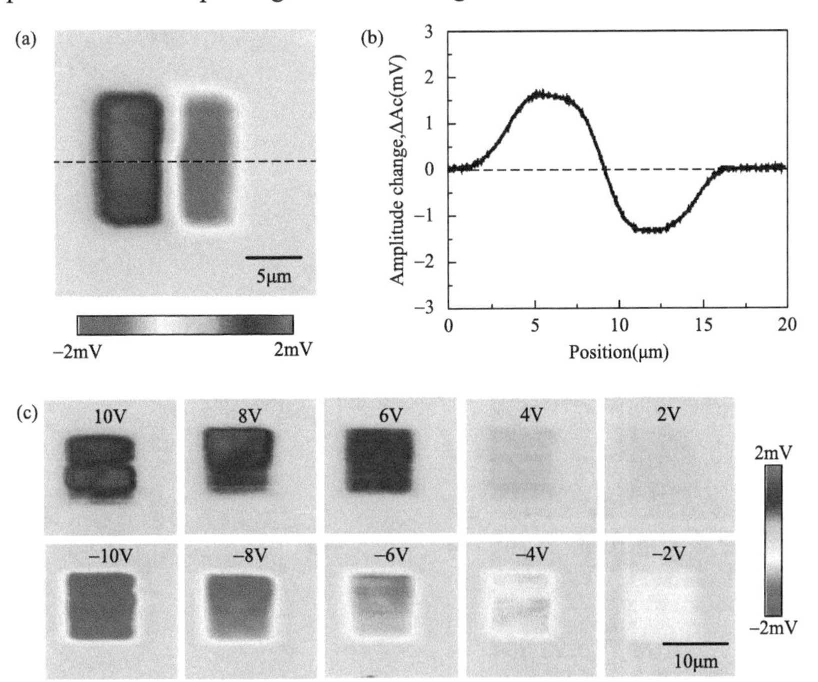

扫码阅全文

http://www.binn.cas.cn/book/202106/W020210610544524930801.pdf

2. Triboelectric Nanogenerator as a Probe for Measuring the Charge Transfer between Liquid and Solid Surfaces

Jinyang Zhang, Shiquan Lin, Mingli Zheng, Zhong Lin Wang*

ACS Nano, 15 (2021) 14830-14837.

简介：本文首次把单电极式摩擦纳米发电机作为一个探针来研究液体-固体接触起电。

ABSTRACT The phenomenon of triboelectricity involves the flow of charged species across an interface, but conclusively establishing the nature of the charge transfer has proven extremely difficult, especially for the liquid-solid cases. Herein, we developed a self-powered droplet triboelectric nanogenerator (droplet-TENG) with spatially arranged electrodes as a probe for measuring the charge transfer process between liquid and solid interfaces. The information on the electric signal on spatially arranged electrodes shows that the charge transfer between droplets and the solid is an accumulation process during the dropping and that the electron is the dominant charge-transfer species. Such a droplet-TENG showed a high sensitivity to the ratio of solvents in the mixed organic solution, and we postulated this is due to the possibility of generation of a hydrogen bond, affecting the electric signal on the spatially arranged electrodes. This work demonstrated a chemical sensing application based on the self-powered droplet triboelectric nanogenerator.

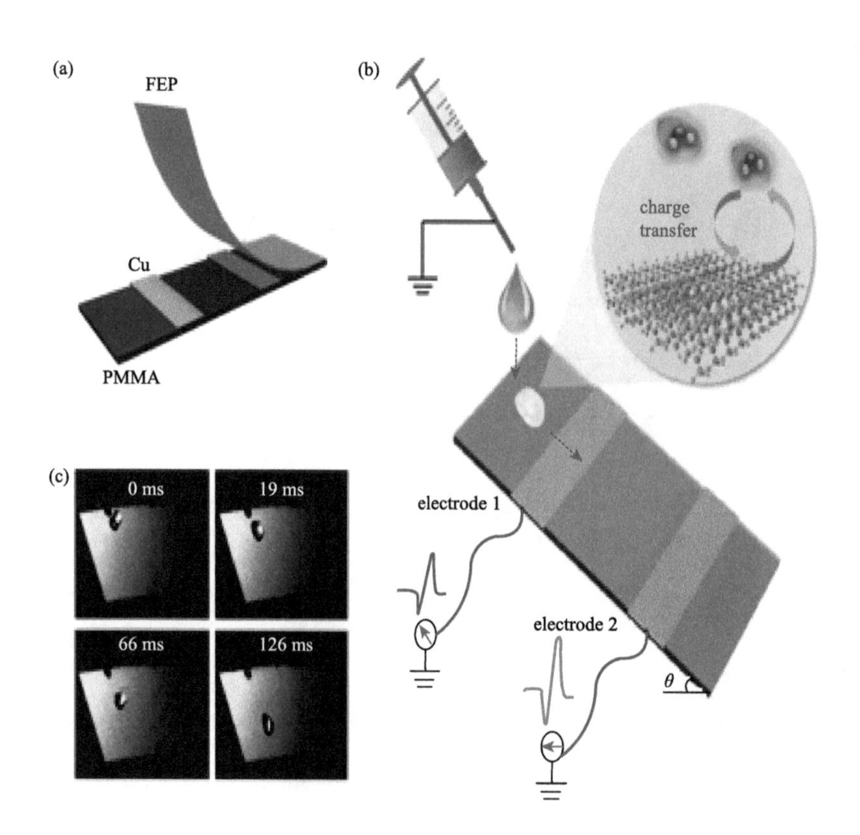

扫码阅全文

http://www.binn.cas.cn/book/202106/W020211210519358384499.pdf

第三部分

Part Ⅲ

Ⅲ.1 摩擦纳米发电机：微小功率源及其应用
Triboelectric Nanogenerators—As a Power Source

1. Nanoscale Triboelectric-Effect-Enabled Energy Conversion for Sustainably Powering Portable Electronics

Sihong Wang, Long Lin *and* Zhong Lin Wang*

Nano Letters, 12 (2012) 6339-6346.

简介：首个报道利用摩擦纳米发电机驱动手提式电子器件。

ABSTRACT Harvesting energy from our living environment is an effective approach for sustainable, maintenance-free, and green power source for wireless, portable, or implanted electronics. Mechanical energy scavenging based on triboelectric effect has been proven to be simple, cost-effective, and robust. However, its output is still insufficient for sustainably driving electronic devices/systems. Here, we demonstrated a rationally designed arch-shaped triboelectric nanogenerator (TENG) by utilizing the contact electrification between a polymer thin film and a metal thin foil. The working mechanism of the TENG was studied by finite element simulation. The output voltage, current density, and energy volume density reached 230 V, 15.5 μA/cm^2, and 128 mW/cm^3, respectively, and an energy conversion efficiency as high as 10%-39% has been demonstrated. The TENG was systematically studied and demonstrated as a sustainable power source that can not only drive instantaneous operation of light-emitting diodes (LEDs) but also charge a lithium ion battery as a regulated power module for powering a wireless sensor system and a commercial cell phone, which is the first demonstration of the nanogenerator for driving personal mobile electronics, opening the chapter of impacting general people's life by nanogenerators.

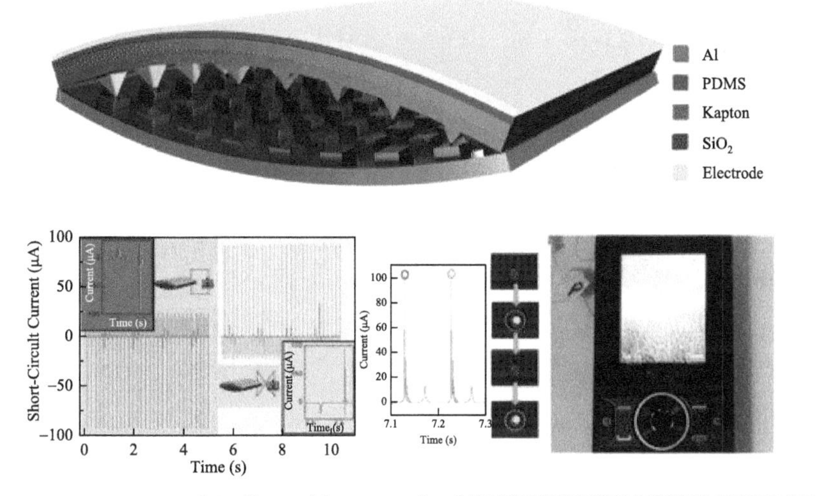

扫码阅全文

http://www.binn.cas.cn/book/202106/W020210623351451773675.pdf

2. Hybrid Energy Cell for Degradation of Methyl Orange by Self-Powered Electrocatalytic Oxidation

Ya Yang, Hulin Zhang, Sangmin Lee, Dongseob Kim, Woonbong Hwang *and* Zhong Lin Wang*

Nano Letters, 13 (2013) 803-808.

简介：首个报道利用复合式纳米发电机进行污染物分解。

ABSTRACT In general, methyl orange (MO) can be degraded by an electrocatalytic oxidation process driven by a power source due to the generation of superoxidative hydroxyl radical on the anode. Here, we report a hybrid energy cell that is used for a self-powered electrocatalytic process for the degradation of MO without using an external power source. The hybrid energy cell can simultaneously or individually harvest mechanical and thermal energies. The mechanical energy was harvested by the triboelectric nanogenerator (TENG) fabricated at the top by using a flexible polydimethysiloxane (PDMS) nanowire array with diameters of about 200nm. A pyroelectric nanogenerator (PENG) was fabricated below the TENG to harvest thermal energy. The power output of the device can be directly used for electrodegradation of MO, demonstrating a self-powered electrocatalytic oxidation process.

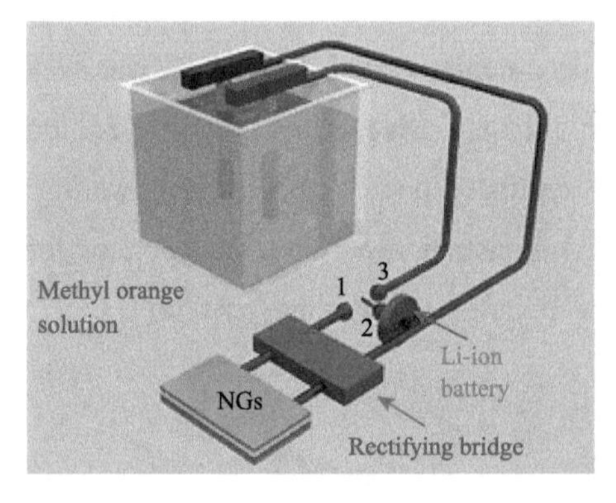

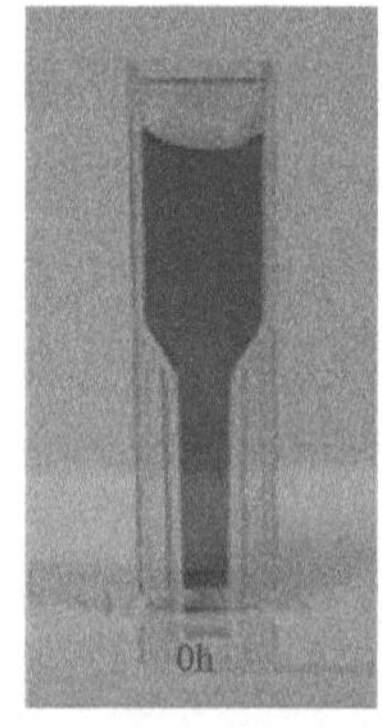

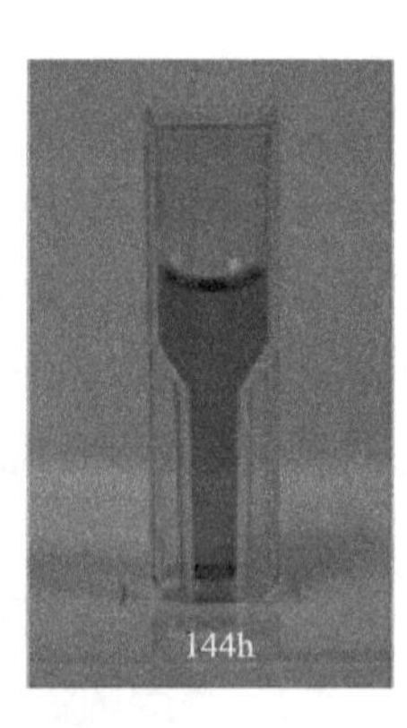

扫码阅全文

http://www.binn.cas.cn/book/202106/W020210623351452284545.pdf

3. Power-Generating Shoe Insole Based on Triboelectric Nanogenerators for Self-Powered Consumer Electronics

Guang Zhu, Peng Bai, Jun Chen, Zhong Lin Wang*

Nano Energy, 2 (2013) 688-692.

简介：首个报道利用基于摩擦纳米发电机的鞋垫来驱动消费产品。

ABSTRACT A major application of energy-harvesting technology is to power portable and wearable consumer electronics. We report a packaged power-generating insole with built-in flexible multi-layered triboelectric nanogenerators that enable harvesting mechanical pressure during normal walking. Using the insole as a direct power source, we develop a fully packaged self-lighting shoe that has broad applications for display and entertainment purposes. Furthermore, a prototype of a wearable charging gadget is introduced for charging portable consumer electronics, such as cellphones. This work presents a successful initial attempt in applying energy-harvesting technology for self-powered electronics in our daily life, which will have broad impact on people's living style in the near future.

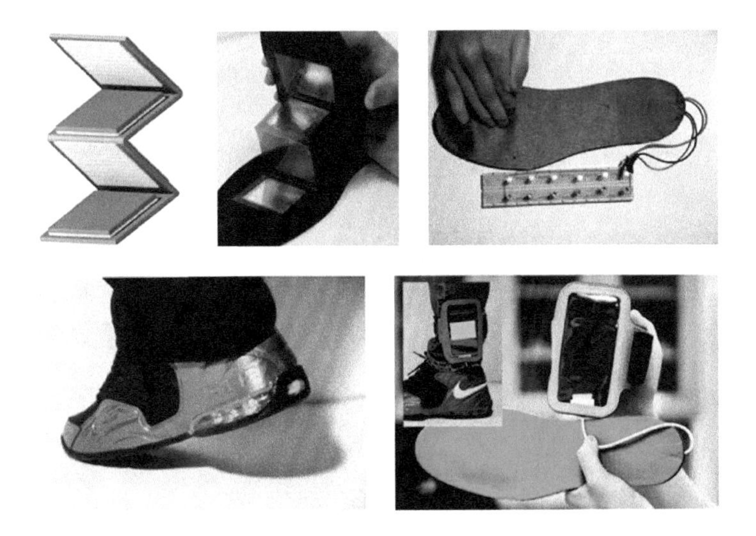

扫码阅全文

http://www.binn.cas.cn/book/202106/W020210623351453309251.pdf

4. Triboelectric Nanogenerator for Harvesting Wind Energy and as Self-Powered Wind Vector Sensor System

Ya Yang, Guang Zhu, Hulin Zhang, Jun Chen, Xiandai Zhong, Zong-Hong Lin, Yuanjie Su, Peng Bai, Xiaonan Wen *and* Zhong Lin Wang*

ACS Nano,7 (2013) 9461-9468.

简介：首个报道利用摩擦纳米发电机收集风能。

ABSTRACT We report a triboelectric nanogenerator (TENG) that plays dual roles as a sustainable power source by harvesting wind energy and as a self-powered wind vector sensor system for wind speed and direction detection. By utilizing the wind-induced resonance vibration of a fluorinated ethylene-propylene film between two aluminum foils, the integrated TENGs with dimensions of 2.5 cm×2.5 cm×22 cm deliver an output voltage up to 100 V, an output current of 1.6 μA, and a corresponding output power of 0.16 mW under an external load of 100 MΩ, which can be used to directly light up tens of commercial light-emitting diodes. Furthermore, a self-powered wind vector sensor system has been developed based on the rationally designed TENGs, which is capable of detecting the wind direction and speed with a sensitivity of 0.09 μA/(m/s). This work greatly expands the applicability of TENGs as power sources for self-sustained electronics and also self-powered sensor systems for ambient wind detection.

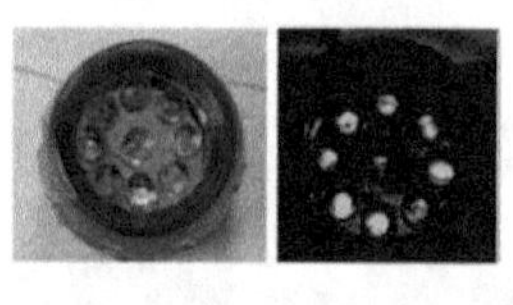

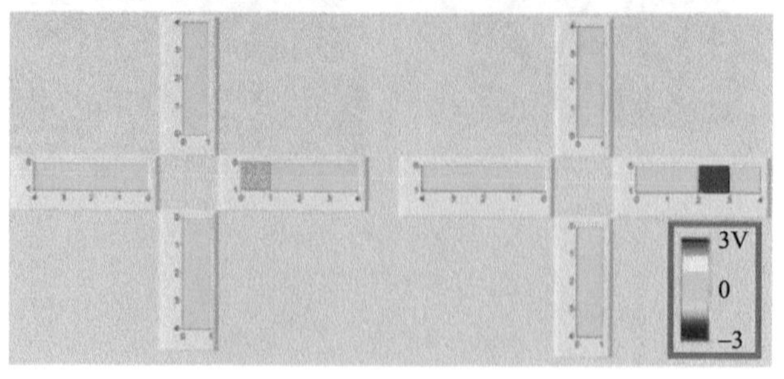

扫码阅全文

http://www.binn.cas.cn/book/202106/W020210623351453970724.pdf

5. Radial-Arrayed Rotary Electrification for High Performance Triboelectric Generator

Guang Zhu, Jun Chen, Tiejun Zhang, Qingshen Jing *and* Zhong Lin Wang*

Nature Communications, 5 (2014) 3456.

简介：首个报道高性能摩擦纳米发电机及其潜在的商业应用。

ABSTRACT Harvesting mechanical energy is an important route in obtaining cost-effective, clean and sustainable electric energy. Here we report a two-dimensional planar-structured triboelectric generator on the basis of contact electrification. The radial arrays of micro-sized sectors on the contact surfaces enable a high output power of 1.5 W (area power density of 19 $mW{\cdot}cm^{-2}$) at an efficiency of 24%. The triboelectric generator can effectively harness various ambient motions, including light wind, tap water flow and normal body movement. Through a power management circuit, a triboelectric-generator-based power-supplying system can provide a constant direct-current source for sustainably driving and charging commercial electronics, immediately demonstrating the feasibility of the triboelectric generator as a practical power source. Given exceptional power density, extremely low cost and unique applicability resulting from distinctive mechanism and structure, the triboelectric generator can be applied not only to self-powered electronics but also possibly to power generation at a large scale.

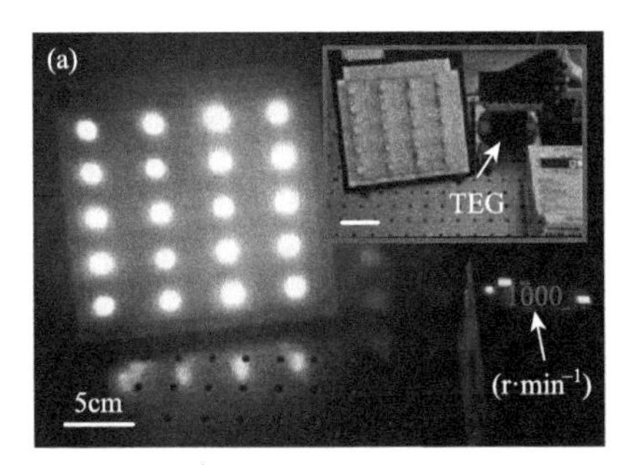

扫码阅全文

http://www.binn.cas.cn/book/202106/W020210623351454317982.pdf

6. *In Vivo* Powering of Pacemaker by Breathing-Driven Implanted Triboelectric Nanogenerator

Qiang Zheng, Bojing Shi, Fengru Fan, Xinxin Wang, Ling Yan, Weiwei Yuan, Sihong Wang, Hong Liu , Zhou Li* *and* Zhong Lin Wang*

Advanced Materials, 26 (2014) 5851-5856.

简介：首个报道利用呼吸带动的摩擦纳米发电机来驱动模拟的心脏起搏器。

ABSTRACT The first application of an implanted triboelectric nanogenerator (iTENG) that enables harvesting energy from in vivo mechanical movement in breathing to directly drive a pacemaker is reported. The energy harvested by iTENG from animal breathing is stored in a capacitor and successfully drives a pacemaker prototype to regulate the heart rate of a rat. This research shows a feasible approach to scavenge biomechanical energy, and presents a crucial step forward for lifetime-implantable self-powered medical devices.

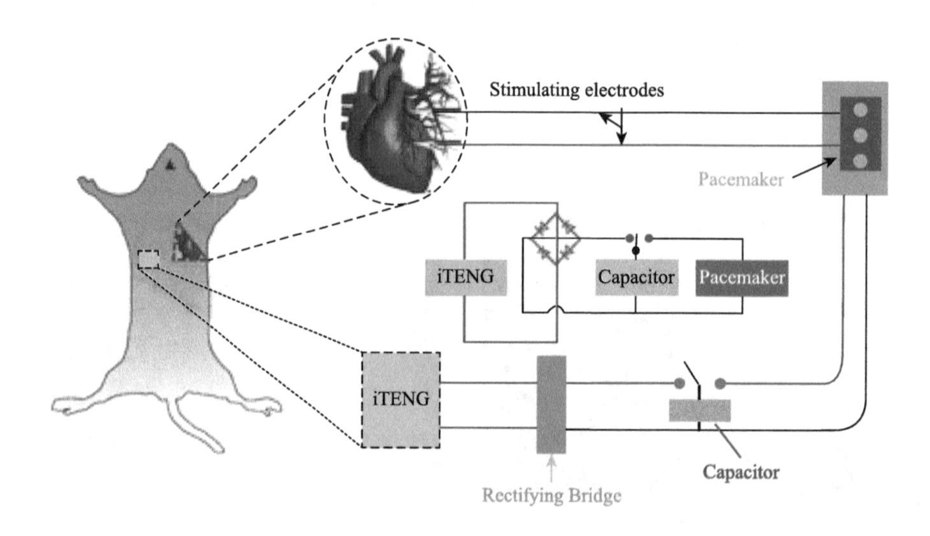

扫码阅全文

http://www.binn.cas.cn/book/202106/W020210623351454824313.pdf

7. Biodegradable Triboelectric Nanogenerator as a Life-Time Designed Implantable Power Source

Qiang Zheng, Yang Zou, Yalan Zhang, Zhuo Liu, Bojing Shi, Xinxin Wang, Yiming Jin, Han Ouyang, Zhou Li*, Zhong Lin Wang*

Science Advances, 2 (2016) e1501478.

简介：首个报道利用生物可降解的材料制作的体内植入式摩擦纳米发电机。

ABSTRACT Transient electronics built with degradable organic and inorganic materials is an emerging area and has shown great potential for in vivo sensors and therapeutic devices. However, most of these devices require external power sources to function, which may limit their applications for in vivo cases. We report a biodegradable triboelectric nanogenerator (BD-TENG) for in vivo biomechanical energy harvesting, which can be degraded and resorbed in an animal body after completing its work cycle without any adverse long-term effects. Tunable electrical output capabilities and degradation features were achieved by fabricated BD-TENG using different materials. When applying BD-TENG to power two complementary micrograting electrodes, a DC-pulsed electrical field was generated, and the nerve cell growth was successfully orientated, showing its feasibility for neuron-repairing process. Our work demonstrates the potential of BD-TENG as a power source for transient medical devices.

扫码阅全文

http://www.binn.cas.cn/book/202106/W020210623351455202161.pdf

8. Symbiotic Cardiac Pacemaker

Han Ouyang, Zhuo Liu, Ning Li, Bojing Shi, Yang Zou, Feng Xie, Ye Ma, Zhe Li, Hu Li, Qiang Zheng, Xuecheng Qu, Yubo Fan, Zhong Lin Wang*, Hao Zhang* *and* Zhou Li*

Nature Communications, 10 (2019) 1821.

简介：首个报道利用生物可降解的材料制作的摩擦纳米发电机利用动物呼吸来驱动一个商用心脏起搏器。这是世界上首次，也是王中林团队努力了十年的结果。

ABSTRACT Self-powered implantable medical electronic devices that harvest biomechanical energy from cardiac motion, respiratory movement and blood flow are part of a paradigm shift that is on the horizon. Here, we demonstrate a fully implanted symbiotic pacemaker based on an implantable triboelectric nanogenerator, which achieves energy harvesting and storage as well as cardiac pacing on a large-animal scale. The symbiotic pacemaker successfully corrects sinus arrhythmia and prevents deterioration. The open circuit voltage of an implantable triboelectric nanogenerator reaches up to 65.2 V. The energy harvested from each cardiac motion cycle is 0.495 μJ, which is higher than the required endocardial pacing threshold energy (0.377 μJ). Implantable triboelectric nanogenerators for implantable medical devices offer advantages of excellent output performance, high power density, and good durability, and are expected to find application in fields of treatment and diagnosis as in vivo symbiotic bioelectronics.

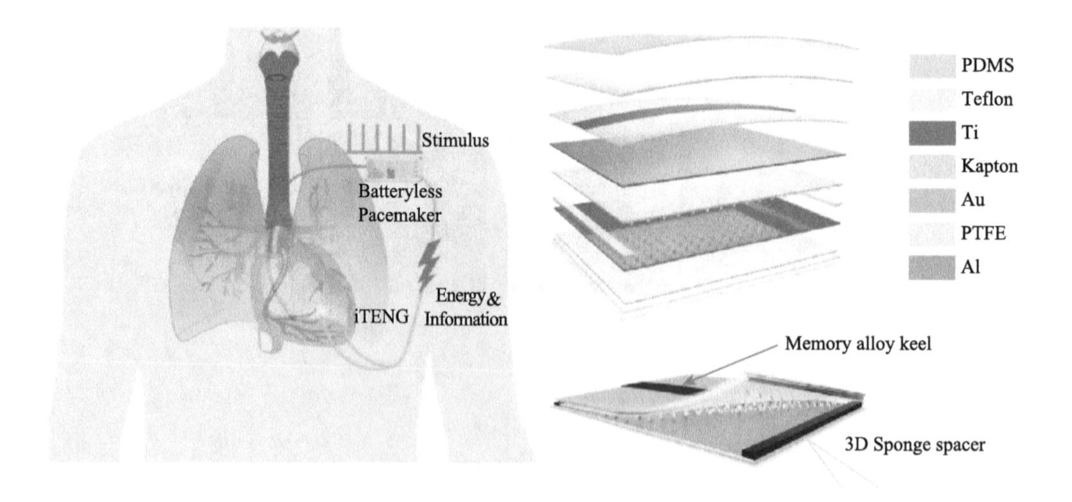

扫码阅全文

http://www.binn.cas.cn/book/202106/W020210623351456700634.pdf

9. Sustainably Powering Wearable Electronics Solely by Biomechanical Energy

Jie Wang, Shengming Li, Fang Yi, Yunlong Zi, Jun Lin, Xiaofeng Wang, Youlong Xu *and* **Zhong Lin Wang***

Nature Communications, 7 (2016) 12744.

简介：首个报道完全利用生物机械能带动的摩擦纳米发电机来驱动穿戴式电子产品。

ABSTRACT Harvesting biomechanical energy is an important route for providing electricity to sustainably drive wearable electronics, which currently still use batteries and therefore need to be charged or replaced/disposed frequently. Here we report an approach that can continuously power wearable electronics only by human motion, realized through a triboelectric nanogenerator (TENG) with optimized materials and structural design. Fabricated by elastomeric materials and a helix inner electrode sticking on a tube with the dielectric layer and outer electrode, the TENG has desirable features including flexibility, stretchability, isotropy, weavability, water-resistance and a high surface charge density of 250 μC·m^{-2}. With only the energy extracted from walking or jogging by the TENG that is built in outsoles, wearable electronics such as an electronic watch and fitness tracker can be immediately and continuously powered.

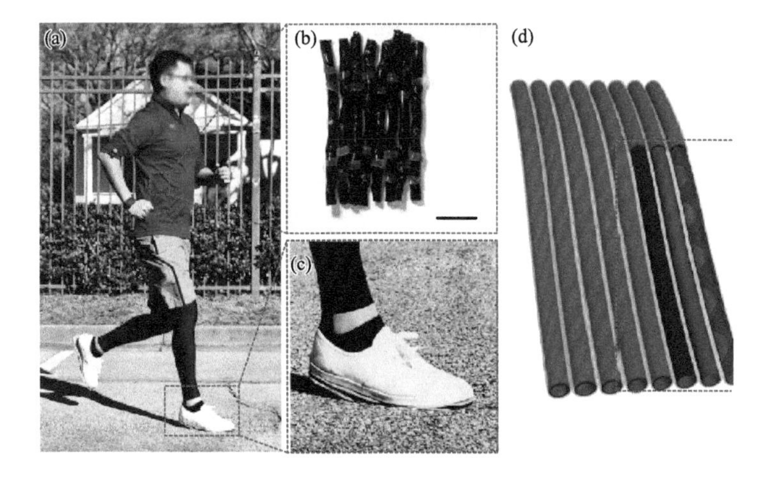

扫码阅全文

http://www.binn.cas.cn/book/202106/W020210623351457444129.pdf

10. Electrochemical Cathodic Protection Powered by Triboelectric Nanogenerator

Wenxi Guo, Xiaoyi Li, Mengxiao Chen, Lu Xu, Lin Dong, Xia Cao, Wei Tang, Jing Zhu, Changjian Lin*, Caofeng Pan* *and* Zhong Lin Wang*

Advanced Functional Materials, 24 (2014) 6691-6699.

简介：首个报道利用摩擦纳米发电机输出来做电化学阴极保护。

ABSTRACT Metal corrosion is universal in the nature and the corrosion prevention for metals plays an important role everywhere in national economic development and daily life. Here a disk triboelectric nanogenerator (TENG) with segmental structures is introduced as power source to achieve a special cathodic protection effect for steels. The output transferred charges and short-circuit current density of the TENG achieve 1.41 mC/min and 10.1 mA/m^2, respectively, when the rotating speed is 1000 revolutions per minute (rpm). The cathodic protection potential, Tafel polarization curves and electrochemical impedance spectra (EIS) measurements are measured to evaluate the corrosion protection effect for the 403 stainless steel (403SS). The cathodic protection potential range from −320 mV to −5320 mV is achieved by changing rotation speeds and external resistance when the steel is coupled in a 0.5 m NaCl solution to the negative pole of the disk TENG. The corrosion tests results indicate that the TENG can produce 59.1% degree of protection for Q235 steels in 0.5 m NaCl solution. Furthermore, an application of marine corrosion prevention is presented by mounting the TENG onto a buoy. This work demonstrates a versatile, cost-effect and self-powered system to scavenging mechanical energy from environment, leading to effectively protect the metal corrosion without additional power sources.

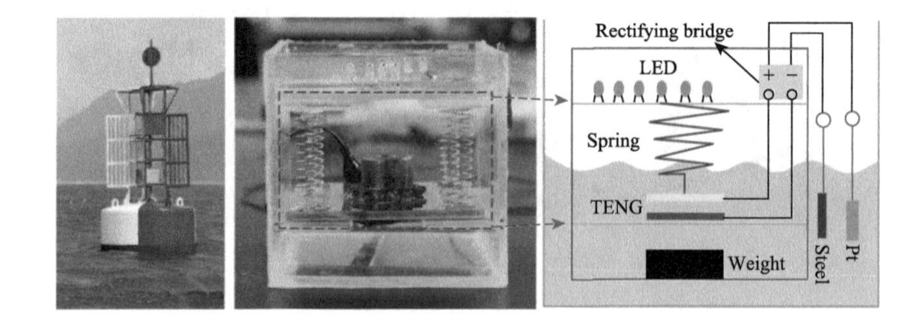

扫码阅全文

http://www.binn.cas.cn/book/202106/W020210623351458028085.pdf

11. Ultrathin, Rollable, Paper-Based Triboelectric Nanogenerator for Acoustic Energy Harvesting and Self-Powered Sound Recording

Xing Fan, Jun Chen, Jin Yang, Peng Bai, Zhaoling Li *and* **Zhong Lin Wang***

ACS Nano, 9 (2015) 4236-4243.

简介：首个报道利用纸质摩擦纳米发电机来收集声音能量，并作为自驱动麦克风。

ABSTRACT A 125 μm thickness, rollable, paper-based triboelectric nanogenerator (TENG) has been developed for harvesting sound wave energy, which is capable of delivering a maximum power density of 121 mW/m^2 and 968 W/m^3 under a sound pressure of 117 dB$_{SPL}$. The TENG is designed in the contact-separation mode using membranes that have rationally designed holes at one side. The TENG can be implemented onto a commercial cell phone for acoustic energy harvesting from human talking; the electricity generated can be used to charge a capacitor at a rate of 0.144 V/s. Additionally, owing to the superior advantages of a broad working bandwidth, thin structure, and flexibility, a self-powered microphone for sound recording with rolled structure is demonstrated for all-sound recording without an angular dependence. The concept and design presented in this work can be extensively applied to a variety of other circumstances for either energy-harvesting or sensing purposes, for example, wearable and flexible electronics, military surveillance, jet engine noise reduction, low-cost implantable human ear, and wireless technology applications.

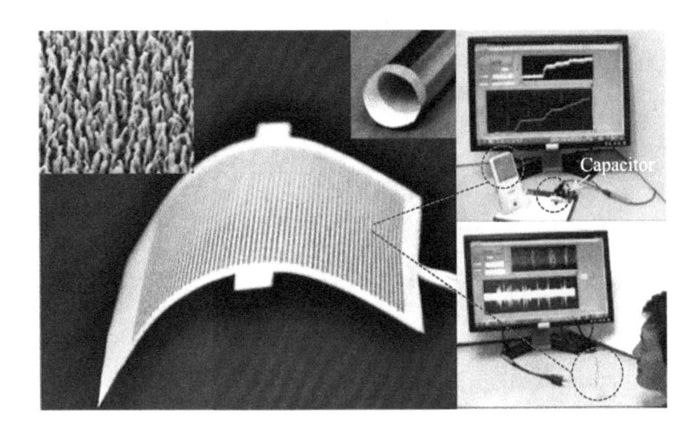

扫码阅全文

http://www.binn.cas.cn/book/202106/W020210623351458594816.pdf

12. Triboelectrification-Enabled Self-Powered Detection and Removal of Heavy Metal Ions in Wastewater

Zhaoling Li, Jun Chen, Hengyu Guo, Xing Fan, Zhen Wen, Min-Hsin Yeh, Chongwen Yu, Xia Cao* *and* **Zhong Lin Wang***

Advanced Materials, 28 (2016) 2983-2991.

简介：首个报道利用摩擦纳米发电机的原理来探测和清除废水中的重金属离子。

ABSTRACT A fundamentally new working principle into the field of self-powered heavy-metal-ion detection and removal using the triboelectrification effect is introduced. The as-developed tribo-nanosensors can selectively detect common heavy metal ions. The water-driven triboelectric nanogenerator is taken as a sustainable power source for heavy-metal-ion removal by recycling the kinetic energy from flowing wastewater.

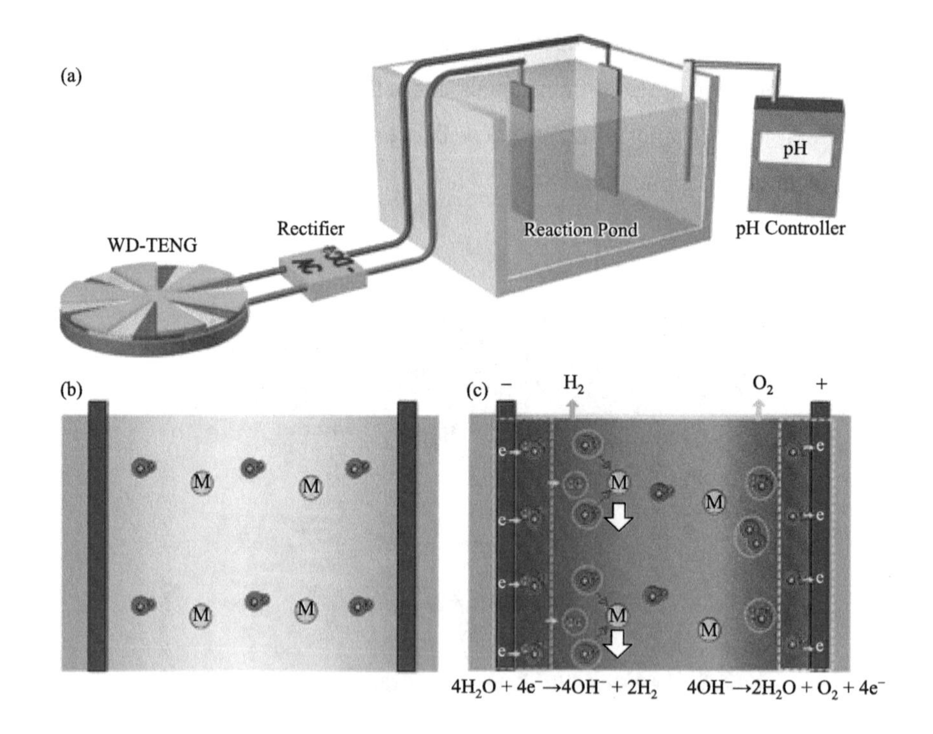

扫码阅全文

http://www.binn.cas.cn/book/202106/W020210623351459789227.pdf

13. Self-Powered Water Splitting Using Flowing Kinetic Energy

Wei Tang, Yu Han, Chang Bao Han, Cai Zhen Gao, Xia Cao* and Zhong Lin Wang*

Advanced Materials, 27(2015) 272-276.

简介：首个报道利用水流驱动的摩擦纳米发电机进行水分解。

ABSTRACT By utilizing a water-flow-driven triboelectric nanogenerator, a fully self-powered water-splitting process is demonstrated using the electricity converted from a water flow without additional energy costs. Considering the extremely low costs, the demonstrated approach is universally applicable and practically usable for future water electrolysis, which may initiate a research direction in the field of triboelectrolysis and possibly impacts energy science in general.

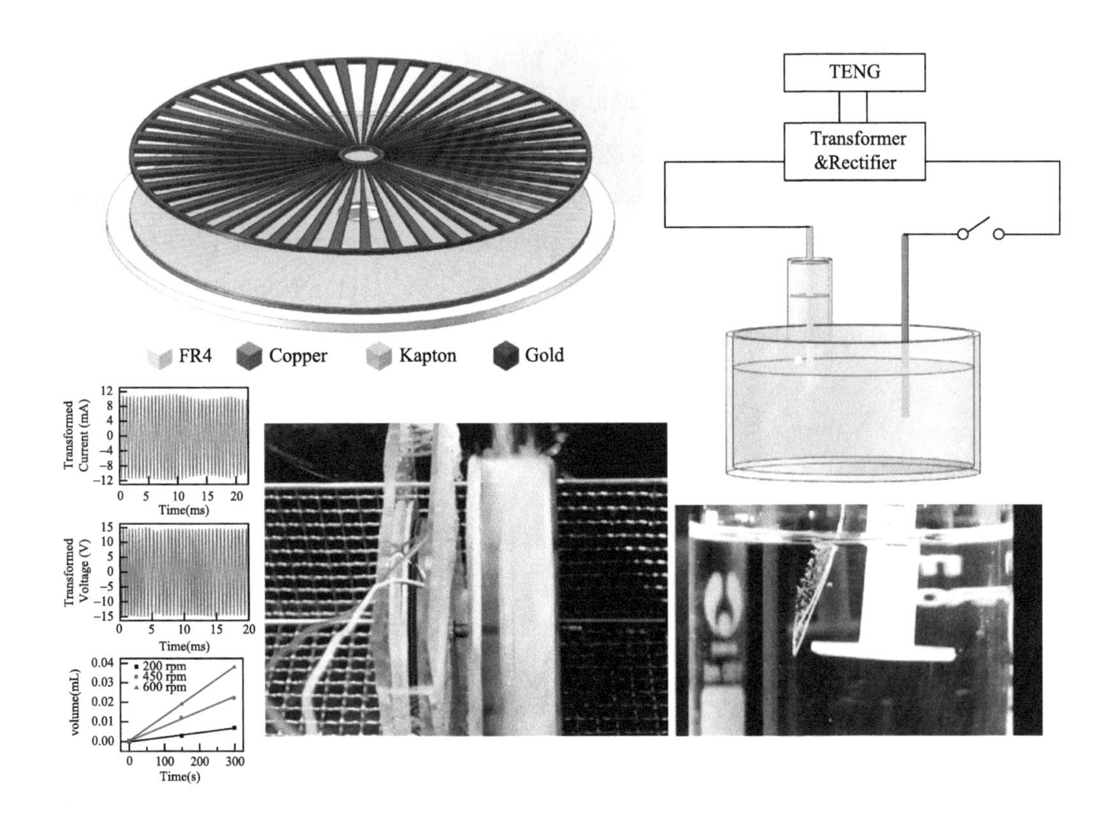

扫码阅全文

http://www.binn.cas.cn/book/202106/W020210623351460275717.pdf

14. Triboelectric Nanogenerator for Sustainable Wastewater Treatment via a Self-Powered Electrochemical Process

Shuwen Chen, Ning Wang, Long Ma, Tao Li , Magnus Willander, Yang Jie, Xia Cao* *and* Zhong Lin Wang*

Advanced Energy Materials, 6 (2016) 1501778.

简介：首个报道利用摩擦纳米发电机的自驱动电化学进行污水处理。

ABSTRACT By harvesting the flowing kinetic energy of water using a rotating triboelectric nanogenerator (R-TENG), this study demonstrates a self-powered wastewater treatment system that simultaneously removes rhodamine B (RhB) and copper ions through an advanced electrochemical unit. With the electricity generated by R-TENG, the removal efficiency (RE) of RhB can reach the vicinity of 100% within just 15 min when the initial concentration of RhB is around 100 ppm at optimized conditions. The removal efficiency of copper ions can reach 97.3% after 3 h within an initial concentration of 150 ppm at an optimized condition. Importantly, a better performance and higher treating efficiency are found by using the pulsed output of R-TENG than those using direct current (DC) supply for pollutant removal when consuming equal amount of energy. The recovered copper layer on the cathode through R-TENG is much denser, more uniform, and with smaller grain size (d = 20 nm) than those produced by DC process, which also hints at very promising applications of the R-TENG in electroplating industry. In light of the merits such as easy portability, low cost, and effectiveness, this R-TENG-based self-powered electrochemical system holds great potential in wastewater treatment and electroplating industry.

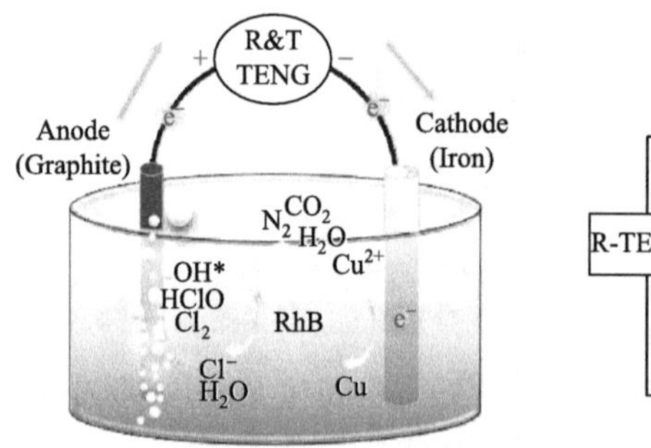

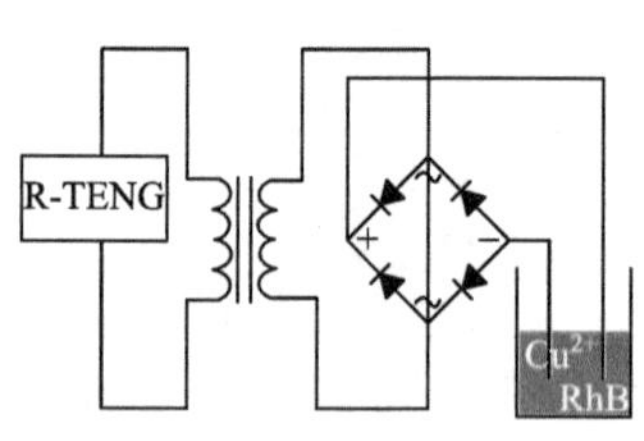

扫码阅全文

http://www.binn.cas.cn/book/202106/W020210623351460725531.pdf

15. Electric Eel-Skin-Inspired Mechanically Durable and Super-Stretchable Nanogenerator for Deformable Power Source and Fully Autonomous Conformable Electronic-Skin Applications

Ying-Chih Lai, Jianan Deng, Simiao Niu, Wenbo Peng, Changsheng Wu, Ruiyuan Liu, Zhen Wen *and* Zhong Lin Wang*

Advanced Materials, 28 (2016) 10024-10032.

简介：首个报道利用摩擦纳米发电机做仿生电子皮肤。

ABSTRACT Electric eel-skin-inspired mechanically durable and super-stretchable nanogenerator is demonstrated for the first time by using triboelectric effect. This newly designed nanogenerator can produce electricity by touch or tapping despite under various extreme mechanical deformations or even after experiencing damage. This device can be used not only as deformable and wearable power source but also as fully autonomous and self-sufficient adaptive electronic skin system.

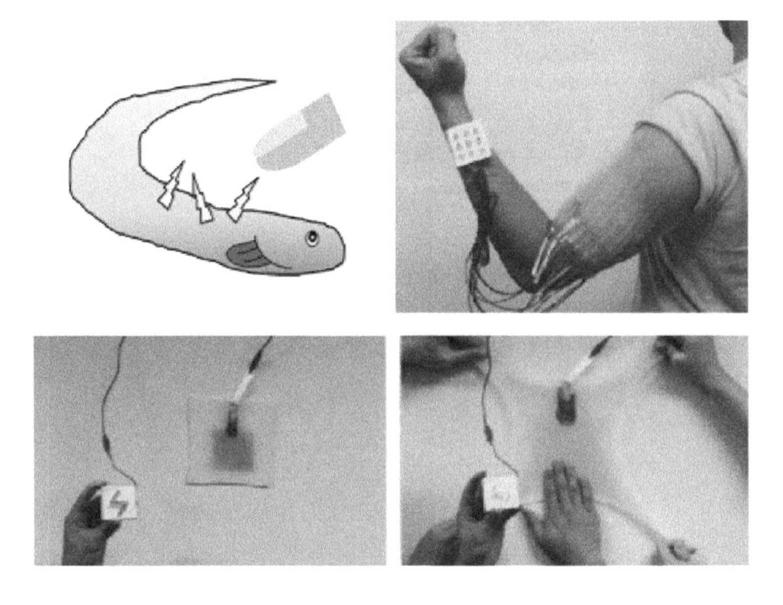

扫码阅全文

http://www.binn.cas.cn/book/202106/W020210623351461875901.pdf

16. A Self-Powered Implantable Drug-Delivery System Using Biokinetic Energy

Peiyi Song, Shuangyang Kuang, Nishtha Panwar, Guang Yang, Danny Jian Hang Tng, Swee Chuan Tjin, Wun Jern Ng, Maszenan Bin Abdul Majid, Guang Zhu*, Ken-Tye Yong* *and* Zhong Lin Wang*

Advanced Materials, 29 (2017) 1605668.

简介：首个报道利用摩擦纳米发电机来驱动药物投递。

ABSTRACT The first triboelectric-nanogenerator (TENG)-based self-powered implantable drug-delivery system is presented. Pumping flow rates from 5.3 to 40 μL min^{-1} under different rotating speeds of the TENG are realized. The implantable drug-delivery system can be powered with a TENG device rotated by human hand motion. *Ex vivo* trans-sclera drug delivery in porcine eyes is demonstrated by utilizing the biokinetic energies of human hands.

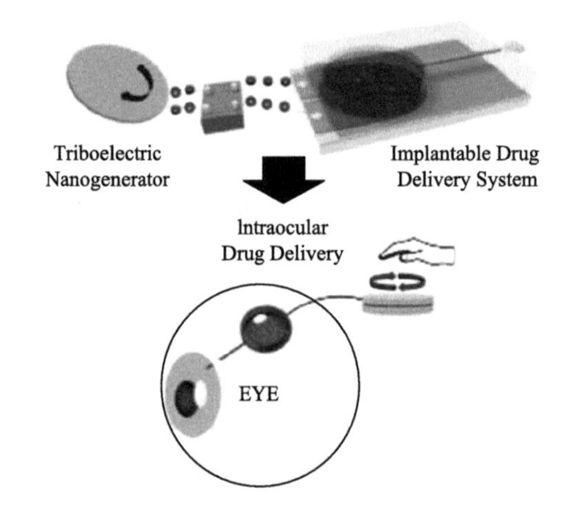

扫码阅全文

http://www.binn.cas.cn/book/202106/W020210623351462833928.pdf

17. High Efficient Harvesting of Underwater Ultrasonic Wave Energy by Triboelectric Nanogenerator

Yi Xi, Jie Wang, Yunlong Zi, Xiaogan Li, Changbao Han, Xia Cao, Chenguo Hu, Zhong Lin Wang*

Nano Energy, 38 (2017) 101-108.

简介：首个报道利用水下超声波来驱动滚珠式摩擦纳米发电机来发电，能量转换效率达到13%。

ABSTRACT Ultrasonic waves are existing in water and in our living environment, but harvesting the sonic wave energy especially in water is rather challenging, simply because of the high pressure generated by water. In this work, a triboelectric nanogenerator (TENG) has been designed using spherical pellets as the media for performing the contact-separation operation during ultrasonic wave excitation for energy harvesting. The fabricated TENG can attain an instantaneous output current from several milliamps to about one hundred milliamps and achieve an output power of 0.362 W/cm^2 at an ultrasonic wave frequency of 80 kHz. The average power conversion efficiency of the TENG has reached 13.1%. The equivalent output galvanostatic current is 1.43 mA, which is the highest value reported so far. The developed TENG pellets has been demonstrated to continuously power up to 12 lamps with 0.75 W each, and can continuously drive a temperature-humidity meter, an electronic watch, and directly drive a health monitor. This study presents the outstanding potential of TENG for underwater applications.

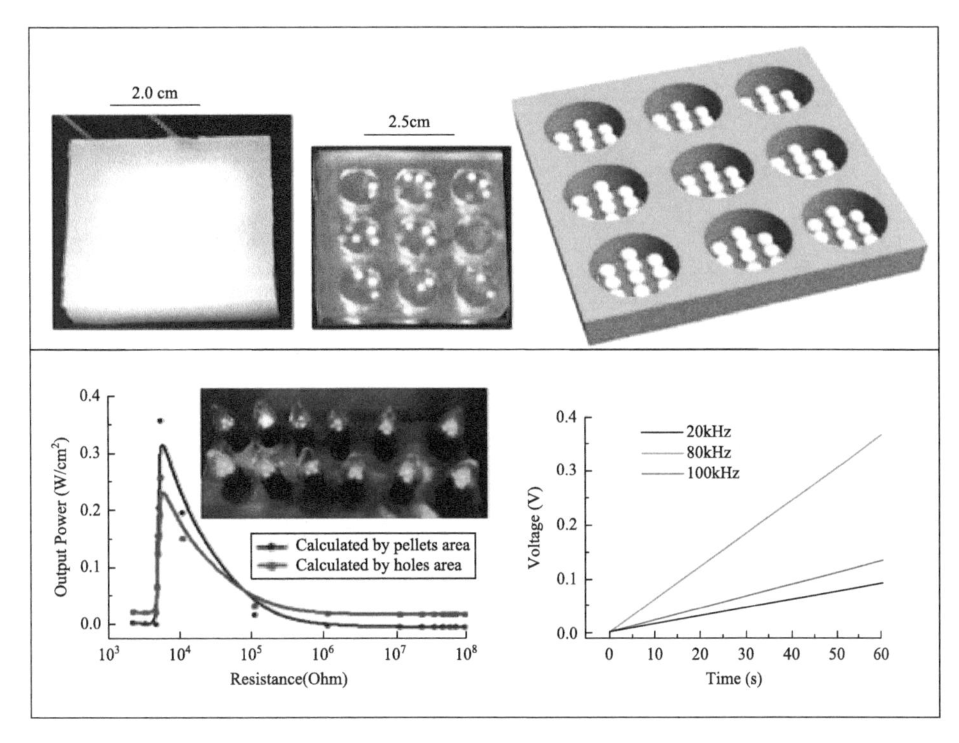

扫码阅全文

http://www.binn.cas.cn/book/202106/W020210623351464273717.pdf

18. A Bionic Stretchable Nanogenerator for Underwater Sensing and Energy Harvesting

Yang Zou, Puchuan Tan, Bojing Shi, Han Ouyang, Dongjie Jiang, Zhuo Liu, Hu Li, Min Yu, Chan Wang, Xuecheng Qu, Luming Zhao, Yubo Fan*, Zhong Lin Wang* *and* Zhou Li*

Nature Communications, 10 (2019) 2695.

简介：首个报道可拉伸摩擦纳米发电机进行水下传感与能量回收。

ABSTRACT Soft wearable electronics for underwater applications are of interest, but depend on the development of a waterproof, long-term sustainable power source. In this work, we report a bionic stretchable nanogenerator for underwater energy harvesting that mimics the structure of ion channels on the cytomembrane of electrocyte in an electric eel. Combining the effects of triboelectrification caused by flowing liquid and principles of electrostatic induction, the bionic stretchable nanogenerator can harvest mechanical energy from human motion underwater and output an open-circuit voltage over 10 V. Underwater applications of a bionic stretchable nanogenerator have also been demonstrated, such as human body multi-position motion monitoring and an undersea rescue system. The advantages of excellent flexibility, stretchability, outstanding tensile fatigue resistance (over 50,000 times) and underwater performance make the bionic stretchable nanogenerator a promising sustainable power source for the soft wearable electronics used underwater.

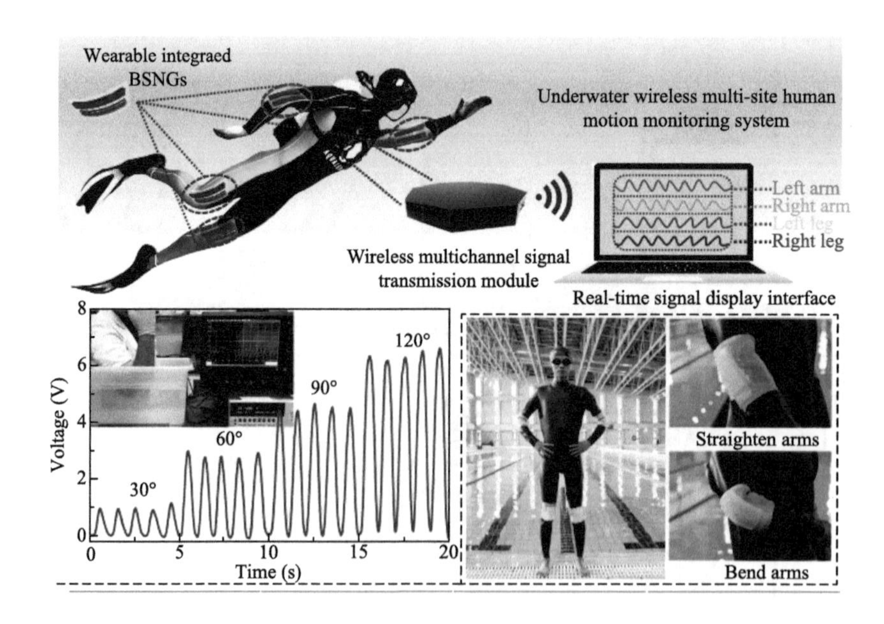

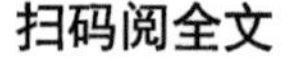

http://www.binn.cas.cn/book/202106/W020210623351465388340.pdf

19. Wind-Driven Self-Powered Wireless Environmental Sensors for Internet of Things at Long Distance

Di Liu, Baodong Chen, Jie An, Chengyu Li, Guoxu Liu, Jiajia Shao, Wei Tang*, Chi Zhang*, Zhong Lin Wang*

Nano Energy, 73 (2020) 104819.

简介：首个报道利用风力带动的摩擦纳米发电机来进行自驱动环境检测与长距离无线信号发射，为其在物联网中的应用做出了示范。

ABSTRACT As the technology of the Internet of Things (IoT) develops, trillions of wireless nodes need to be powered. It's of significant importance to design self-powered systems by utilizing tirboelectric nanogenerators (TENG) to harness energy from environment. In this paper, a wind-driven self-powered wireless environmental sensing system is presented. A TENG working in the contact-separation mode with high output performance is designed and optimized to convert wind energy into electricity. Nine TENG units are then assembled into a 3D printed shell and demonstrated as a power source for two environmental monitoring systems. At a wind speed of 4.5 m/s, a Carbon monoxide (CO) monitor works sustainably and sends out a signal from 1.5 km away every 18 min, by using the radio transmission with 433 MHz ultrahigh frequency (Sub-1G). Another temperature and humidity hybrid sensor can send out the monitoring data every 9 min through Bluetooth (2.4 GHz) in a range of 50 m. Above all, this wind-driven self-powered system constructs a platform, which can be widely used in any other wireless remote environmental sensing scene, requiring sustainability and free of maintenance.

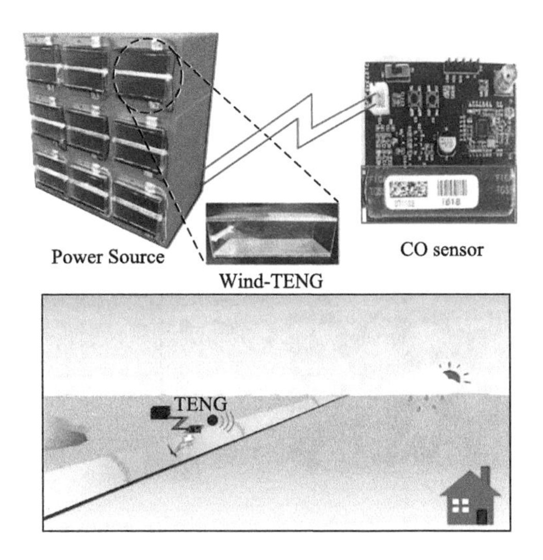

扫码阅全文

http://www.binn.cas.cn/book/202106/W020210623351466410552.pdf

Ⅲ.2 摩擦纳米发电机：自驱动传感器
Triboelectric Nanogenerators—As a Self-Powered Sensor

1. Transparent Triboelectric Nanogenerators and Self-Powered Pressure Sensors Based on Micropatterned Plastic Films

Feng-Ru Fan, Long Lin, Guang Zhu, Wenzhuo Wu, Rui Zhang *and* **Zhong Lin Wang***

Nano Letters, 12 (2012) 3109-3114.

简介：首个报道透明的摩擦纳米发电机作为自驱动压力传感。

ABSTRACT Transparent, flexible and high efficient power sources are important components of organic electronic and optoelectronic devices. In this work, based on the principle of the previously demonstrated triboelectric generator, we demonstrate a new high-output, flexible and transparent nanogenerator by using transparent polymer materials. We have fabricated three types of regular and uniform polymer patterned arrays (line, cube, and pyramid) to improve the efficiency of the nanogenerator. The power generation of the pyramid-featured device far surpassed that exhibited by the unstructured films and gave an output voltage of up to 18 V at a current density of ～0.13 μA/cm^2. Furthermore, the as-prepared nanogenerator can be applied as a self-powered pressure sensor for sensing a water droplet (8 mg, ～3.6 Pa in contact pressure) and a falling feather (20 mg, ～0.4 Pa in contact pressure) with a low-end detection limit of ～13 mPa.

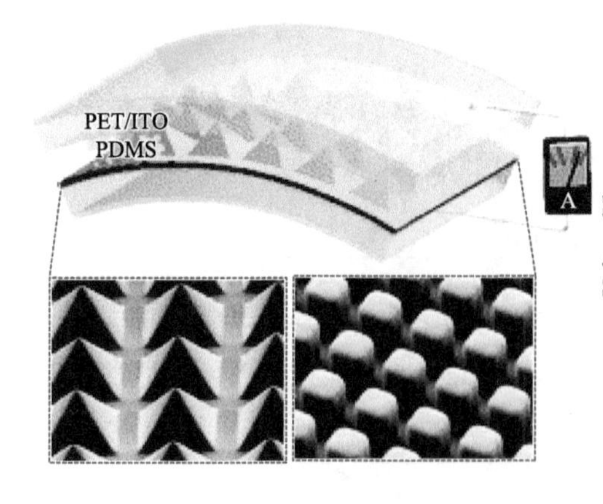

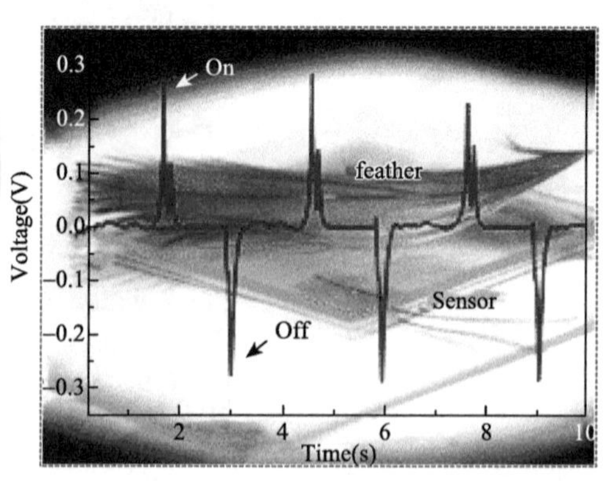

扫码阅全文

http://www.binn.cas.cn/book/202106/W020210623367049861026.pdf

2. A Self-Powered Triboelectric Nanosensor for Mercury Ion Detection

Zong-Hong Lin, Guang Zhu, Yu Sheng Zhou, Ya Yang, Peng Bai, Jun Chen and **Zhong Lin Wang***

Angewandte Chemie, 52 (2013) 5065-5069.

简介：首个报道利用摩擦纳米发电机的原理来选择性地探测汞离子。

ABSTRACT Moving to mercury: The first triboelectric effect-based sensor for the detection of Hg^{2+} ions by using Au nanoparticles (see picture; red) as electrical performance enhancer and recognition element has been successfully demonstrated. This self-powered and stand-alone triboelectric nanosensor has the advantages of simplicity, low cost, high selectivity, and sensitivity.

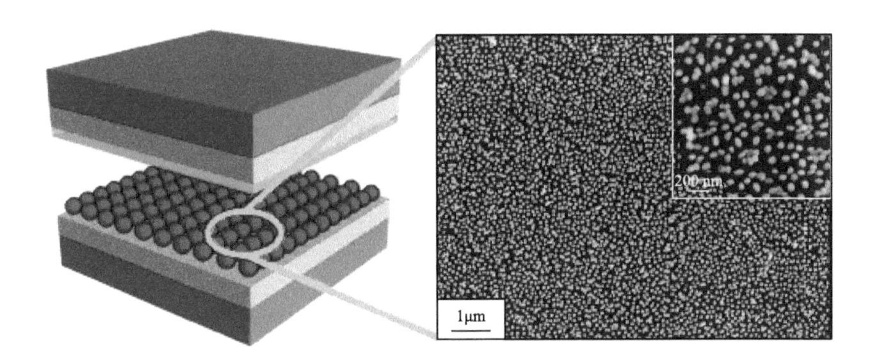

扫码阅全文

http://www.binn.cas.cn/book/202106/W020210623367050218002.pdf

3. A Highly Efficient Triboelectric Negative Air Ion Generator

Hengyu Guo, Jie Chen, Longfei Wang, Aurelia Chi Wang, Yafeng Li, Chunhua An, Jr-Hau He, Chenguo Hu, Vincent K. S. Hsiao* *and* Zhong Lin Wang*

Nature Sustainability, 4 (2021) 147-153.

简介：首个报道利用人体活动所驱动的摩擦纳米发电机来产生负离子。

ABSTRACT Negative air ions (NAIs) have been widely harnessed in recent technologies for air pollutant removal and their beneficial effects on human health, including allergy relief and neurotransmitter modulation. Herein, we report a corona-type, mechanically stimulated triboelectric NAI generator. Using the high output voltage from a triboelectric nanogenerator, air molecules can be locally ionized from carbon fibre electrodes through various movements, with the electron-ion transformation efficiency reaching up to 97%. Using a palm-sized device, 1×10^{13} NAIs (theoretically 1×10^{5} ions·cm^{-3} in 100 m^{3} space) are produced in one sliding motion, and particulate matter (PM 2.5) can be rapidly reduced from 999 to 0 μg·m^{-3} in 80 s (in a 5,086 cm^{3} glass chamber) under an operation frequency of 0.25 Hz. This triboelectric NAI generator is simple, safe and effective, providing an appealing alternative, sustainable avenue to improving health and contributing to a cleaner environment.

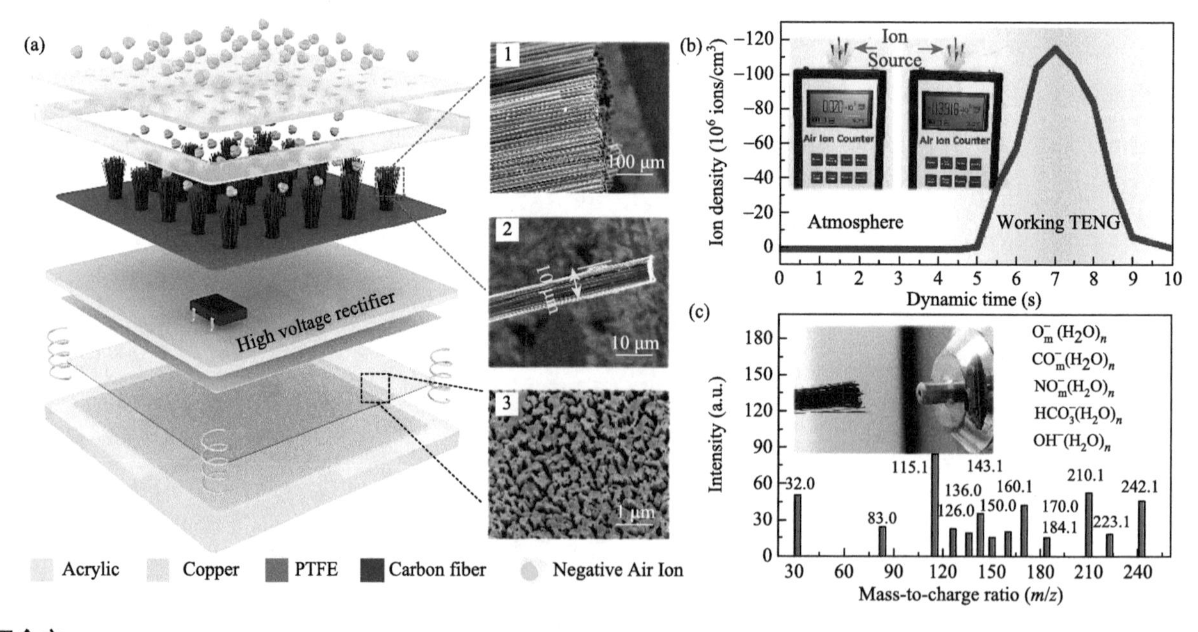

扫码阅全文

http://www.binn.cas.cn/book/202106/W020210628574205685987.pdf

4. Harmonic-Resonator-Based Triboelectric Nanogenerator as a Sustainable Power Source and a Self-Powered Active Vibration Sensor

Jun Chen, Guang Zhu, Weiqing Yang, Qingshen Jing, Peng Bai, Ya Yang, Te-Chien Hou *and* Zhong Lin Wang*

Advanced Materials, 25 (2013) 6094-6099.

简介：首个报道利用振动驱动的摩擦纳米发电机。

ABSTRACT A harmonic-resonator-based triboelectric nanogenerator (TENG) is presented as a sustainable power source and an active vibration sensor. It can effectively respond to vibration frequencies ranging from 2 to 200 Hz with a considerably wide working bandwidth of 13.4 Hz. This work not only presents a new principle in the field of vibration energy harvesting but also greatly expands the applicability of TENGs.

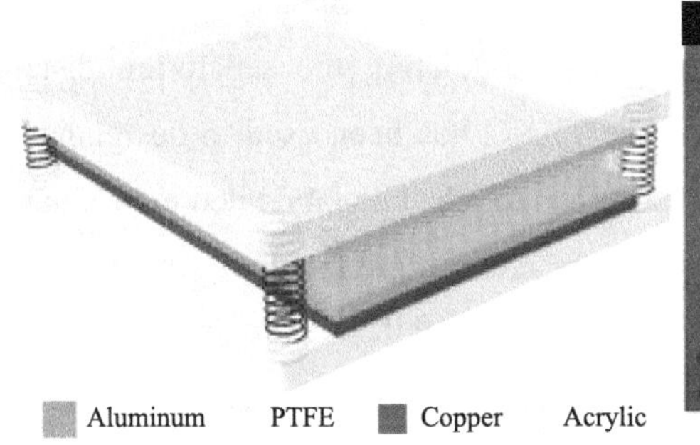

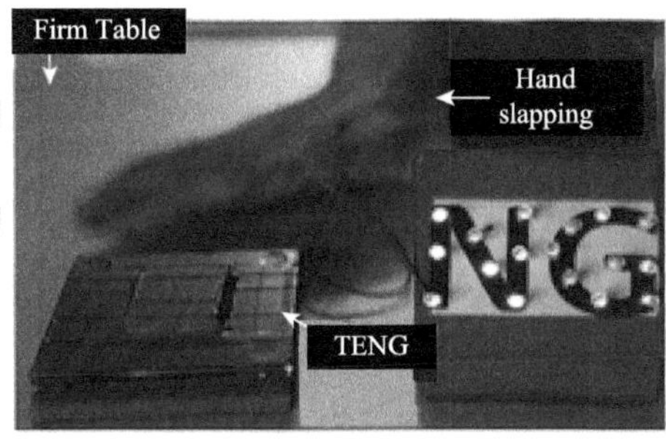

扫码阅全文

http://www.binn.cas.cn/book/202106/W020210623367050663643.pdf

5. Human Skin Based Triboelectric Nanogenerators for Harvesting Biomechanical Energy and as Self-Powered Active Tactile Sensor System

Ya Yang, Hulin Zhang, Zong-Hong Lin, Yusheng Zhou, Qingshen Jing, Yuanjie Su, Jin Yang, Jun Chen, Chenguo Hu, *and* **Zhong Lin Wang**[*]

ACS Nano, 7 (2013) 9213-9222.

简介：首个报道基于人皮肤的摩擦纳米发电机。

ABSTRACT We report human skin based triboelectric nanogenerators (TENG) that can either harvest biomechanical energy or be utilized as a self-powered tactile sensor system for touch pad technology. We constructed a TENG utilizing the contact/separation between an area of human skin and a polydimethylsiloxane (PDMS) film with a surface of micropyramid structures, which was attached to an ITO electrode that was grounded across a loading resistor. The fabricated TENG delivers an open-circuit voltage up to −1000 V, a short-circuit current density of 8 mA/m^2, and a power density of 500 mW/m^2 on a load of 100 MΩ, which can be used to directly drive tens of green light-emitting diodes. The working mechanism of the TENG is based on the charge transfer between the ITO electrode and ground via modulating the separation distance between the tribo-charged skin patch and PDMS film. Furthermore, the TENG has been used in designing an independently addressed matrix for tracking the location and pressure of human touch. The fabricated matrix has demonstrated its self-powered and high-resolution tactile sensing capabilities by recording the output voltage signals as a mapping figure, where the detection sensitivity of the pressure is about (0.29±0.02) V/kPa and each pixel can have a size of 3mm×3mm. The TENGs may have potential applications in human-machine interfacing, micro/nano-electromechanical systems, and touch pad technology.

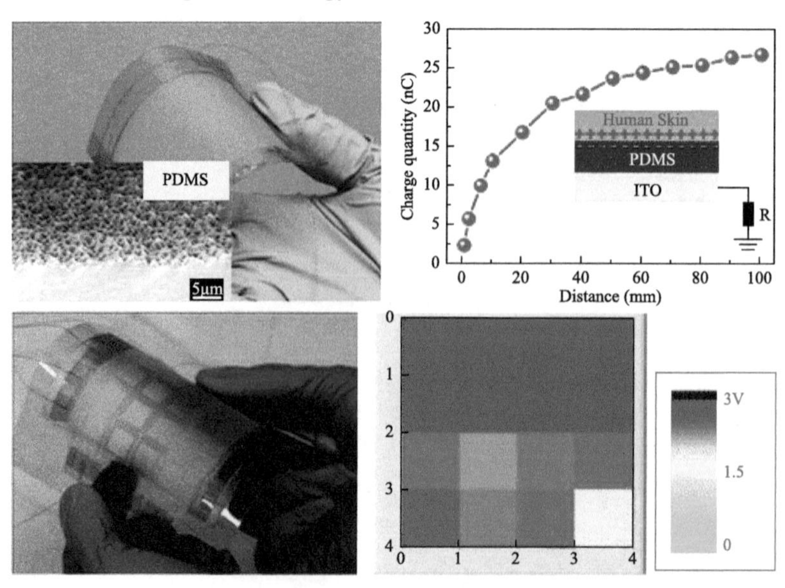

扫码阅全文

http://www.binn.cas.cn/book/202106/W020210623367051019305.pdf

6. Triboelectric Active Sensor Array for Self-Powered Static and Dynamic Pressure Detection and Tactile Imaging

Long Lin, Yannan Xie, Sihong Wang, Wenzhuo Wu, Simiao Niu, Xiaonan Wen *and* **Zhong Lin Wang***

ACS Nano, 7 (2013) 8266-8274.

简介：首个报道利用摩擦纳米发电机做静态和动态压力传感。

ABSTRACT We report an innovative, large-area, and self-powered pressure mapping approach based on the triboelectric effect, which converts the mechanical stimuli into electrical output signals. The working mechanism of the triboelectric active sensor (TEAS) was theoretically studied by both analytical method and numerical calculation to gain an intuitive understanding of the relationship between the applied pressure and the responsive signals. Relying on the unique pressure response characteristics of the open-circuit voltage and short-circuit current, we realize both static and dynamic pressure sensing on a single device for the first time. A series of comprehensive investigations were carried out to characterize the performance of the TEAS, and high sensitivity (0.31 kPa^{-1}), ultrafast response time (<5ms), long-term stability (30,000 cycles), as well as low detection limit (2.1 Pa) were achieved. The pressure measurement range of the TEAS was adjustable, which means both gentle pressure detection and large-scale pressure sensing were enabled. Through integrating multiple TEAS units into a sensor array, the as-fabricated TEAS matrix was capable of monitoring and mapping the local pressure distribution applied on the device with distinguishable spatial profiles. This work presents a technique for tactile imaging and progress toward practical applications of nanogenerators, providing potential solutions for accomplishment of artificial skin, human-electronic interfacing, and self-powered systems.

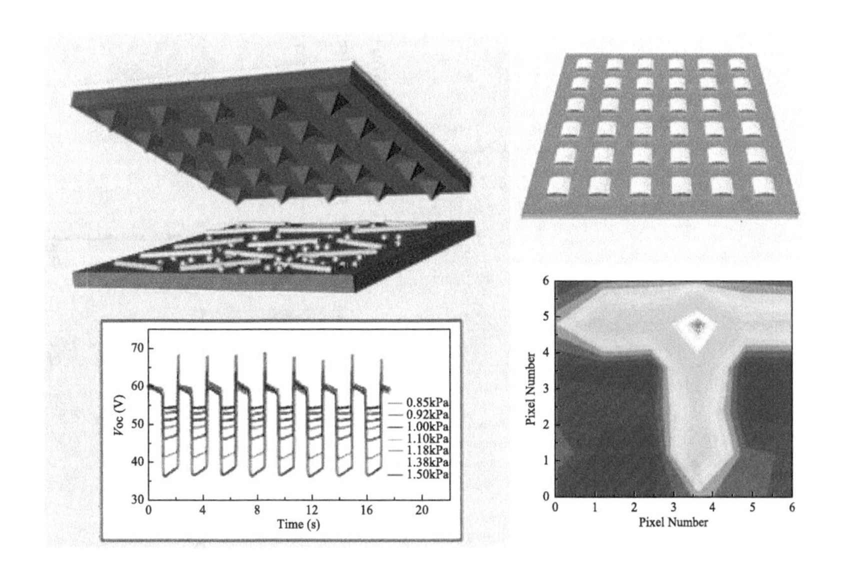

扫码阅全文

http://www.binn.cas.cn/book/202106/W020210623367051354476.pdf

7. Triboelectrification-Based Organic Film Nanogenerator for Acoustic Energy Harvesting and Self-Powered Active Acoustic Sensing

Jin Yang, Jun Chen, Ying Liu, Weiqing Yang, Yuanjie Su *and* Zhong Lin Wang*

ACS Nano, 8 (2014) 2649-2657.

简介：首个报道利用摩擦纳米发电机收集声波振动能。

ABSTRACT As a vastly available energy source in our daily life, acoustic vibrations are usually taken as noise pollution with little use as a power source. In this work, we have developed a triboelectrification-based thin-film nanogenerator for harvesting acoustic energy from ambient environment. Structured using a polytetrafluoroethylene thin film and a holey aluminum film electrode under carefully designed straining conditions, the nanogenerator is capable of converting acoustic energy into electric energy via triboelectric transduction. With an acoustic sensitivity of 9.54 $V\cdot Pa^{-1}$ in a pressure range from 70 to 110 dB and a directivity angle of 52°, the nanogenerator produced a maximum electric power density of 60.2 $mW\cdot m^{-2}$, which directly lit 17 commercial light-emitting diodes (LEDs). Furthermore, the nanogenerator can also act as a self-powered active sensor for automatically detecting the location of an acoustic source with an error less than 7 cm. In addition, an array of devices with varying resonance frequencies was employed to widen the overall bandwidth from 10 to 1700 Hz, so that the nanogenerator was used as a superior self-powered microphone for sound recording. Our approach presents an adaptable, mobile, and cost-effective technology for harvesting acoustic energy from ambient environment, with applications in infrastructure monitoring, sensor networks, military surveillance, and environmental noise reduction.

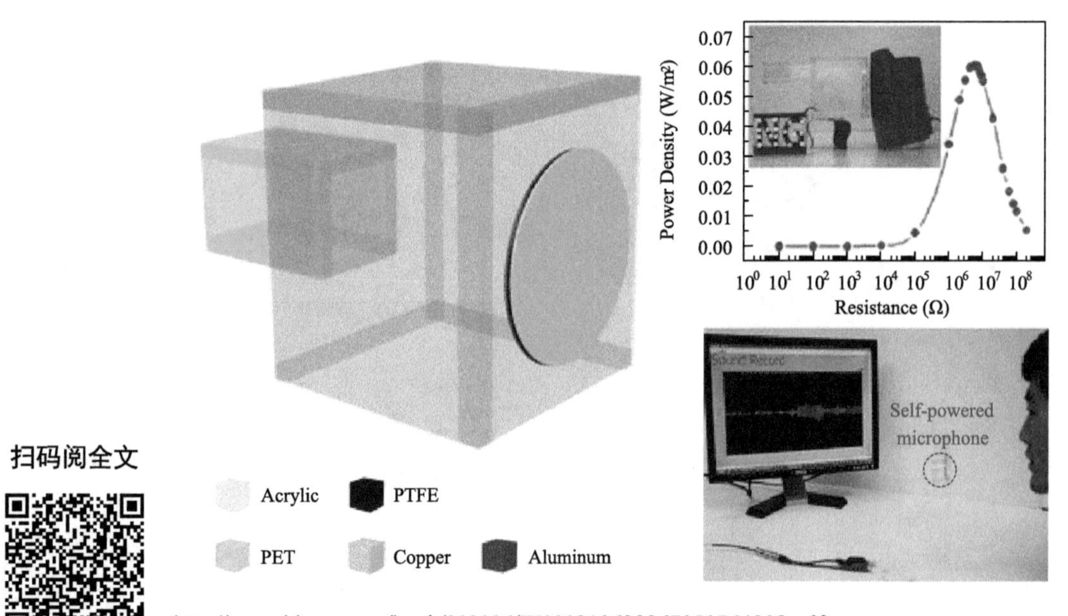

扫码阅全文

http://www.binn.cas.cn/book/202106/W020210623367051764908.pdf

8. Nanometer Resolution Self-Powered Static and Dynamic Motion Sensor Based on Micro-Grated Triboelectrification

Yu Sheng Zhou, Guang Zhu, Simiao Niu, Ying Liu, Peng Bai, Qingsheng Jing *and* Zhong Lin Wang*

Advanced Materials, 26 (2014) 1719-1724.

简介：首个报道利用摩擦纳米发电机来做纳米尺度的自驱动位移传感。

ABSTRACT A one-dimensional displacement and speed sensing technology that consists of a pair of micro-grating structures and utilizes the coupling between the triboelectric effect and electrostatic induction is demonstrated. Its distinct advantages, including being self-powered, high resolution, large dynamic range, and long detecting distance, show extensive potential applications in automation, manufacturing, process control, and portable devices.

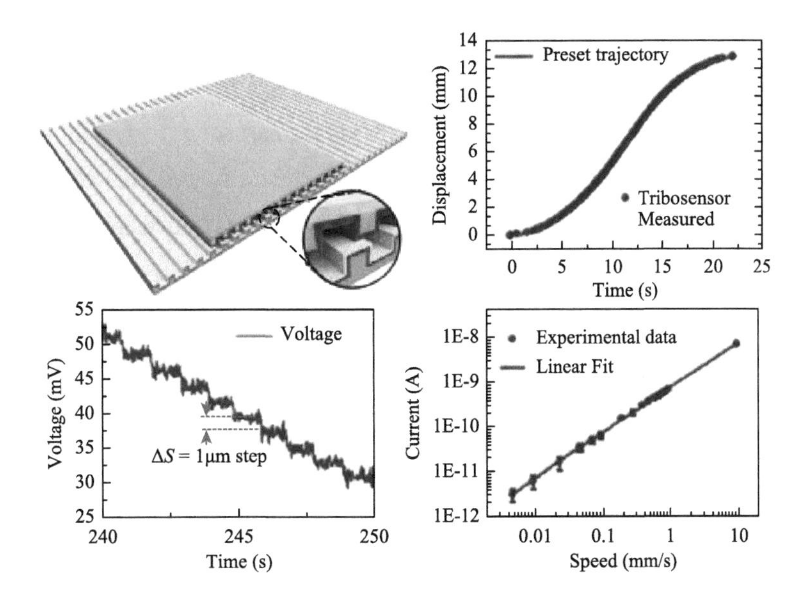

扫码阅全文

http://www.binn.cas.cn/book/202106/W020210623367052569244.pdf

9. Personalized Keystroke Dynamics for Self-Powered Human-Machine Interfacing

Jun Chen, Guang Zhu, Jin Yang, Qingshen Jing, Peng Bai, Weiqing Yang, Xuewei Qi, Yuanjie Su *and* Zhong Lin Wang*

ACS Nano, 9 (2015) 105-116.

简介：首个报道利用摩擦纳米发电机做人机界面的个性化打字输入键盘。

ABSTRACT The computer keyboard is one of the most common, reliable, accessible, and effective tools used for human-machine interfacing and information exchange. Although keyboards have been used for hundreds of years for advancing human civilization, studying human behavior by keystroke dynamics using smart keyboards remains a great challenge. Here we report a self-powered, non-mechanical-punching keyboard enabled by contact electrification between human fingers and keys, which converts mechanical stimuli applied to the keyboard into local electronic signals without applying an external power. The intelligent keyboard (IKB) can not only sensitively trigger a wireless alarm system once gentle finger tapping occurs but also trace and record typed content by detecting both the dynamic time intervals between and during the inputting of letters and the force used for each typing action. Such features hold promise for its use as a smart security system that can realize detection, alert, recording, and identification. Moreover, the IKB is able to identify personal characteristics from different individuals, assisted by the behavioral biometric of keystroke dynamics. Furthermore, the IKB can effectively harness typing motions for electricity to charge commercial electronics at arbitrary typing speeds greater than 100 characters per min. Given the above features, the IKB can be potentially applied not only to self-powered electronics but also to artificial intelligence, cyber security, and computer or network access control.

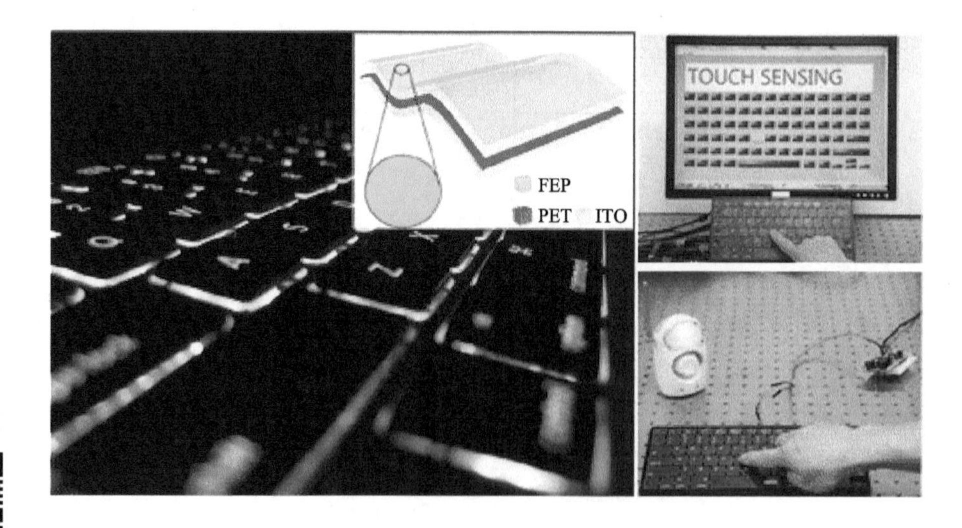

扫码阅全文

http://www.binn.cas.cn/book/202106/W020210623367052942371.pdf

10. Eardrum-Inspired Active Sensors for Self-Powered Cardiovascular System Characterization and Throat-Attached Anti-Interference Voice Recognition

Jin Yang, Jun Chen, Yuanjie Su, Qingshen Jing, Zhaoling Li, Fang Yi, Xiaonan Wen, Zhaona Wang *and* **Zhong Lin Wang***

Advanced Materials, 27 (2015) 1316-1326.

简介：首个报道利用摩擦纳米发电机做反干扰的声音识别装置和耳蜗助听器。

ABSTRACT The first bionic membrane sensor based on triboelectrification is reported for self-powered physiological and behavioral measurements such as local internal body pressures for non-invasive human health assessment. The sensor can also be for self-powered anti-interference throat voice recording and recognition, as well as high-accuracy multimodal biometric authentication, thus potentially expanding the scope of applications in self-powered wearable medical/health monitoring, interactive input/control devices as well as accurate, reliable, and less intrusive biometric authentication systems.

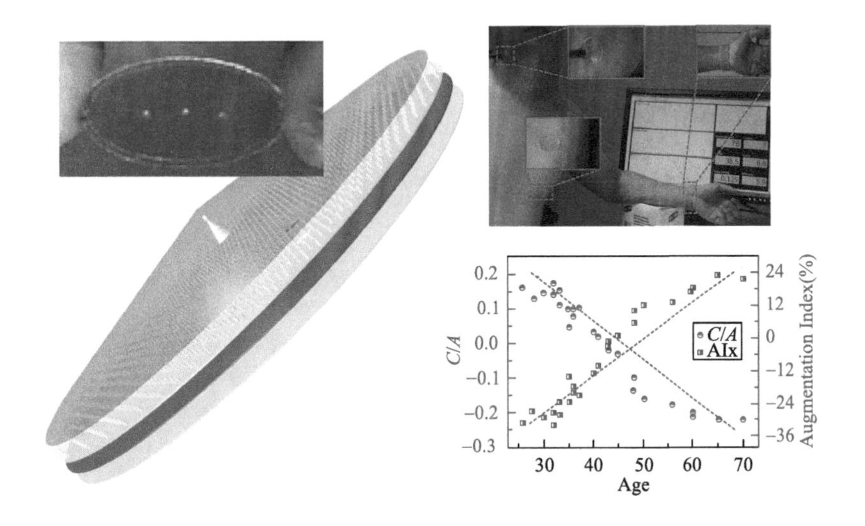

扫码阅全文

http://www.binn.cas.cn/book/202106/W020210623367053908735.pdf

11. A Self-Powered Angle Measurement Sensor Based on Triboelectric Nanogenerator

Ying Wu, Qingshen Jing, Jun Chen, Peng Bai, Junjie Bai, Guang Zhu, Yuanjie Su *and* Zhong Lin Wang*

Advanced Functional Materials, 25 (2015) 2166-2174.

简介：首个报道利用摩擦纳米发电机做自驱动角度传感器。

ABSTRACT A self-powered, sliding electrification based quasi-static triboelectric sensor (QS-TES) for detecting angle from rotating motion is reported. This innovative, cost-effective, simply-designed QS-TES has a two-dimensional planar structure, which consists of a rotator coated with four channel coded Cu foil material and a stator with a fluorinated ethylenepropylene film. On the basis of coupling effect between triboelectrification and electrostatic induction, the sensor generates electric output signals in response to mechanical rotating motion of an object mounted with the sensor. The sensor can read and remember the absolute angular position, angular velocity, and acceleration regardless being continuously monitored or segmented monitored. Under the rotation speed of 100 $r \cdot min^{-1}$, the output voltage of the sensor reaches as high as 60 V. Given a relatively low threshold voltage of ± 0.5 V for data processing, the robustness of the device is guaranteed. The resolution of the sensor is 22.5° and can be further improved by increasing the number of channels. Triggered by the output voltage signal, the rotating characteristics of the steering wheel can be real-time monitored and mapped by being mounted to QS-TES. This work not only demonstrates a new principle in the field of angular measurement but also greatly expands the applicability of triboelectric nanogenerator as self-powered sensors.

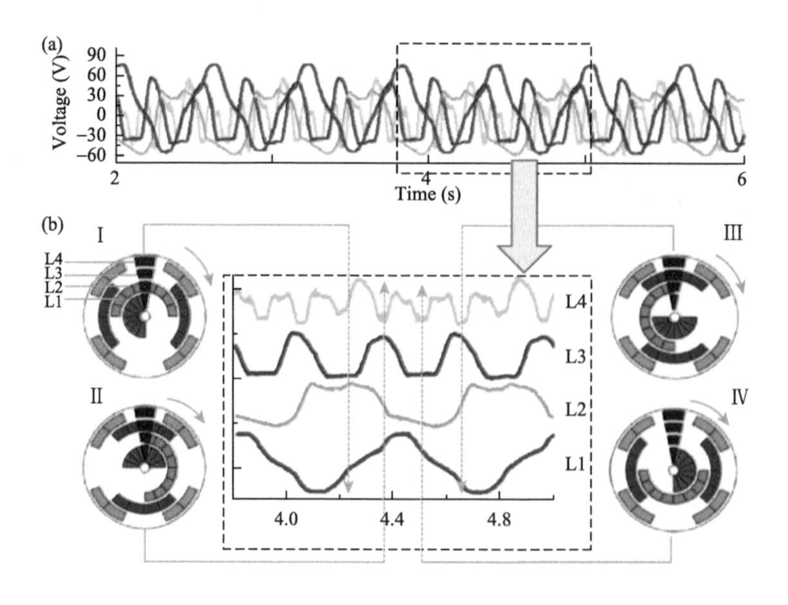

扫码阅全文

http://www.binn.cas.cn/book/202106/W020210623367054489909.pdf

12. β-Cyclodextrin Enhanced Triboelectrification for Self-Powered Phenol Detection and Electrochemical Degradation

Zhaoling Li, Jun Chen, Jin Yang, Yuanjie Su, Xing Fan, Ying Wu, Chongwen Yu *and* **Zhong Lin Wang***

Energy & Environmental Science, 8 (2015) 887-896.

简介：首个报道利用摩擦纳米发电机做自驱动的污染物传感器与清理装置。

ABSTRACT We report a unique route that creatively harnessed β-cyclodextrin enhanced triboelectrification for self-powered phenol detection as well as electrochemical degradation. A detection sensitivity of 0.01 μM^{-1} was demonstrated in the sensing range of 10 μM to 100 μM. In addition, β-cyclodextrin enhanced triboelectrification was designed to harvest kinetic impact energy from wastewater waves to electrochemically degrade the phenol in a self-powered manner without using an external power source.

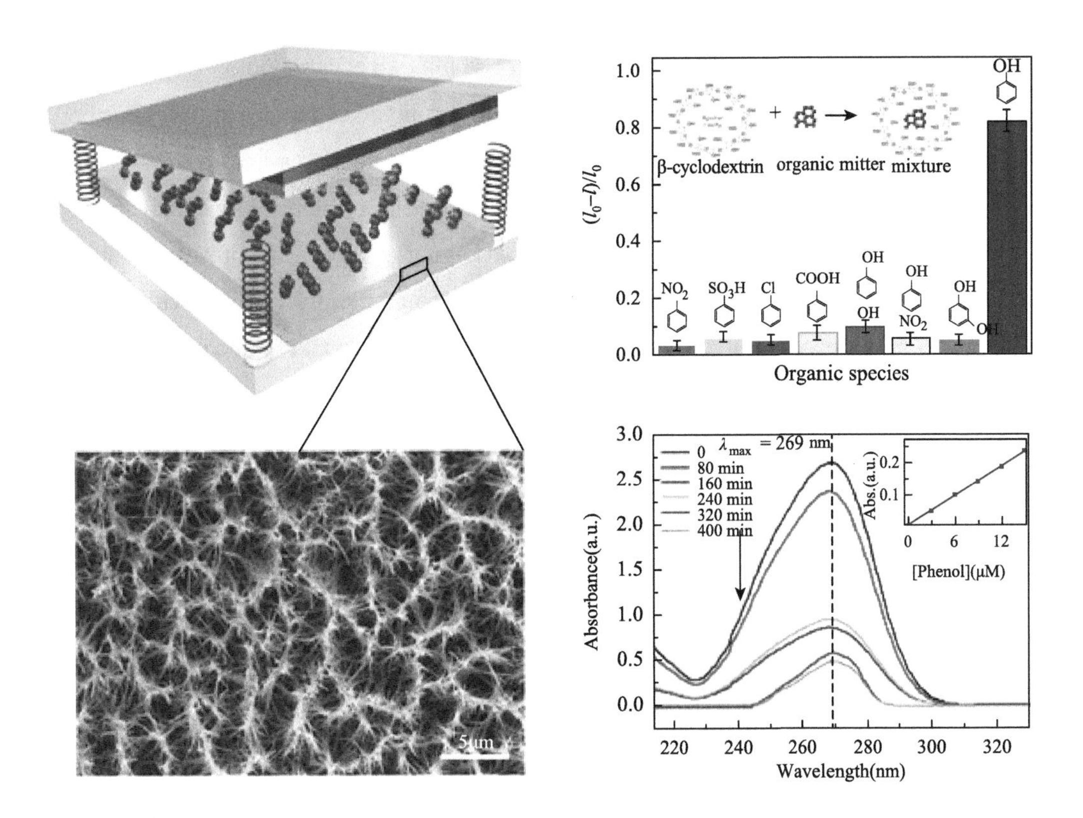

扫码阅全文

http://www.binn.cas.cn/book/202106/W020210623367055075674.pdf

13. Self-Powered, One-Stop and Multifunctional Implantable Triboelectric Active Sensor for Real-Time Biomedical Monitoring

Ye Ma, Qiang Zheng, Yang Liu, Bojin Shi, Xiang Xue, Weiping Ji, Zhuo Liu, Yiming Jin, Yang Zou, Zhao An, Wei Zhang, Xinxin Wang, Wen Jiang, Zhiyun Xu, Zhong Lin Wang*, Zhou Li* *and* **Hao Zhang***

Nano Letters, 16 (2016) 6042-6051.

简介：首个报道利用摩擦纳米发电机做多功能、自驱动、植入式生物传感。

ABSTRACT Operation time of implantable electronic devices is largely constrained by the lifetime of batteries, which have to be replaced periodically by surgical procedures once exhausted, causing physical and mental suffering to patients and increasing healthcare costs. Besides the efficient scavenging of the mechanical energy of internal organs, this study proposes a self-powered, flexible, and one-stop implantable triboelectric active sensor (iTEAS) that can provide continuous monitoring of multiple physiological and pathological signs. As demonstrated in human-scale animals, the device can monitor heart rates, reaching an accuracy of ~99%. Cardiac arrhythmias such as atrial fibrillation and ventricular premature contraction can be detected in real-time. Furthermore, a novel method of monitoring respiratory rates and phases is established by analyzing variations of the output peaks of the iTEAS. Blood pressure can be independently estimated and the velocity of blood flow calculated with the aid of a separate arterial pressure catheter. With the core-shell packaging strategy, monitoring functionality remains excellent during 72 h after closure of the chest. The in vivo biocompatibility of the device is examined after 2 weeks of implantation, proving suitability for practical use. As a multifunctional biomedical monitor that is exempt from needing an external power supply, the proposed iTEAS holds great potential in the future of the healthcare industry.

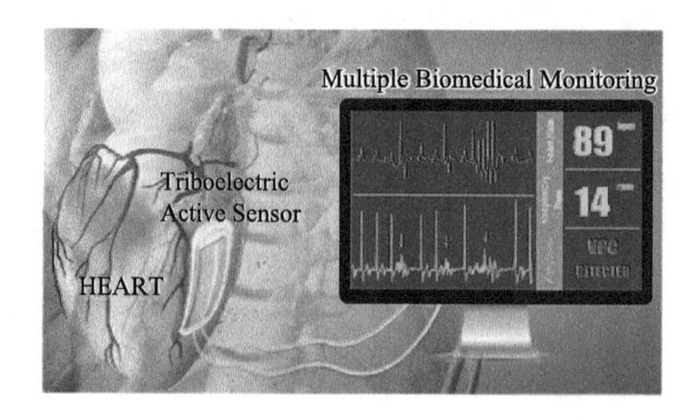

扫码阅全文

http://www.binn.cas.cn/book/202106/W020210623367055435763.pdf

14. Self-Powered Pulse Sensor for Antidiastole of Cardiovascular Disease

Han Ouyang, Jingjing Tian, Guanglong Sun, Yang Zou, Zhuo Liu, Hu Li, Luming Zhao, Bojing Shi, Yubo Fan, Yifan Fan*, Zhong Lin Wang* *and* **Zhou Li***

Advanced Materials, 29 (2017) 1703456.

简介：首个报道利用摩擦纳米发电机做心血管监测自驱动传感。

ABSTRACT Cardiovascular diseases are the leading cause of death globally; fortunately, 90% of cardiovascular diseases are preventable by long-term monitoring of physiological signals. Stable, ultralow power consumption, and high-sensitivity sensors are significant for miniaturized wearable physiological signal monitoring systems. Here, this study proposes a flexible self-powered ultrasensitive pulse sensor (SUPS) based on triboelectric active sensor with excellent output performance (1.52 V), high peak signal-noise ratio (45 dB), long-term performance (10^7 cycles), and low cost price. Attributed to the crucial features of acquiring easy-processed pulse waveform, which is consistent with second derivative of signal from conventional pulse sensor, SUPS can be integrated with a bluetooth chip to provide accurate, wireless, and real-time monitoring of pulse signals of cardiovascular system on a smart phone/PC. Antidiastole of coronary heart disease, atrial septal defect, and atrial fibrillation are made, and the arrhythmia (atrial fibrillation) is indicative diagnosed from health, by characteristic exponent analysis of pulse signals accessed from volunteer patients. This SUPS is expected to be applied in self-powered, wearable intelligent mobile diagnosis of cardiovascular disease in the future.

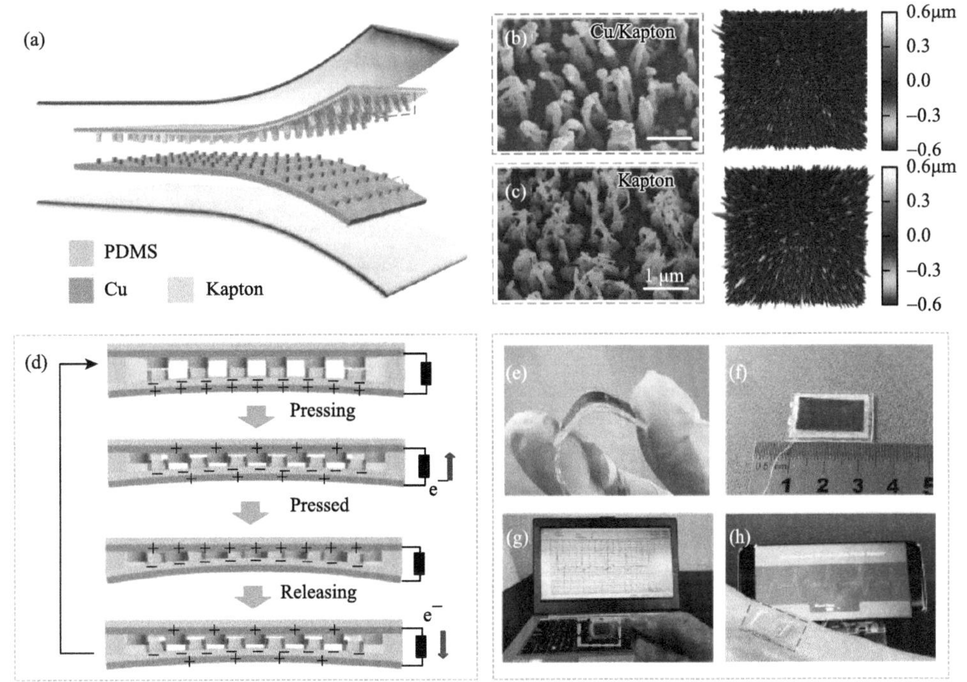

扫码阅全文

15. 3D Orthogonal Woven Triboelectric Nanogenerator for Effective Biomechanical Energy Harvesting and as Self-Powered Active Motion Sensors

Kai Dong, Jianan Deng, Yunlong Zi, Yi-cheng Wang, Cheng Xu, Haiyang Zou, Wenbo Ding, Yejing Dai, Bohong Gu, Baozhong Sun *and* Zhong Lin Wang*

Advanced Materials, 29 (2017) 1702648.

简介：首个报道基于纺织品的摩擦纳米发电机，并应用于能量回收和自驱动人动传感。

ABSTRACT The development of wearable and large-area energy-harvesting textiles has received intensive attention due to their promising applications in next-generation wearable functional electronics. However, the limited power outputs of conventional textiles have largely hindered their development. Here, in combination with the stainless steel/polyester fiber blended yarn, the polydimethylsiloxane-coated energy-harvesting yarn, and nonconductive binding yarn, a high-power-output textile triboelectric nanogenerator (TENG) with 3D orthogonal woven structure is developed for effective biomechanical energy harvesting and active motion signal tracking. Based on the advanced 3D structural design, the maximum peak power density of 3D textile can reach 263.36 mW·m^{-2} under the tapping frequency of 3 Hz, which is several times more than that of conventional 2D textile TENGs. Besides, its collected power is capable of lighting up a warning indicator, sustainably charging a commercial capacitor, and powering a smart watch. The 3D textile TENG can also be used as a self-powered active motion sensor to constantly monitor the movement signals of human body. Furthermore, a smart dancing blanket is designed to simultaneously convert biomechanical energy and perceive body movement. This work provides a new direction for multifunctional self-powered textiles with potential applications in wearable electronics, home security, and personalized healthcare.

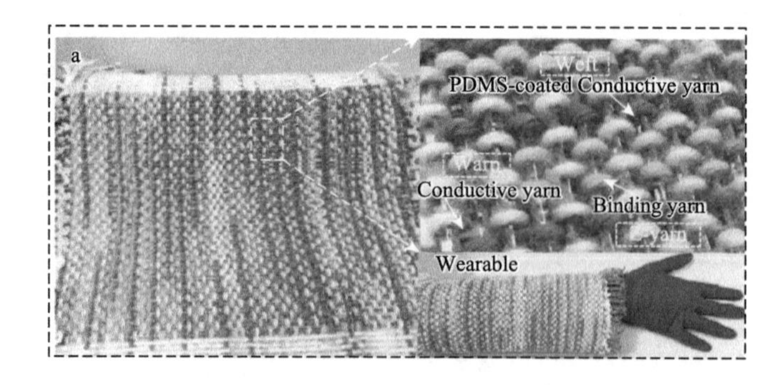

扫码阅全文

http://www.binn.cas.cn/book/202106/W020210623367057433550.pdf

16. A Self-Powered Angle Sensor at Nanoradian-Resolution for Robotic Arms and Personalized Medicare

Ziming Wang, Jie An, Jinhui Nie, Jianjun Luo, Jiajia Shao, Tao Jiang, Baodong Chen, Wei Tang* *and* Zhong Lin Wang*

Advanced Materials, 32 (2020) 2001466.

简介：首个报道利用摩擦纳米发电机做自驱动角度传感器，其角度分辨率达到 10^{-9} 弧度。

ABSTRACT As the dominant component for precise motion measurement, angle sensors play a vital role in robotics, machine control, and personalized rehabilitation. Various forms of angle sensors have been developed and optimized over the past decades, but none of them would function without an electric power. Here, a highly sensitive triboelectric self-powered angle sensor (SPAS) exhibiting the highest resolution (2.03 nano-radian) after a comprehensive optimization is reported. In addition, the SPAS holds merits of light weight and thin thickness, which enables its extensive integrated applications with minimized energy consumption: a palletizing robotic arm equipped with the SPAS can precisely reproduce traditional Chinese calligraphy via angular data it collects. In addition, the SPAS can be assembled in a medicare brace to record the flexion/extension of joints, which may benefit personalized orthopedic recuperation. The SPAS paves a new approach for applications in the emerging fields of robotics, sensing, personalized medicare, and artificial intelligence.

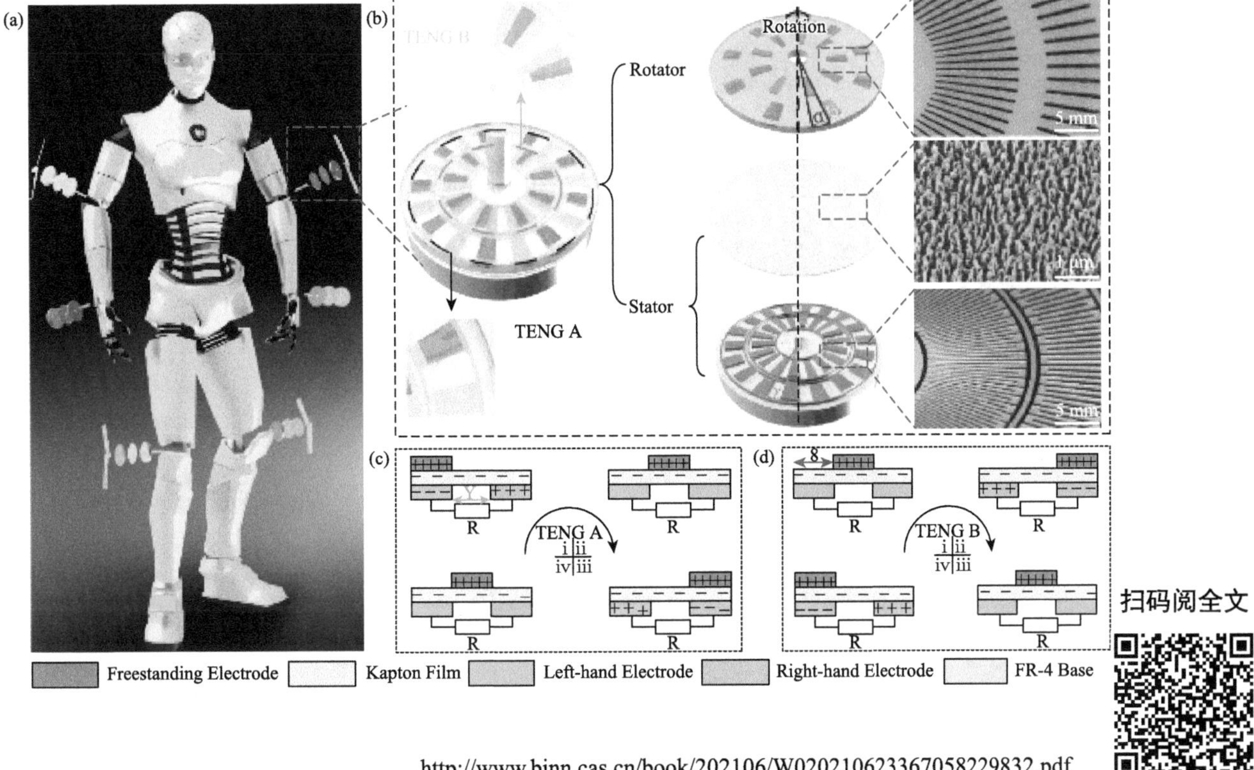

扫码阅全文

http://www.binn.cas.cn/book/202106/W020210623367058229832.pdf

17. All-in-One 3D Acceleration Sensor Based on Coded Liquid-Metal Triboelectric Nanogenerator for Vehicle Restraint System

Binbin Zhang, Zhiyi Wu, Zhiming Lin, Hengyu Guo, Fengjun Chun, Weiqing Yang, Zhong Lin Wang*

Materials Today, 43 (2021) 37-44.

简介：摩擦纳米发电机作为加速度传感器的一篇文章。

ABSTRACT Vehicle restraint systems play an irreplaceable role to limit passenger injuries when an accident occurs, in which, the 3D acceleration sensor (AS) is an essential component to detect the collision position and force. However, there are some defects for commercial sensors such as passive sensing, low sensitivity and high manufacturing cost. Here, we report a lightweight, high-sensitivity, low-cost and self-powered 3D AS based on a liquid-metal triboelectric nanogenerator (LM-TENG). In view of the coded strategy of the electrodes, the 3D AS retains the smallest size, lowest weight and highest integration compared to the currently reported self-powered AS. The fabricated sensor possesses wide detection range from 0 to 100 m/s^2 in the horizontal direction and 0 to 50 m/s^2 in the vertical direction at a sensitivity of 800 mV/g. The open-circuit voltage shows a negligible decrease after continuously operating for 100,000 times, showing excellent stability and durability. Furthermore, the 3D AS is demonstrated as a part of the airbag system to spot the collision position and force of the car simultaneously. This work will further promote the commercialization of TENG-based sensor and exhibits a prospective application in the vehicle restraint system.

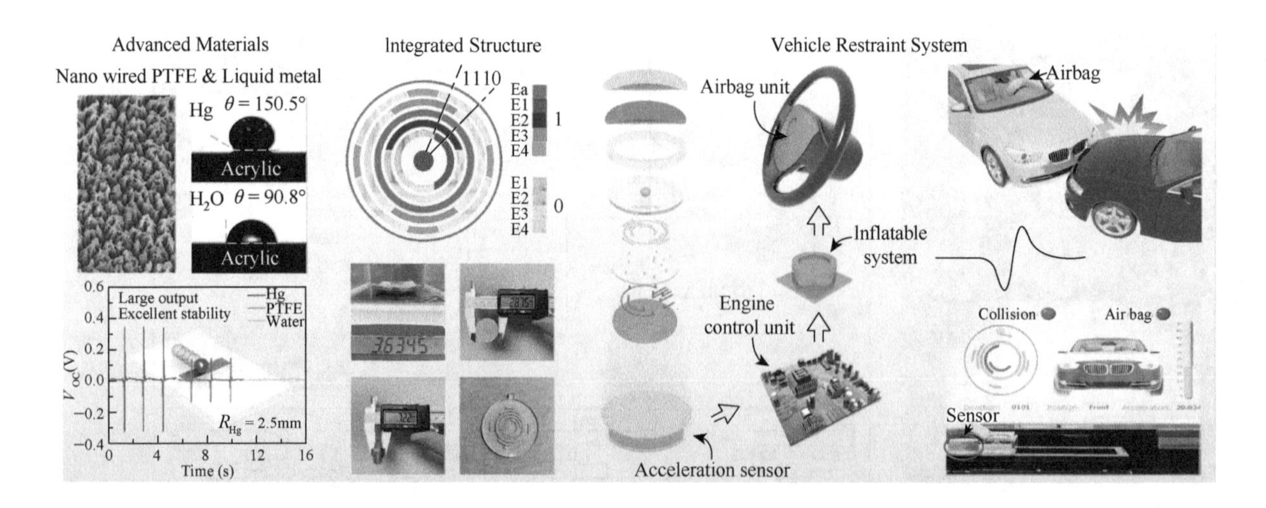

扫码阅全文

http://www.binn.cas.cn/book/202106/W020210623367060435292.pdf

18. Multilanguage-Handwriting Self-Powered Recognition Based on Triboelectric Nanogenerator Enabled Machine Learning

Weiqiang Zhang1, Linfeng Deng1, Lei Yang, Ping Yang, Dongfeng Diao, Pengfei Wang, Zhong Lin Wang

Nano Energy, 77 (2020) 105174.

简介：摩擦纳米发电机应用于人机界面的手写识别。

ABSTRACT Handwriting signature is widely used and the main challenge for handwriting recognition is how to obtain comprehensive handwriting information. Triboelectric nanogenerator is sensitive to external triggering force and can be used to record personal handwriting signals and associated characteristics. In this work, micro/nano structure textured TENG acting as a smart self-powered handwriting pad is developed and its effectiveness for handwriting recognition is demonstrated. Three individuals' handwriting signals of English words, Chinese characters and Arabic numerals are acquired by leaf-inspired TENG, and the other three people's handwriting signals of English sentences and the corresponding Chinese sentences are obtained by cylindrical microstructured PDMS based TENG, and these signals exhibit unique features. Combined with the machine learning method, the people's handwriting was successfully identified. The classification accuracies of 99.66%, 93.63%, 91.36%, 99.05%, and 97.73% were reached for English words, Arabic numerals, Chinese characters, English sentences, and the corresponding Chinese sentences, respectively. The results strongly suggested that the textured TENG exhibited great potential in personal handwriting signature identification, security defense, and private information protection applications.

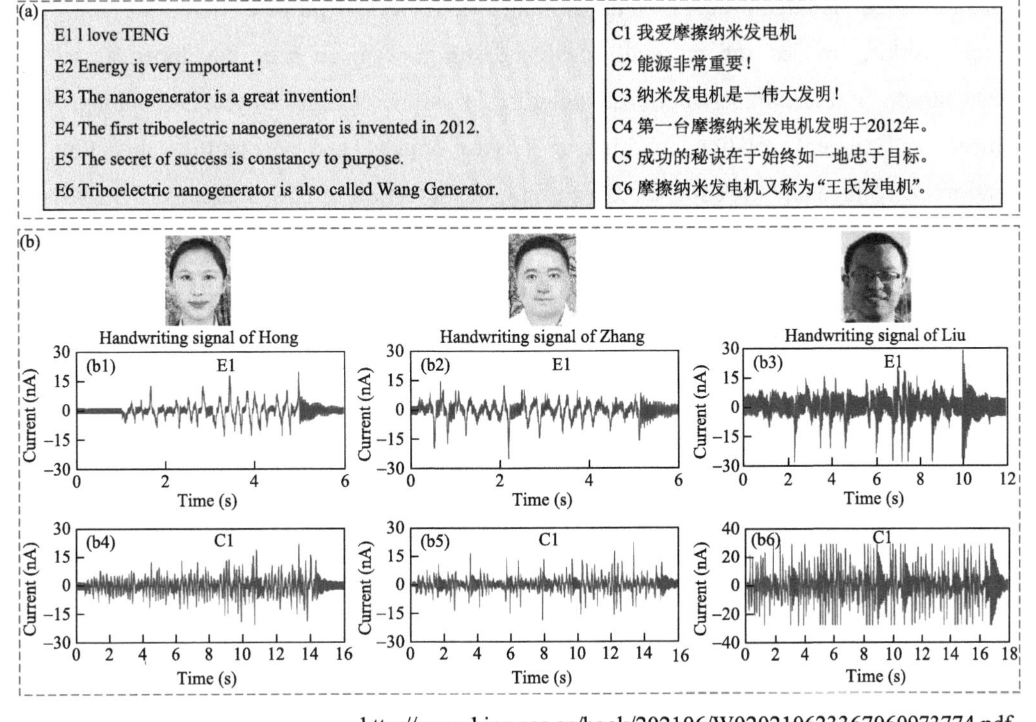

扫码阅全文

19. Triboelectric Nanogenerators as Self-Powered Active Sensors

Sihong Wang, Long Lin, Zhong Lin Wang*

Nano Energy, 11 (2015) 436-462.

简介：摩擦纳米发电机作为自驱动传感器的第一篇综述。

ABSTRACT The development of internet of things and the related sensor technology have been a key driving force for the rapid development of industry and information technology. The requirement of wireless, sustainable and independent operation is becoming increasingly important for sensor networks that currently could include thousands even to millions of sensor nodes with different functionalities. For these purposes, developing technologies of self-powered sensors that can utilize the ambient environmental energy to drive the operation themselves is highly desirable and mandatory. The realization of self-powered sensors generally has two approaches: the first approach is to develop environmental energy harvesting devices for driving the traditional sensors; the other is to develop a new category of sensors-self-powered active sensors-that can actively generate electrical signal itself as a response to a stimulation/triggering from the ambient environment. The recent invention and intensive development of triboelectric nanogenerators (TENGs) as a new technology for mechanical energy harvesting can be utilized as self-powered active mechanical sensors, because the parameters (magnitude, frequency, number of periods, etc.) of the generated electrical signal are directly determined by input mechanical behaviors. In this review paper, we first briefly introduce the fundamentals of TENGs, including the four basic working modes. Then, the most updated progress of developing TENGs as self-powered active sensors is reviewed. TENGs with different working modes and rationally designed structures have been developed as self-powered active sensors for a variety of mechanical motions, including pressure change, physical touching, vibrations, acoustic waves, linear displacement, rotation, tracking of moving objects, and acceleration detection. Through combining the open-circuit voltage and the short-circuit current, the detection of both static and dynamic processes has been enabled. The integration of individual sensor elements into arrays or matrixes helps to realize the mapping or parallel detection for multiple points. On the other hand, the relationship between the amplitude of TENG-generated electrical signal and the chemical state of its triboelectric surface enables TENGs to function as self-powered active chemical sensors. Through continuous research on the TENG-based self-powered active sensors in the coming years to further improve the sensitivity and realize the self-powered operation for the entire sensor node systems, they will soon have broad applications in touch screens, electronic skins, healthcare, environmental/infrastructure monitoring, national security, and more.

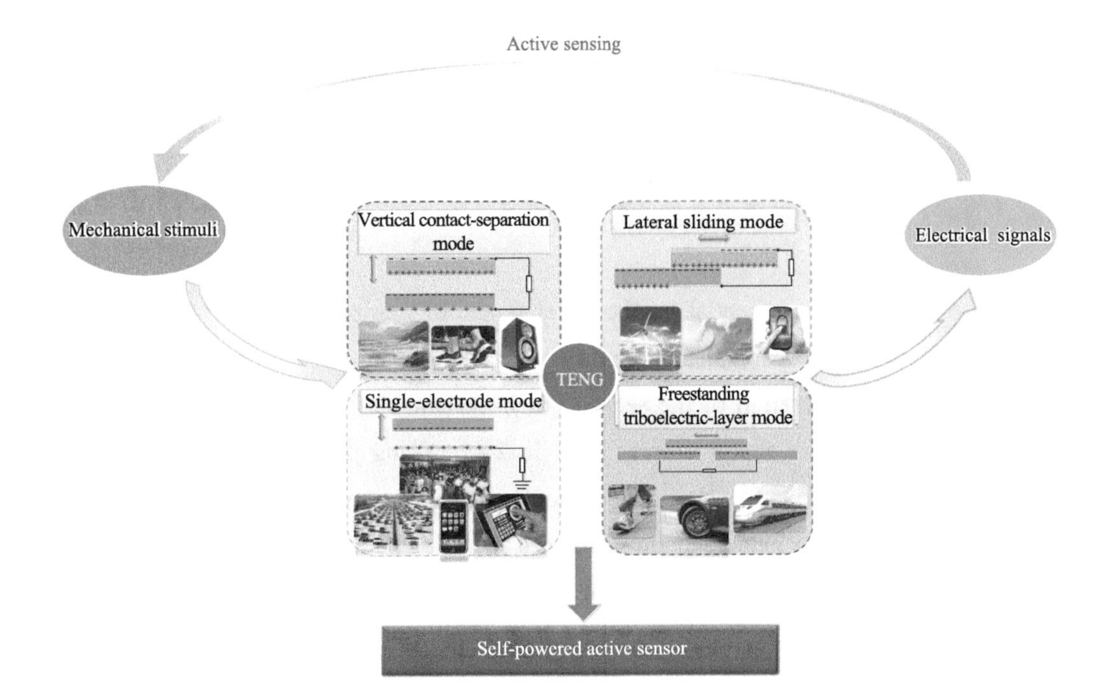

扫码阅全文

http://www.binn.cas.cn/book/202106/W020210623367063321034.pdf

20. Self-Powered Sensors and Systems Based on Nanogenerators

Zhiyi Wu, Tinghai Cheng *and* **Zhong Lin Wang***

Sensors, 20 (2020) 2925.

简介：摩擦纳米发电机作为自驱动传感器的最新一篇综述。

ABSTRACT Sensor networks are essential for the development of the Internet of Things and the smart city. A general sensor, especially a mobile sensor, has to be driven by a power unit. When considering the high mobility, wide distribution and wireless operation of the sensors, their sustainable operation remains a critical challenge owing to the limited lifetime of an energy storage unit. In 2006, Wang proposed the concept of self-powered sensors/system, which harvests ambient energy to continuously drive a sensor without the use of an external power source. Based on the piezoelectric nanogenerator (PENG) and triboelectric nanogenerator (TENG), extensive studies have focused on self-powered sensors. TENG and PENG, as effective mechanical-to-electricity energy conversion technologies, have been used not only as power sources but also as active sensing devices in many application fields, including physical sensors, wearable devices, biomedical and health care, human-machine interface, chemical and environmental monitoring, smart traffic, smart cities, robotics, and fiber and fabric sensors. In this review, we systematically summarize the progress made by TENG and PENG in those application fields. A perspective will be given about the future of self-powered sensors.

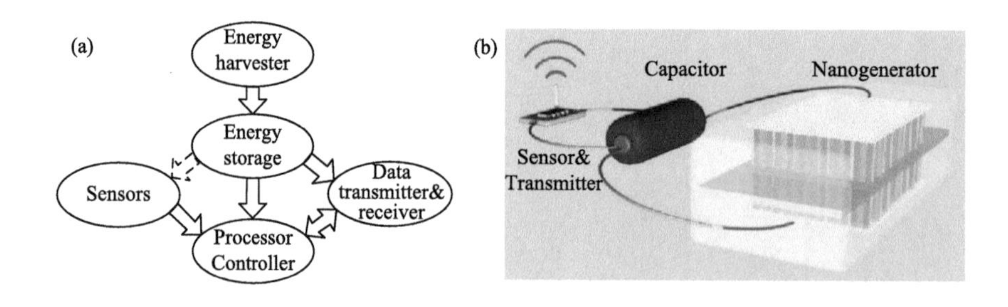

扫码阅全文

http://www.binn.cas.cn/book/202106/W020210623367064944709.pdf

Ⅲ.3 摩擦纳米发电机：高压电压源

Triboelectric Nanogenerators—As a High Voltage Source

1. Triboelectric-Generator-Driven Pulse Electrodeposition for Micropatterning

Guang Zhu, Caofeng Pan, Wenxi Guo, Chih-Yen Chen, Yusheng Zhou, Ruomeng Yu *and* Zhong Lin Wang*

Nano Letters, 12 (2012) 4960-4965.

简介：首次报道了摩擦纳米发电机作为高压电源进行电镀。

ABSTRACT By converting ambient energy into electricity, energy harvesting is capable of at least offsetting, or even replacing, the reliance of small portable electronics on traditional power supplies, such as batteries. Here we demonstrate a novel and simple generator with extremely low cost for efficiently harvesting mechanical energy that is typically present in the form of vibrations and random displacements/deformation. Owing to the coupling of contact charging and electrostatic induction, electric generation was achieved with a cycled process of contact and separation between two polymer films. A detailed theory is developed for understanding the proposed mechanism. The instantaneous electric power density reached as high as 31.2 mW/cm^3 at a maximum open circuit voltage of 110 V. Furthermore, the generator was successfully used without electric storage as a direct power source for pulse electrodeposition (PED) of micro/nanocrystalline silver structure. The cathodic current efficiency reached up to 86.6%. Not only does this work present a new type of generator that is featured by simple fabrication, large electric output, excellent robustness, and extremely low cost, but also extends the application of energy-harvesting technology to the field of electrochemistry with further utilizations including, but not limited to, pollutant degradation, corrosion protection, and water splitting.

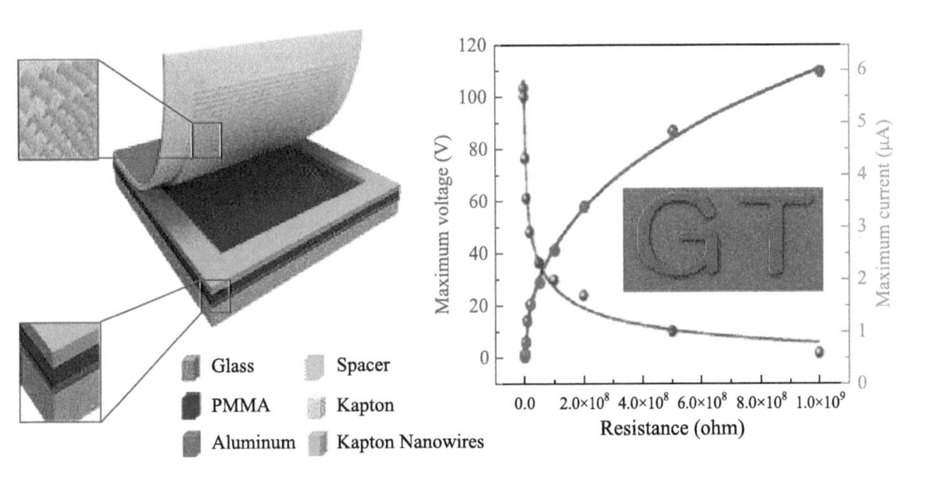

扫码阅全文

http://www.binn.cas.cn/book/202106/W020210623369412884846.pdf

2. Self-Powered Cleaning of Air Pollution by Wind Driven Triboelectric Nanogenerator

Shu wen Chen, Cai zhen Gao, Wei Tang, Hua rui Zhu, Yu Han, Qian wen Jiang, Tao Li, Xia Cao*, Zhong Lin Wang*

Nano Energy, 14 (2015) 217-225.

简介：首次报道了利用风力驱动的摩擦纳米发电机作为高压电源来进行空气净化。该研究后来经过几年的不断更新，最后孵化出了科技公司——中科纳清股份有限公司。

ABSTRACT Air pollution is one of the major challenges faced by the human kind, but cleaning of air is a horrendous task and hugely expensive, because of its large scope and the cost of energy. Up to now, all of the air cleaning systems are generally driven by external power, making it rather expensive and infeasible. Here, we introduce the first self-powered air cleaning system focusing on sulfur dioxide (SO_2) and dust removal as driven by the electricity generated by natural wind, with the use of rotating triboelectric nanogenerator (R-TENG). Distinguished from traditional approach of electrostatic precipitation by applying a voltage of thousand volt, our technology takes the advantages of high output voltage of R-TENG, typically in the order of a few hundreds volt. This self-powered air cleaning system not only adsorbs dust particles in air, but also oxidizes SO_2 without producing byproducts. Therefore, it could be potential for easing the haze-fog situation, which is one of the most important directions in self-powered electrochemistry.

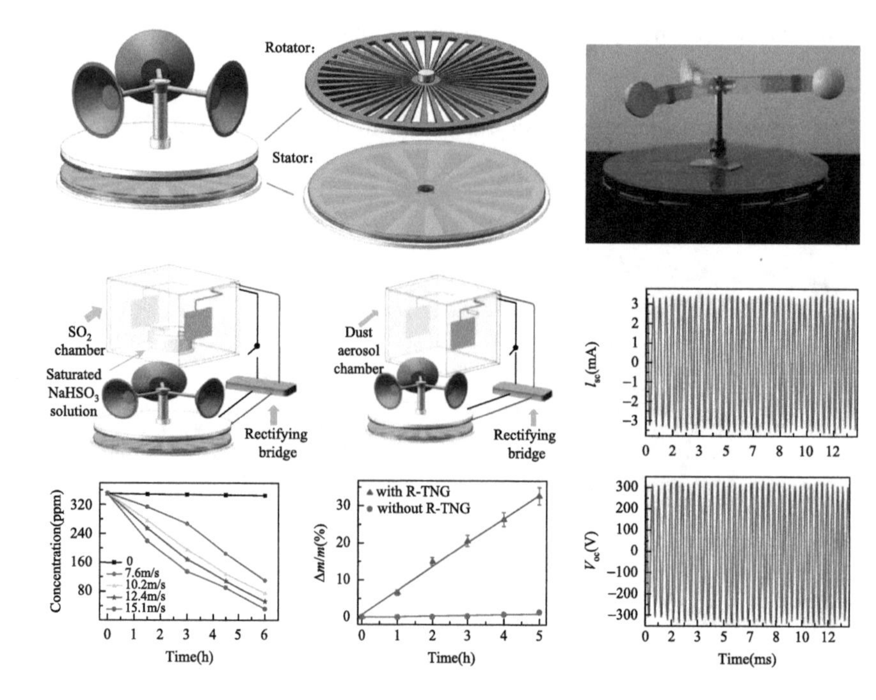

扫码阅全文

http://www.binn.cas.cn/book/202106/W020210623369413640016.pdf

3. Triboelectric Nanogenerators for Sensitive Nano-Coulomb Molecular Mass Spectrometry

Anyin Li, Yunlong Zi, Hengyu Guo, Zhong Lin Wang* *and* **Facundo M. Fernández***

Nature Nanotechnology, 12 (2017) 481-487.

简介：首次报道了摩擦纳米发电机作为高压电源应用在高分辨质谱仪上。由于每次输入的总电荷数一定，该技术比传统质谱仪在分析微量化学或生物成分中更具优势。

ABSTRACT Ion sources for molecular mass spectrometry are usually driven by direct current power supplies with no user control over the total charges generated. Here, we show that the output of triboelectric nanogenerators (TENGs) can quantitatively control the total ionization charges in mass spectrometry. The high output voltage of TENGs can generate single-or alternating-polarity ion pulses, and is ideal for inducing nanoelectrospray ionization (nanoESI) and plasma discharge ionization. For a given nanoESI emitter, accurately controlled ion pulses ranging from 1.0 to 5.5 nC were delivered with an onset charge of 1.0 nC. Spray pulses can be generated at a high frequency of 17 Hz (60 ms in period) and the pulse duration is adjustable on-demand between 60 ms and 5.5 s. Highly sensitive (～0.6 zeptomole) mass spectrometry analysis using minimal sample (18 pl per pulse) was achieved with a 10 pg·ml^{-1} cocaine sample. We also show that native protein conformation is conserved in TENG-ESI, and that patterned ion deposition on conductive and insulating surfaces is possible.

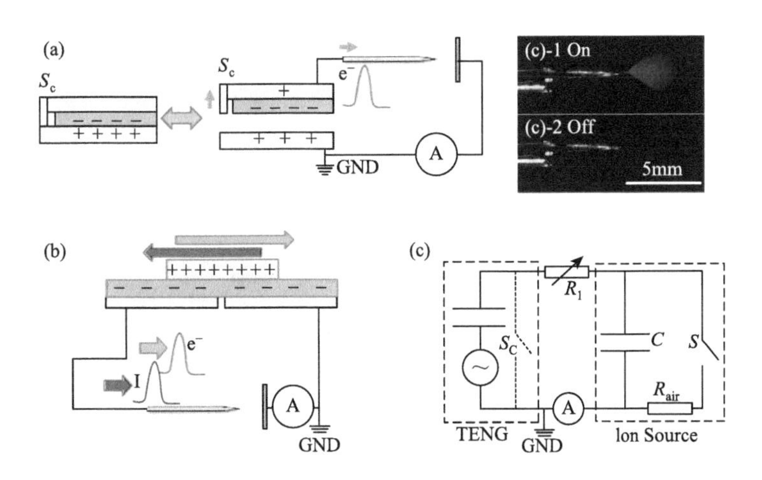

扫码阅全文

http://www.binn.cas.cn/book/202106/W020210623369414815361.pdf

4. Field Emission of Electrons Powered by a Triboelectric Nanogenerator

Yunlong Zi, Changsheng Wu, Wenbo Ding, Xingfu Wang, Yejing Dai, Jia Cheng, Jiyu Wang, Zhengjun Wang *and* Zhong Lin Wang*

Advanced Functional Materials, 28 (2018) 1800610.

简介：首次报道了摩擦纳米发电机作为高压电源进行电子的场致发射。

ABSTRACT Field emission of electrons usually requires high voltage (HV) of at least 100 V, which limits its applications due to the high cost, instability, portability issues, etc., of the HV instrument. Triboelectric nanogenerators (TENGs) have been developed to provide an HV of at least several kV for portable/mobile instrument, with controllability already demonstrated. Here, the field emission of electrons driven by TENG, namely, tribo-field emission, is presented for the first time. The emission voltage and the charge transfer per cycle can be tuned and controlled by TENG. The current peak generated from TENG with limited charge transfer is demonstrated to be more favorable than the direct current HV source in terms of emitter protection. A unidirectional continuous emission current is achieved through the tribo-field emission. A cathode-ray tube can be powered by TENG, with hours of illumination demonstrated through only one sliding motion. Such approach can provide a potential solution for controllable, stable, and portable field-emission devices without any additional external power sources.

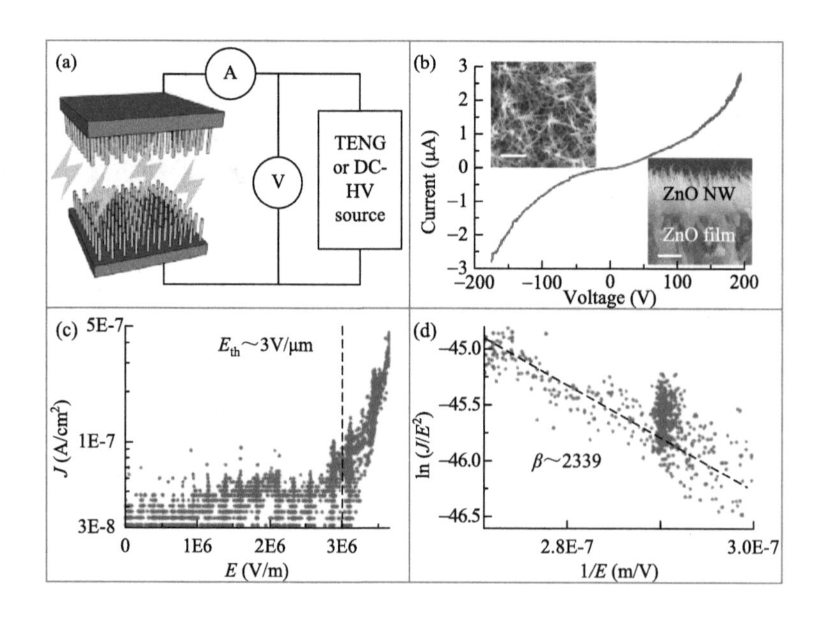

扫码阅全文

http://www.binn.cas.cn/book/202106/W020210623369415156039.pdf

5. Triboelectric Microplasma Powered by Mechanical Stimuli

Jia Cheng, Wenbo Ding, Yunlong Zi, Yijia Lu, Linhong Ji, Fan Liu, Changsheng Wu *and* Zhong Lin Wang*

Nature Communications, 9 (2018) 3733.

简介：首次报道了摩擦纳米发电机作为高压电源实现微型等离子体激发。

ABSTRACT Triboelectric nanogenerators (TENGs) naturally have the capability of high voltage output to breakdown gas easily. Here we present a concept of triboelectric microplasma by integrating TENGs with the plasma source so that atmospheric-pressure plasma can be powered only by mechanical stimuli. Four classical atmospheric-pressure microplasma sources are successfully demonstrated, including dielectric barrier discharge (DBD), atmospheric-pressure non-equilibrium plasma jets (APNP-J), corona discharge, and microspark discharge. For these types of microplasma, analysis of electric characteristics, optical emission spectra, COMSOL simulation and equivalent circuit model are carried out to explain transient process of different discharge. The triboelectric microplasma has been applied to patterned luminescence and surface treatment successfully as a first-step evaluation as well as to prove the system feasibility. This work offers a promising, facile, portable and safe supplement to traditional plasma sources, and will enrich the diversity of plasma applications based on the reach of existing technologies.

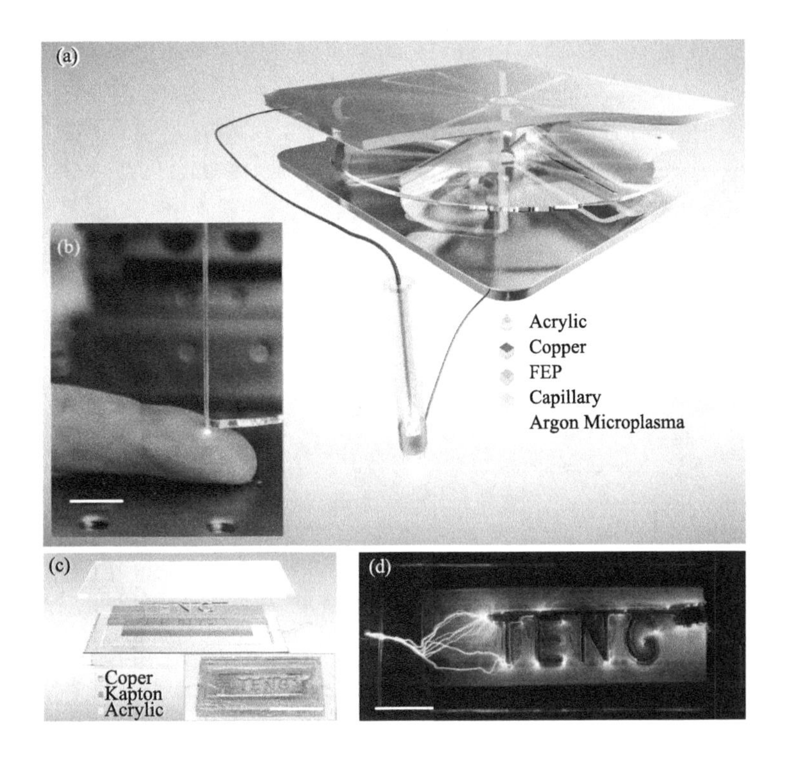

扫码阅全文

http://www.binn.cas.cn/book/202106/W020210623369415758862.pdf

6. Triboelectric Micromotors Actuated by Ultralow Frequency Mechanical Stimuli

Hang Yang, Yaokun Pang, Tianzhao Bu, Wenbo Liu, Jianjun Luo, Dongdong Jiang, Chi Zhang* *and* Zhong Lin Wang*

Nature Communications, 10 (2019) 2309.

简介：首次报道了摩擦纳米发电机作为高压电源驱动静电马达。

ABSTRACT A high-speed micromotor is usually actuated by a power source with high voltage and frequency. Here we report a triboelectric micromotor by coupling a micromotor and a triboelectric nanogenerator, in which the micromotor can be actuated by ultralow-frequency mechanical stimuli. The performances of the triboelectric micromotor are exhibited at various structural parameters of the micromotor, as well as at different mechanical stimuli of the triboelectric nanogenerator. With a sliding range of 50 mm at 0.1 Hz, the micromotor can start to rotate and reach over $1000 \cdot \mathrm{r\,min}^{-1}$ at 0.8 Hz. The maximum operation efficiency of the triboelectric micromotor can reach 41%. Additionally, the micromotor is demonstrated in two scanning systems for information recognition. This work has realized a high-speed micromotor actuated by ultralow frequency mechanical stimuli without an external power supply, which has extended the application of triboelectric nanogenerator in micro/nano electromechanical systems, intelligent robots and autonomous driving.

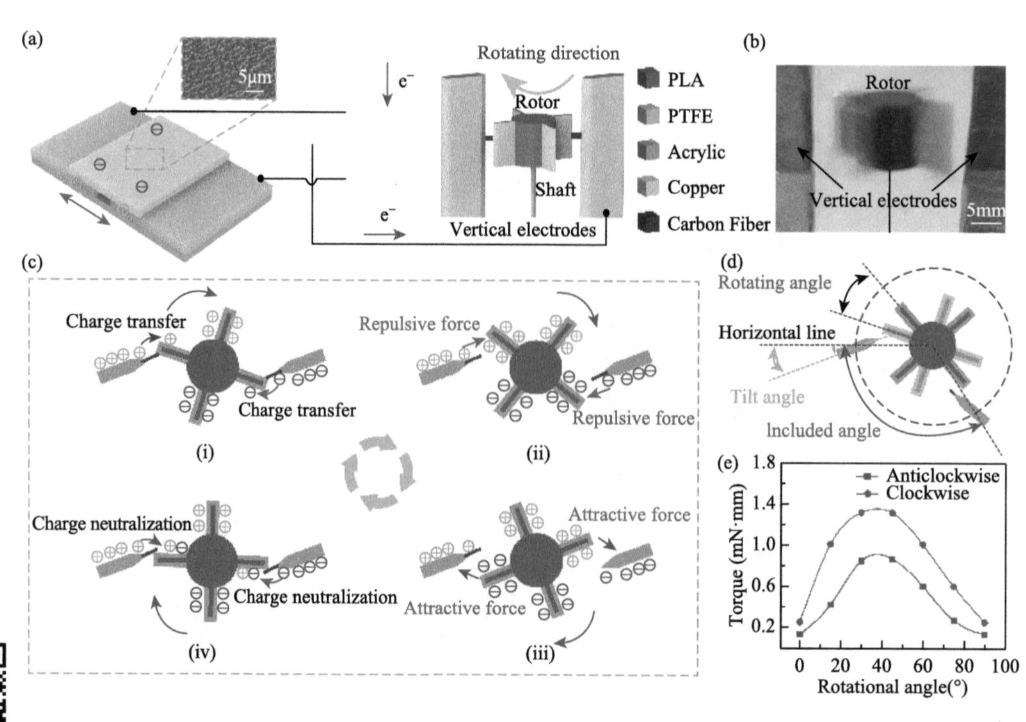

扫码阅全文

http://www.binn.cas.cn/book/202106/W020210623369416382582.pdf

Ⅲ.4 摩擦纳米发电机：蓝色能源
Triboelectric Nanogenerators—For Blue Energy

1. Networks of Triboelectric Nanogenerators for Harvesting Water Wave Energy: A Potential Approach toward Blue Energy

Jun Chen, Jin Yang, Zhaoling Li, Xing Fan, Yunlong Zi, Qingshen Jing, Hengyu Guo, Zhen Wen, Ken C. Pradel, Simiao Niu *and* Zhong Lin Wang[*]

ACS Nano, 9 (2015) 3324-3331.

简介：首次报道了利用摩擦纳米发电机收集水波能量，迈向蓝色能源。

ABSTRACT With 70% of the earth's surface covered with water, wave energy is abundant and has the potential to be one of the most environmentally benign forms of electric energy. However, owing to lack of effective technology, water wave energy harvesting is almost unexplored as an energy source. Here, we report a network design made of triboelectric nanogenerators (TENGs) for large-scale harvesting of kinetic water energy. Relying on surface charging effect between the conventional polymers and very thin layer of metal as electrodes for each TENG, the TENG networks (TENG-NW) that naturally float on the water surface convert the slow, random, and high-force oscillatory wave energy into electricity. On the basis of the measured output of a single TENG, the TENG-NW is expected to give an average power output of 1.15 MW from 1 km^2 surface area. Given the compelling features, such as being lightweight, extremely cost-effective, environmentally friendly, easily implemented, and capable of floating on the water surface, the TENG-NW renders an innovative and effective approach toward large-scale blue energy harvesting from the ocean.

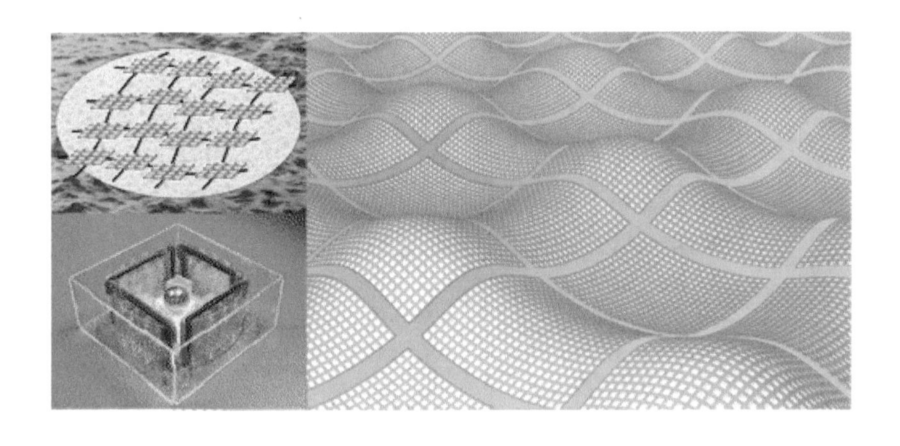

扫码阅全文

http://www.binn.cas.cn/book/202106/W020210623371004431908.pdf

2. Triboelectric Nanogenerator Based on Fully Enclosed Rolling Spherical Structure for Harvesting Low-Frequency Water Wave Energy

Xiaofeng Wang, Simiao Niu, Yajiang Yin, Fang Yi, Zheng You *and* Zhong Lin Wang*

Advanced Energy Materials, (2015) 1501467.

简介：首次报道了球形滚动式摩擦纳米发电机收集水波能量，迈向蓝色能源。

ABSTRACT Water waves are increasingly regarded as a promising source for large-scale energy applications. Triboelectric nanogenerators (TENGs) have been recognized as one of the most promising approaches for harvesting wave energy. This work examines a freestanding, fully enclosed TENG that encloses a rolling ball inside a rocking spherical shell. Through the optimization of materials and structural parameters, a spherical TENG of 6 cm in diameter actuated by water waves can provide a peak current of 1 μA over a wide load range from a short-circuit condition to 10 GΩ, with an instantaneous output power of up to 10 mW. A multielectrode arrangement is also studied to improve the output of the TENG under random wave motions from all directions. Moreover, at a frequency of 1.43 Hz, the wave-driven TENG can directly drive tens of LEDs and charge a series of supercapacitors to rated voltage within several hours. The stored energy can power an electronic thermometer for 20 min. This rolling-structured TENG is extremely lightweight, has a simple structure, and is capable of rocking on or in water to harvest wave energy; it provides an innovative and effective approach toward large-scale blue energy harvesting of oceans and lakes.

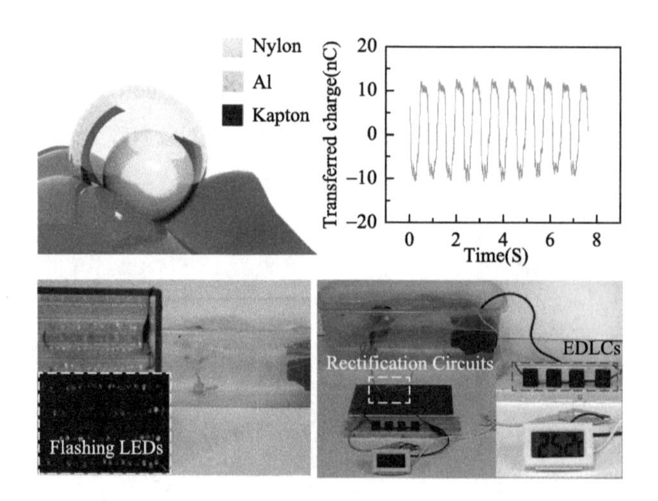

扫码阅全文

http://www.binn.cas.cn/book/202106/W020210623371005730425.pdf

3. Robust Swing-Structured Triboelectric Nanogenerator for Efficient Blue Energy Harvesting

Tao Jiang, Hao Pang, Jie An, Pinjing Lu, Yawei Feng, Xi Liang, Wei Zhong, *and* **Zhong Lin Wang***

Advanced Energy Materials, (2020) 2000064 .

简介：首次报道了摆钟式摩擦纳米发电机高效收集水波能量。

ABSTRACT Ocean wave energy is a promising renewable energy source, but harvesting such irregular, "random," and mostly ultra-low frequency energies is rather challenging due to technological limitations. Triboelectric nanogenerators (TENGs) provide a potential efficient technology for scavenging ocean wave energy. Here, a robust swing-structured triboelectric nanogenerator (SS-TENG) with high energy conversion efficiency for ultra-low frequency water wave energy harvesting is reported. The swing structure inside the cylindrical TENG greatly elongates its operation time, accompanied with multiplied output frequency. The design of the air gap and flexible dielectric brushes enable mininized frictional resistance and sustainable triboelectric charges, leading to enhanced robustness and durability. The TENG performance is controlled by external triggering conditions, with a long swing time of 88 s and a high energy conversion efficiency, as well as undiminished performance after continuous triggering for 400,000 cycles. Furthermore, the SS-TENG is demonstrated to effectively harvest water wave energy. Portable electronic devices are successfully powered for self-powered sensing and environment monitoring. Due to the excellent performance of the distinctive mechanism and structure, the SS-TENG in this work provides a good candidate for harvesting blue energy on a large scale.

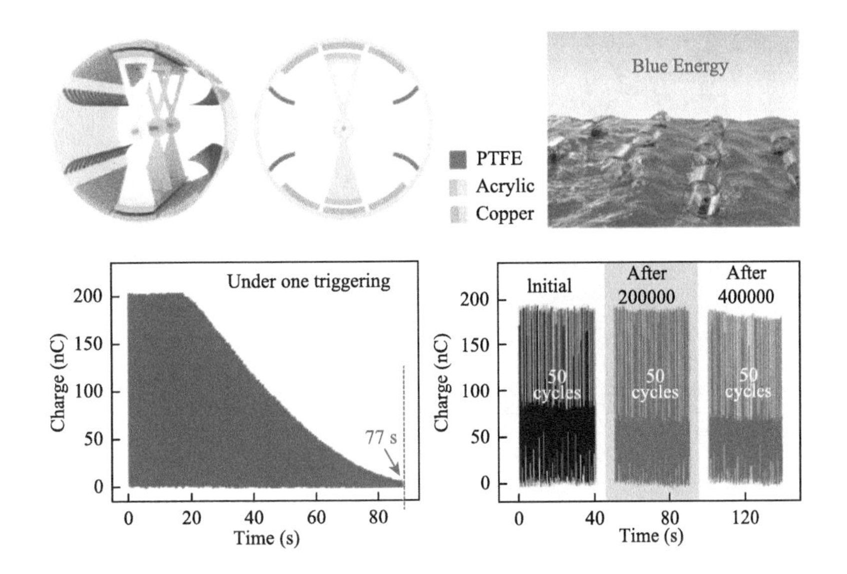

扫码阅全文

http://www.binn.cas.cn/book/202106/W020210623371006401702.pdf

4. New Wave Power

Zhong Lin Wang

Nature, 542 (2017) 159-160.

简介：首次提出了利用摩擦纳米发电机网格式结构来实现蓝色能源。

ABSTRACT Zhong Lin Wang proposes a radically different way to harvest renewable energy from the ocean using floating nets of nanogenerators.

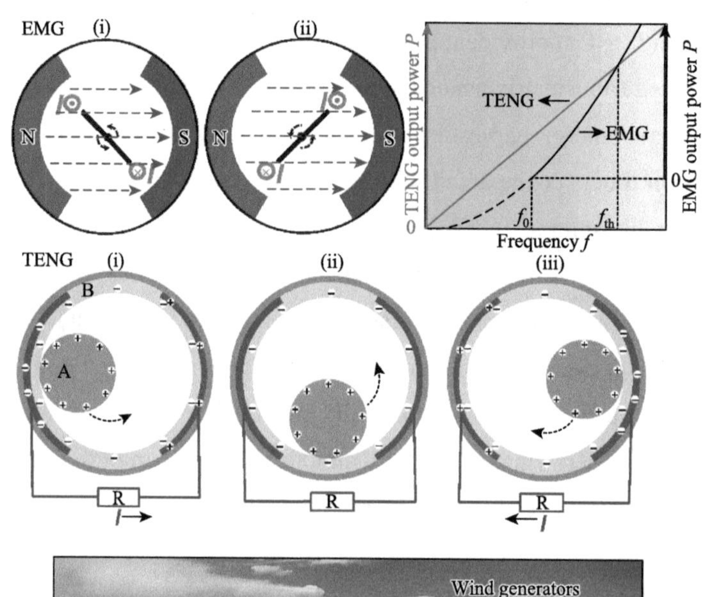

扫码阅全文

http://www.binn.cas.cn/book/202106/W020210623371007112511.pdf

5. Toward the Blue Energy Dream by Triboelectric Nanogenerator Networks

Zhong Lin Wang,* Tao Jiang, Liang Xu

Nano Energy, 39 (2017) 9-23.

简介：蓝色能源的第一篇综述文献。

ABSTRACT Widely distributed across the globe, water wave energy is one of the most promising renewable energy sources, while little has been exploited due to various limitations of current technologies mainly relying on electromagnetic generator (EMG), especially its operation in irregular environment and low frequency (<5 Hz). The newly developed triboelectric nanogenerator (TENG) exhibits obvious advantages over EMG in harvesting energy from low-frequency water wave motions, and the network of TENGs was proposed as a potential approach toward large-scale blue energy harvesting. Here, a review is given for recent progress in blue energy harvesting using TENG technology, starting from a comparison between the EMG and the TENG both in physics and engineering design. The fundamental mechanism of nanogenerators is presented based on Maxwell's displacement current. Approaches of water wave energy harvesting by liquid-solid contact electrification TENG are introduced. For fully enclosed TENGs, the structural designs and performance optimizations are discussed, based on which the TENG network is proposed for large-scale blue energy harvesting from water waves. Furthermore, the energy harvested by TENG from various sources such as water wave, human motion and vibration etc, is not only new energy, but more importantly, the energy for the new era-the era of internet of things.

扫码阅全文

http://www.binn.cas.cn/book/202106/W020210623371007312393.pdf

Ⅲ.5 复合式能量回收系统
Hybridized Energy Systems

1. Nanowire Structured Hybrid Cell for Concurrently Scavenging Solar and Mechanical Energies

Chen Xu, Xudong Wang *and* Zhong Lin Wang*

Journal of the American Chemical Society, 131 (2009) 5866-5872.

简介：首次报道了一个器件同时可以收集太阳能与机械能，是复合能量回收系统的开山作。

ABSTRACT Conversion cells for harvesting solar energy and mechanical energy are usually separate and independent entities that are designed and built following different physical principles. Developing a technology that harvests multiple-type energies in forms such as sun light and mechanical around the clock is desperately desired for fully utilizing the energies available in our living environment. We report a hybrid cell that is intended for simultaneously harvesting solar and mechanical energies. Using aligned ZnO nanowire arrays grown on surfaces of a flat substrate, a dye-sensitized solar cell is integrated with a piezoelectric nanogenerator. The former harvests solar energy irradiating on the top, and the latter harvests ultrasonic wave energy from the surrounding. The two energy harvesting approaches can work simultaneously or individually, and they can be integrated in parallel and serial for raising the output current and voltage, respectively, as well as power. It is found that the voltage output from the solar cell can be used to raise the output voltage of the nanogenerator, providing an effective approach for effectively storing and utilizing the power generated by the nanogenerator. Our study demonstrates a new approach for concurrently harvesting multiple types of energies using an integrated hybrid cell so that the energy resources can be effectively and complementary utilized whenever and wherever one or all of them is available.

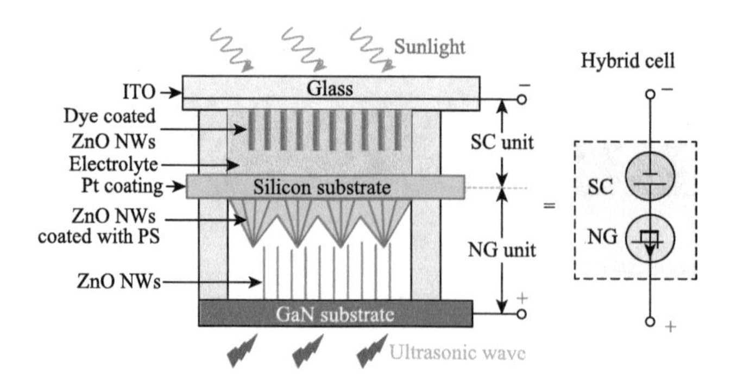

扫码阅全文

http://www.binn.cas.cn/book/202106/W020210623375726039759.pdf

2. Compact Hybrid Cell Based on a Convoluted Nanowire Structure for Harvesting Solar and Mechanical Energy

Chen Xu *and* **Zhong Lin Wang**

Advanced Materials, 23 (2011) 873-877.

简介：首次报道了一个器件同时收集太阳能与机械能的复合能量回收系统。

ABSTRACT A fully integrated, solid-state, compact hybrid cell (CHC) that comprises “convoluted” ZnO nanowire structures for concurrent harvesting of both solar and mechanical energy is demonstrated. The compact hybrid cell is based on a conjunction design of an organic solid-state dye-sensitized solar cell (DSSC) and piezoelectric nanogenerator in one compact structure. The CHC shows a significant increase in output power, clearly demonstrating its potential for simultaneously harvesting multiple types of energy for powering small electronic devices for independent, sustainable, and mobile operation.

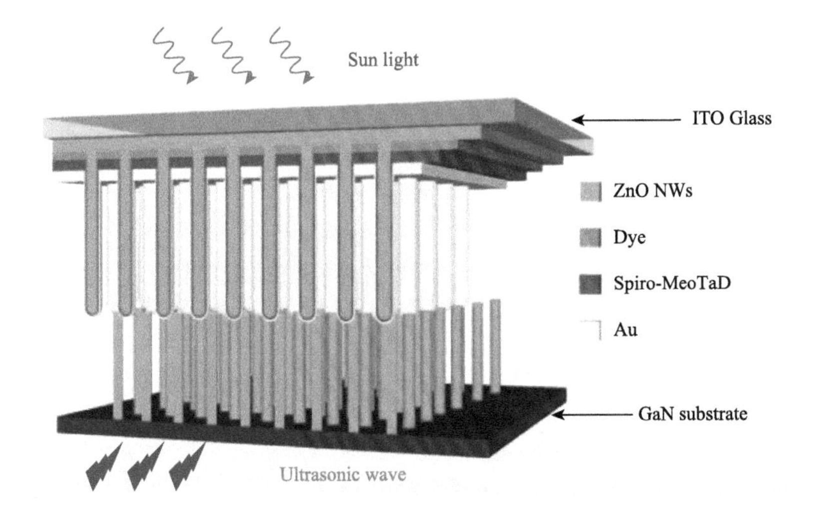

扫码阅全文

http://www.binn.cas.cn/book/202106/W020210623375726711499.pdf

3. Hybrid Nanogenerator for Concurrently Harvesting Biomechanical and Biochemical Energy

Benjamin J. Hansen, Ying Liu, Rusen Yang *and* Zhong Lin Wang*

ACS Nano, 4 (2010) 3647-3652.

简介：首次报道了一个器件同时收集生物机械能与生物化学能的复合能量回收系统。

ABSTRACT Harvesting energy from multiple sources available in our personal and daily environments is highly desirable, not only for powering personal electronics, but also for future implantable sensor-transmitter devices for biomedical and healthcare applications. Here we present a hybrid energy scavenging device for potential *in vivo* applications. The hybrid device consists of a piezoelectric poly(vinylidene fluoride) nanofiber nanogenerator for harvesting mechanical energy, such as from breathing or from the beat of a heart, and a flexible enzymatic biofuel cell for harvesting the biochemical (glucose/O_2) energy in biofluid, which are two types of energy available in vivo. The two energy harvesting approaches can work simultaneously or individually, thereby boosting output and lifetime. Using the hybrid device, we demonstrate a "self-powered" nanosystem by powering a ZnO nanowire UV light sensor.

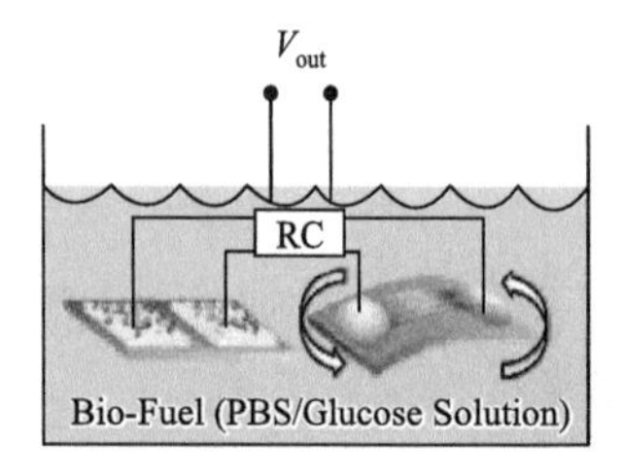

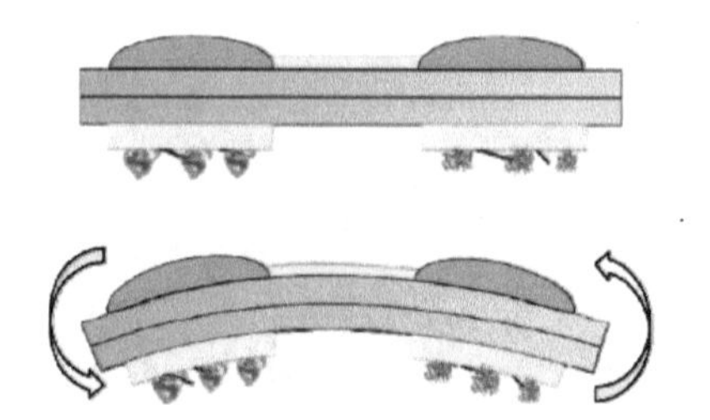

扫码阅全文

http://www.binn.cas.cn/book/202106/W020210623375726908494.pdf

4. Nanowire-Quantum Dot Hybridized Cell for Harvesting Sound and Solar Energies

Minbaek Lee, Rusen Yang, Cheng Li *and* Zhong Lin Wang*

The Journal of Physical Chemistry Letters, 1 (2010) 2929-2935.

简介：首次报道了一个器件同时收集声波振动能与太阳能的复合能源系统。

ABSTRACT We have demonstrated sound-wave-driven nanogenerators using both laterally bonded single wires and vertically aligned nanowire arrays for energy harvesting in the frequency range of 35-1000 Hz. The electricity produced by the single wire generator (SWG) is linearly proportional to the input acoustic energy, while the frequency does not affect the performance of the SWG in this study. By infiltrating CdS/CdTe quantum dots among vertical nanowires, we have fabricated a hybrid cell that simultaneously harvests both sound and solar energies. This demonstrates a possible approach for effectively harvesting available energies in our living environment with and without the presence of light.

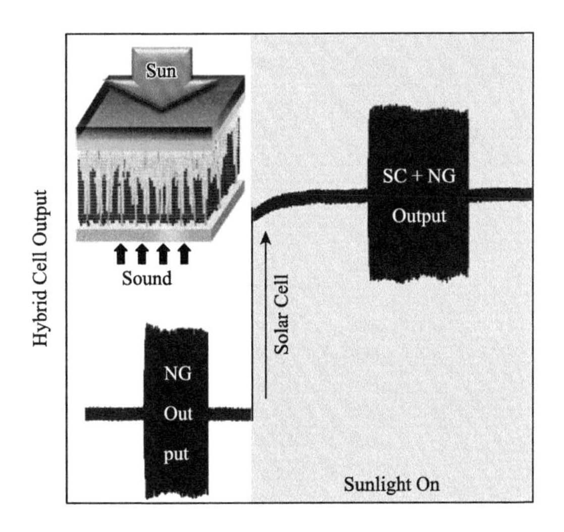

扫码阅全文

http://www.binn.cas.cn/book/202106/W020210623375727067226.pdf

5. Fiber Supercapacitors Made of Nanowire-Fiber Hybrid Structures for Wearable/Flexible Energy Storage

Joonho Bae, Min Kyu Song, Young Jun Park, Jong Min Kim*, Meilin Liu *and* Zhong Lin Wang*

Angewandte Chemie, 123 (2011) 1721-1725.

简介：首次报道了利用纤维制作的超电容。

ABSTRACT Current clothing trends: A wearable and flexible fiber supercapacitor with a fully encapsulated electrolyte is formed by wrapping a plastic wire covered with ZnO nanowires (NWs; see SEM image) around a Kevlar fiber covered with gold-coated ZnO NWs. This supercapacitor shows promise as a highly efficient, wearable energy storage device.

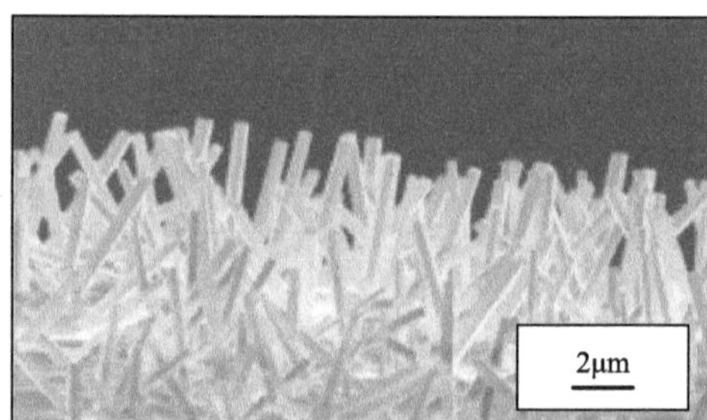

扫码阅全文

http://www.binn.cas.cn/book/202106/W020210623375727390243.pdf

6. Fiber-Based Hybrid Nanogenerators for/as Self-Powered Systems in Biological Liquid

Caofeng Pan, Zetang Li, Wenxi Guo, Jing Zhu *and* **Zhong Lin Wang***

Angewandte Chemie, 50 (2011) 11192-11196.

简介：首次报道了利用纤维制作的复合能源系统。

ABSTRACT Electrifying: A fiber-based hybrid nanogenerator for simultaneous or independent harvesting of mechanical and biochemical energy is presented. The hybrid system consists of a nanogenerator (FNG) and a biofuel cell (FBFC) and provides a peak output of 3.1 V and 200 nA. It also serves as a self-powered sensor for the quantification of pressure variations.

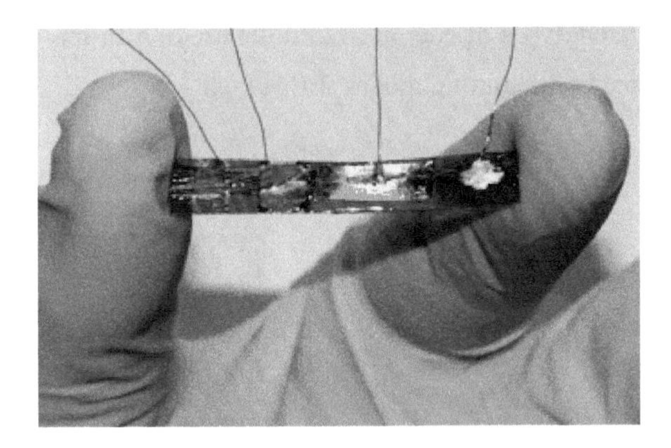

扫码阅全文

http://www.binn.cas.cn/book/202106/W020210623375727596923.pdf

7. Flexible Hybrid Energy Cell for Simultaneously Harvesting Thermal, Mechanical, and Solar Energies

Ya Yang, Hulin Zhang, Guang Zhu, Sangmin Lee, Zong-Hong Lin *and* **Zhong Lin Wang***

ACS Nano, 7 (2013) 785-790.

简介：首次报道了一个器件同时收集热能、机械能与太阳能的复合能源系统。

ABSTRACT We report the first flexible hybrid energy cell that is capable of simultaneously or individually harvesting thermal, mechanical, and solar energies to power some electronic devices. For having both the pyroelectric and piezoelectric properties, a polarized poly(vinylidene fluoride) (PVDF) film-based nanogenerator (NG) was used to harvest thermal and mechanical energies. Using aligned ZnO nanowire arrays grown on the flexible polyester (PET) substrate, a ZnO-poly(3-hexylthiophene) (P3HT) heterojunction solar cell was designed for harvesting solar energy. By integrating the NGs and the solar cells, a hybrid energy cell was fabricated to simultaneously harvest three different types of energies. With the use of a Li-ion battery as the energy storage, the harvested energy can drive four red light-emitting diodes (LEDs).

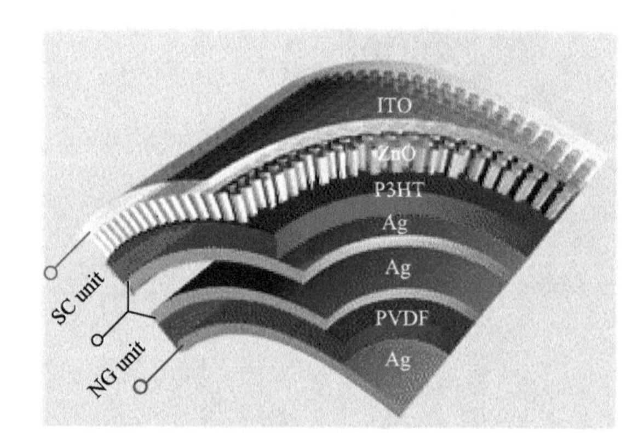

扫码阅全文

http://www.binn.cas.cn/book/202106/W020210623375727944365.pdf

8. Simultaneously Harvesting Mechanical and Chemical Energies by a Hybrid Cell for Self-Powered Biosensors and Personal Electronics

Ya Yang, Hulin Zhang, Jun Chen, Sangmin Lee, Te-Chien Hou *and* Zhong Lin Wang*

Energy & Environmental Science, 6 (2013) 1744-1749.

简介：首次报道了一个器件同时收集化学能与机械能的复合能源系统。

ABSTRACT Electrochemical cells (ECs) are devices that convert chemical energy into electricity through spontaneous oxidation-reduction reactions that occur separately at two electrodes through the transport of protons in the electrolyte solution and the flow of electrons in the external circuit. A triboelectric nanogenerator (TENG) is an effective device that converts mechanical energy into electricity using organic/polymer materials by a contact induced electrification process followed by charge separation. In this paper, we demonstrate the first integration of an EC and a TENG for simultaneously harvesting chemical and mechanical energy, and its application for powering a sensor and even personal electronics. An EC was fabricated using a Cu/NaCl solution/Al structure, on which a thin polydimethylsiloxane (PDMS) film with a micropyramid surface structure was used as the protection layer of the EC for anti-corrosion, anti-contamination and anti-mechanical damage. A TENG was fabricated based on a contact-and-separation process between the PDMS protection layer and the Al electrode layer of the EC. The output performance of the TENG can be increased by embedding $BaTiO_3$ nanoparticles into the PDMS film layer to enhance the dielectric property. Moreover, we also demonstrated that the produced hybrid energies can be stored in a Li-ion battery for lighting up 30 green LEDs.

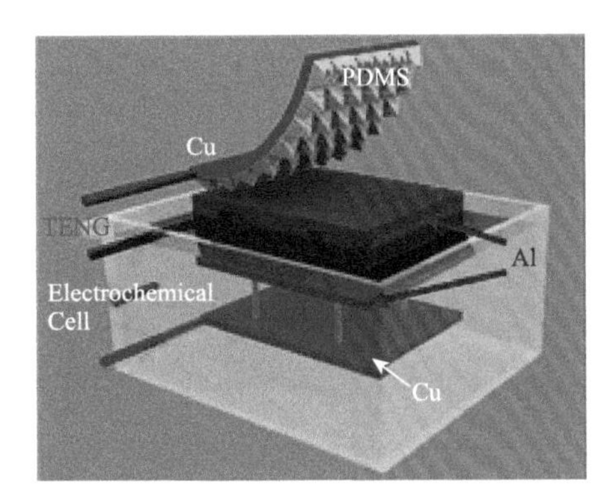

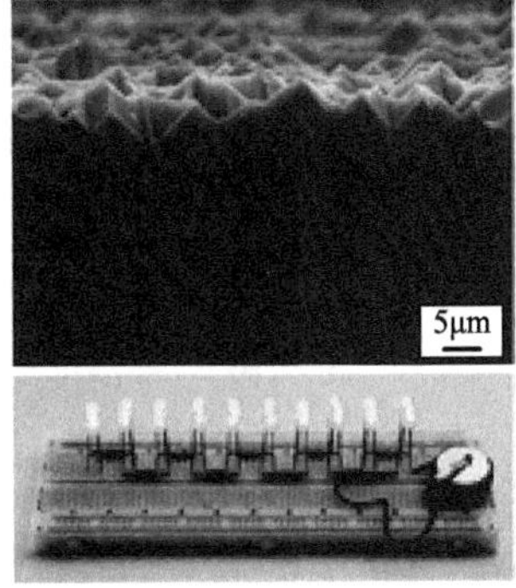

扫码阅全文

http://www.binn.cas.cn/book/202106/W020210623375728262863.pdf

9. Hybridizing Triboelectrification and Electromagnetic Induction Effects for High-Efficient Mechanical Energy Harvesting

Youfan Hu, Jin Yang, Simiao Niu, Wenzhuo Wu *and* Zhong Lin Wang*

ACS Nano, 8 (2014) 7442-7450.

简介：首次报道了把摩擦纳米发电机和传统电磁发电机结合起来的复合能源系统。

ABSTRACT The recently introduced triboelectric nanogenerator (TENG) and the traditional electromagnetic induction generator (EMIG) are coherently integrated in one structure for energy harvesting and vibration sensing/isolation. The suspended structure is based on two oppositely oriented magnets that are enclosed by hollow cubes surrounded with coils, which oscillates in response to external disturbance and harvests mechanical energy simultaneously from triboelectrification and electromagnetic induction. It extends the previous definition of hybrid cell to harvest the same type of energy with multiple approaches. Both the sliding-mode TENG and contact-mode TENG can be achieved in the same structure. In order to make the TENG and EMIG work together, transformers are used to match the output impedance between these two power sources with very different characteristics. The maximum output power of 7.7 and 1.9 mW on the same load of 5 kΩ was obtained for the TENG and EMIG, respectively, after impedance matching. Benefiting from the rational design, the output signal from the TENG and the EMIG are in phase. They can be added up directly to get an output voltage of 4.6 V and an output current of 2.2 mA in parallel connection. A power management circuit was connected to the hybrid cell, and a regulated voltage of 3.3 V with constant current was achieved. For the first time, a logic operation was carried out on a half-adder circuit by using the hybrid cell working as both the power source and the input digit signals. We also demonstrated that the hybrid cell can serve as a vibration isolator. Further applications as vibration dampers, triggers, and sensors are all promising.

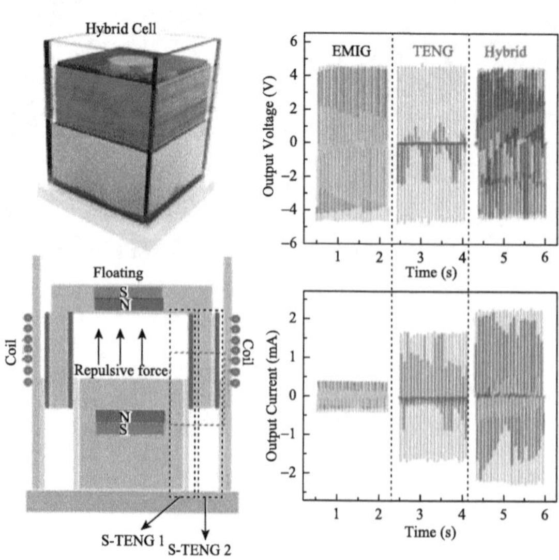

扫码阅全文

http://www.binn.cas.cn/book/202106/W020210623375728535392.pdf

10. Silicon-Based Hybrid Cell for Harvesting Solar Energy and Raindrop Electrostatic Energy

Li Zheng, Zong-Hong Lin, Gang Cheng, Wenzhuo Wu, Xiaonan Wen, Sangmin Lee, Zhong Lin Wang*

Nano Energy, 9 (2014) 291-300.

简介：首次报道了一个器件同时收集太阳能与雨水能的复合能源系统。

ABSTRACT Silicon-based solar cell is by far the most established solar cell technology. The surface of a Si solar cell is usually covered by a layer of transparent material to protect the device from environmental damages/ corrosions. Here, we replaced this protection layer by a transparent triboelectric nanogenerator (TENG), for simultaneously or individually harvesting solar and raindrop energy when either or both of them are available in our living environment. The TENG is made of a specially processed polytetrafluoroethylene (PTFE) film, an indium tin oxide (ITO) and a polyethylene terephthalate (PET) layer. Under solar light irradiation (12 W/m^2) in a rainy day, the fabricated high-efficiency solar cell provides an open-circuit (V_{oc}) of 0.43 V and short-circuit current density (J_{sc}) of 4.2 A/m^2. And the TENG designed for collection of raindrop energy gives an AC V_{oc} of 30 V and J_{sc} of 4.2 mA/m^2 when impacted by water drops at a dripping rate of 0.116 ml/s. In rainy days, the performance of solar cell decreased greatly, while TENG can be a good compensation as for green energy harvesting. From these results, we can see that the hybrid cell formed by a solar cell and a water-drop TENG have great potential for simultaneously/individually harvesting both solar energy and raindrop electrostatic energy under different weather conditions, especially in raining season.

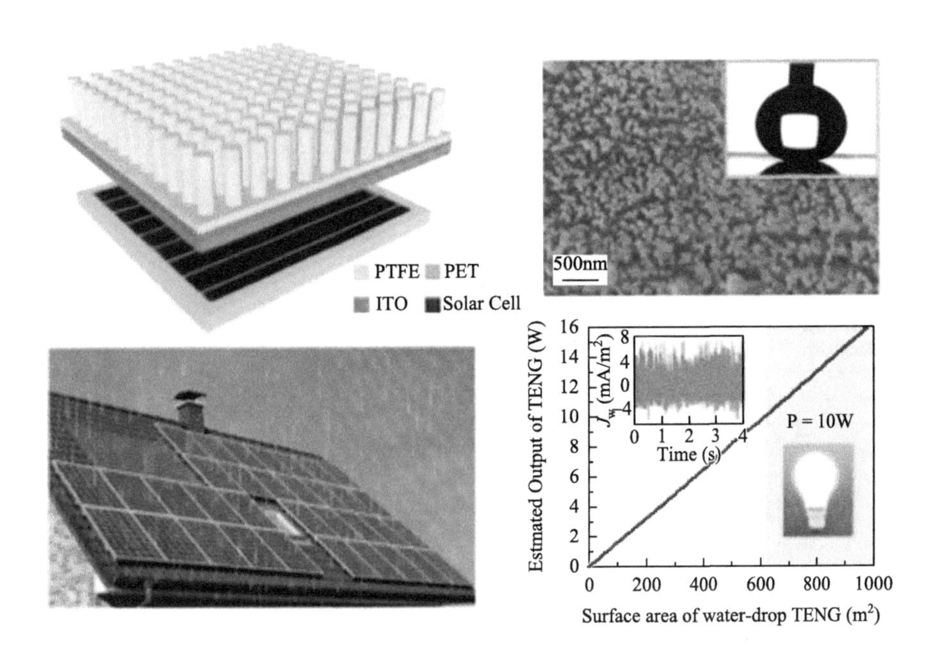

扫码阅全文

http://www.binn.cas.cn/book/202106/W020210623375729356299.pdf

11. A Hybridized Power Panel to Simultaneously Generate Electricity from Sunlight, Raindrops and Wind Around the Clock

Li Zheng, Gang Cheng, Jun Chen, Long Lin, Jie Wang, Yongsheng Liu, Hexing Li *and* Zhong Lin Wang*

Advanced Energy Materials, (2015) 1501152.

简介：首次报道了一个器件同时收集太阳能、雨水能与风能的复合能源系统。

ABSTRACT With the solar panels quickly spreading across the rooftops worldwide, solar power is now very popular. However, the output of the solar cell panels is highly dependent on weather conditions, making it rather unstable. Here, a hybridized power panel that can simultaneously generate power from sunlight, raindrop, and wind is proposed and demonstrated, when any or all of them are available in ambient environment. Without compromising the output performance and conversion efficiency of the solar cell itself, the presented hybrid cell can deliver an average output of 86 $mW \cdot m^{-2}$ from the water drops at a dripping rate of 13.6 $ml \cdot s^{-1}$, and an average output of 8 $mW \cdot m^{-2}$ from wind at a speed of 2.7 $m \cdot s^{-1}$, which is an innovative energy compensation to the common solar cells, especially in rainy seasons or at night. Given the compelling features, such as cost-effectiveness and a greatly expanded working time, the reported hybrid cell renders an innovative way to realize multiple kinds of energy harvesting and as an useful compensation to the currently widely used solar cells. The demonstrated concept here will possibly be adopted in a variety of circumstances and change the traditional way of solar energy harvesting.

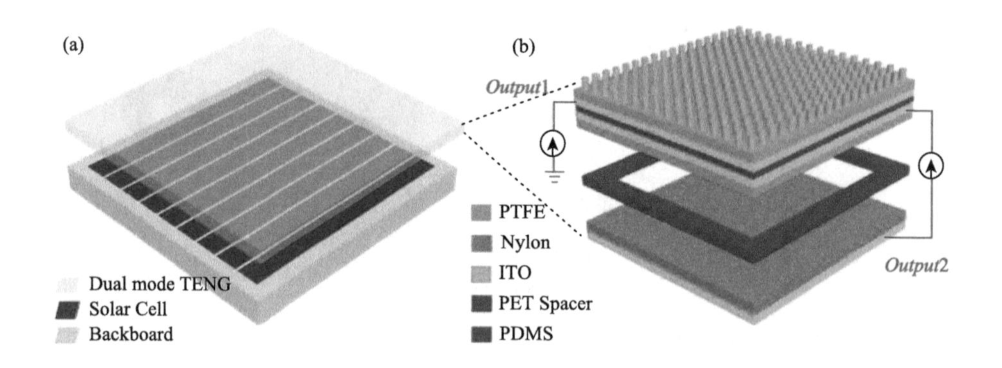

扫码阅全文

http://www.binn.cas.cn/book/202106/W020210623375729946357.pdf

12. Triboelectric-Pyroelectric-Piezoelectric Hybrid Cell for High-Efficient Energy-Harvesting and Self-Powered Sensing

Yunlong Zi, Long Lin, Jie Wang, Sihong Wang, Jun Chen, Xing Fan, Po-Kang Yang, Fang Yi *and* **Zhong Lin Wang***

Advanced Materials, 27 (2015) 2340-2347.

简介：首次报道了把摩擦发电、热势发电与压电发电结合在一起的复合能源系统。

ABSTRACT A triboelectric-pyroelectric-piezoelectric hybrid cell, consisting of a triboelectric nanogenerator and a pyroelectric-piezoelectric nanogenerator, is developed for highly efficient mechanical energy harvesting through multiple mechanisms. The excellent performance of the hybrid cell enhances the energy-harvesting efficiency significantly (by 26.2% at 1 kΩ load resistance), and enables self-powered sensing, which will lead to a variety of advanced applications.

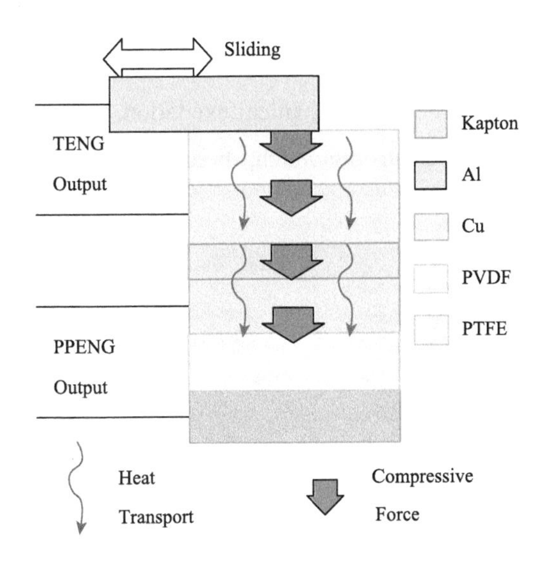

扫码阅全文

http://www.binn.cas.cn/book/202106/W020210623375730419079.pdf

13. Micro-Cable Structured Textile for Simultaneously Harvesting Solar and Mechanical Energy

Jun Chen, Yi Huang, Nannan Zhang, Haiyang Zou, Ruiyuan Liu, Changyuan Tao, Xing Fan* *and* **Zhong Lin Wang***

Nature Energy, (2016) 16138.

简介：首次报道了利用纺织品同时收集风能与太阳能的复合能源系统。

ABSTRACT Developing lightweight, flexible, foldable and sustainable power sources with simple transport and storage remains a challenge and an urgent need for the advancement of next-generation wearable electronics. Here, we report a micro-cable power textile for simultaneously harvesting energy from ambient sunshine and mechanical movement. Solar cells fabricated from lightweight polymer fibres into micro cables are then woven via a shuttle-flying process with fibre-based triboelectric nanogenerators to create a smart fabric. A single layer of such fabric is 320 μm thick and can be integrated into various cloths, curtains, tents and so on. This hybrid power textile, fabricated with a size of 4 cm by 5 cm, was demonstrated to charge a 2 mF commercial capacitor up to 2 V in 1 min under ambient sunlight in the presence of mechanical excitation, such as human motion and wind blowing. The textile could continuously power an electronic watch, directly charge a cell phone and drive water splitting reactions.

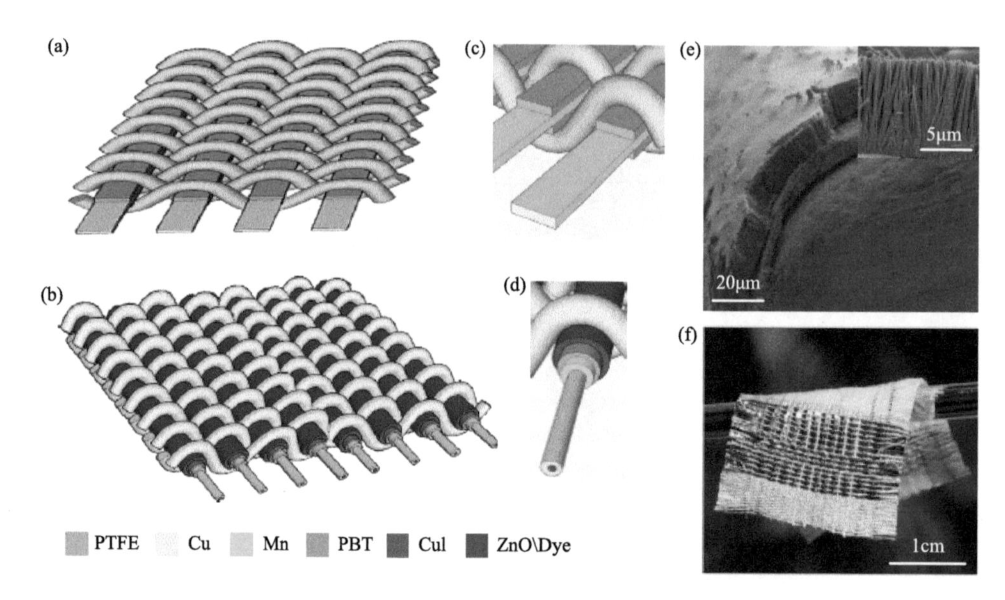

扫码阅全文

http://www.binn.cas.cn/book/202106/W020210623375731222446.pdf

14. Optical Fiber/Nanowire Hybrid Structures for Efficient Three-Dimensional Dye-Sensitized Solar Cells

Benjamin Weintraub, Yaguang Wei *and* Zhong Lin Wang*

Angewandte Chemie, 48 (2009) 8981-8985.

简介：首次报道了利用光纤与径向纳米线阵列相结合来高效收集太阳能的三维结构。

ABSTRACT Wired up: The energy conversion efficiency of three-dimensional dye-sensitized solar cells (DSSCs) in a hybrid structure that integrates optical fibers and nanowire arrays is greater than that of a two-dimensional device. Internal axial illumination enhances the energy conversion efficiency of a rectangular fiber-based hybrid structure (see picture) by a factor of up to six compared to light illumination normal to the fiber axis from outside the device.

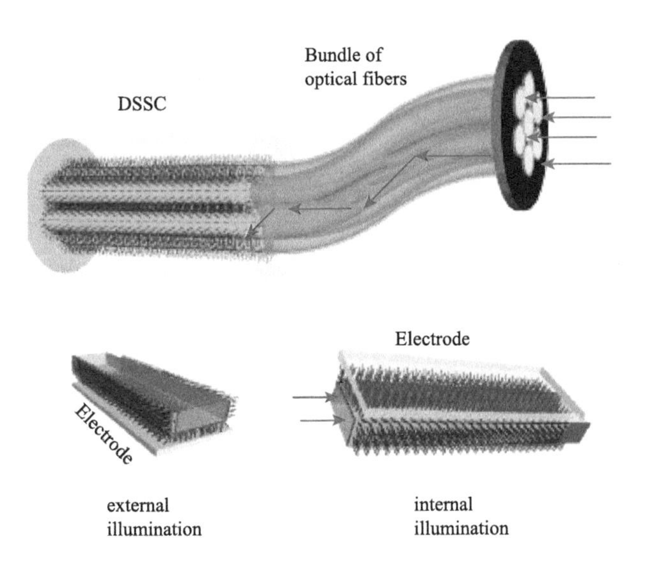

扫码阅全文

http://www.binn.cas.cn/book/202106/W020210623375731950106.pdf

15. Planar Waveguide-Nanowire Integrated Three-Dimensional Dye-Sensitized Solar Cells

Yaguang Wei, Chen Xu, Sheng Xu, Cheng Li, Wenzhuo Wu *and* Zhong Lin Wang*

Nano Letters, 10 (2010) 2092-2096.

简介：首次报道了利用平面光波导与法向纳米线阵列相结合来高效收集太阳能的三维结构。

ABSTRACT We present a new approach to fabricate three-dimensional (3D) dye-sensitized solar cells (DSSCs) by integrating planar optical waveguide and nanowires (NWs). The ZnO NWs are grown normally to the quartz slide. The 3D cell is constructed by alternatively stacking a slide and a planar electrode. The slide serves as a planar waveguide for light propagation. The 3D structure effectively increases the light absorbing surface area due to internal multiple reflections without increasing electron path length to the collecting electrode, resulting in a significant improvement in energy conversion efficiency by a factor of 5.8 on average compared to the planar illumination case. Our approach demonstrates a new methodology for building large scale and high-efficient 3D solar cells that can be expanded to organic-and inorganic-based solar cells.

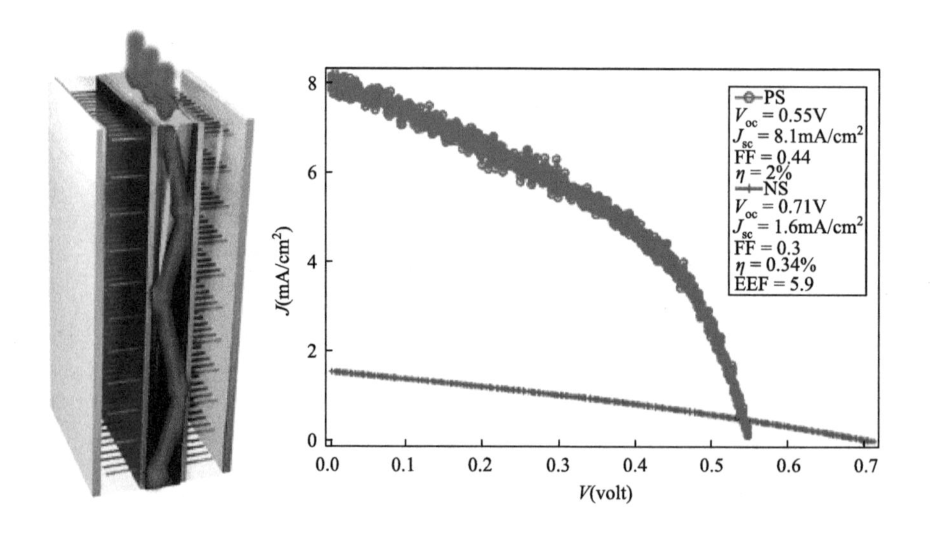

扫码阅全文

http://www.binn.cas.cn/book/202106/W020210623375732185786.pdf

Ⅲ.6　热势纳米发电机与传感
Pyroelectric Nanogenerators and Sensors

1. Pyroelectric Nanogenerators for Harvesting Thermoelectric Energy

Ya Yang, Wenxi Guo, Ken C. Pradel, Guang Zhu, Yusheng Zhou, Yan Zhang, Youfan Hu, Long Lin *and* **Zhong Lin Wang***

Nano Letters, 12 (2012) 2833-2838.

简介：首次报道了利用热势效应来收集热能的方法。

ABSTRACT　Harvesting thermoelectric energy mainly relies on the Seebeck effect that utilizes a temperature difference between two ends of the device for driving the diffusion of charge carriers. However, in an environment that the temperature is spatially uniform without a gradient, the pyroelectric effect has to be the choice, which is based on the spontaneous polarization in certain anisotropic solids due to a time-dependent temperature variation. Using this effect, we experimentally demonstrate the first application of pyroelectric ZnO nanowire arrays for converting heat energy into electricity. The coupling of the pyroelectric and semiconducting properties in ZnO creates a polarization electric field and charge separation along the ZnO nanowire as a result of the time-dependent change in temperature. The fabricated nanogenerator has a good stability, and the characteristic coefficient of heat flow conversion into electricity is estimated to be ~0.05-0.08 Vm^2/W. Our study has the potential of using pyroelectric nanowires to convert wasted energy into electricity for powering nanodevices.

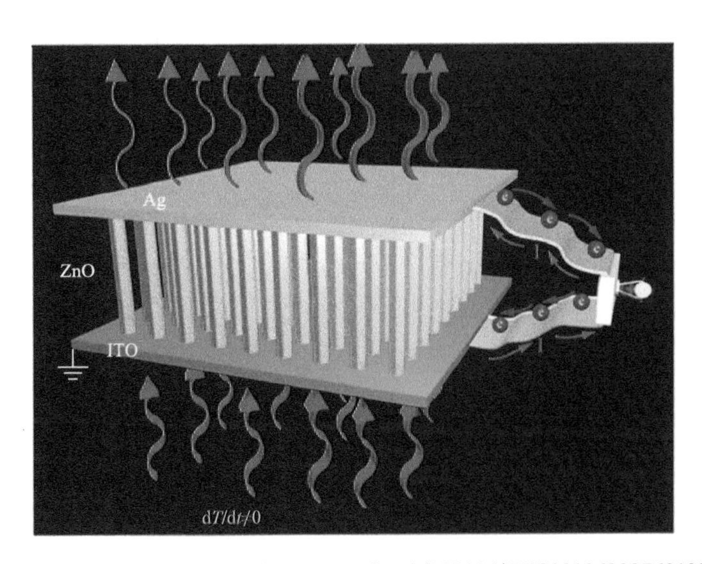

扫码阅全文

http://www.binn.cas.cn/book/202106/W020210623376913319920.pdf

2. Single Micro/Nanowire Pyroelectric Nanogenerators as Self-Powered Temperature Sensors

Ya Yang, Yusheng Zhou, Jyh Ming Wu *and* Zhong Lin Wang*

ACS Nano, 6 (2012) 8456-8461.

简介：首次报道了利用热势纳米发电机做温度传感器。

ABSTRACT We demonstrated the first application of a pyroelectric nanogenerator as a self-powered sensor (or active sensor) for detecting a change in temperature. The device consists of a single lead zirconate titanate (PZT) micro/nanowire that is placed on a thin glass substrate and bonded at its two ends, and it is packaged by polydimethylsiloxane (PDMS). By using the device to touch a heat source, the output voltage linearly increases with an increasing rate of change in temperature. The response time and reset time of the fabricated sensor are about 0.9 and 3 s, respectively. The minimum detecting limit of the change in temperature is about 0.4 K at room temperature. The sensor can be used to detect the temperature of a finger tip. The electricity generated under a large change in temperature can light up a liquid crystal display (LCD).

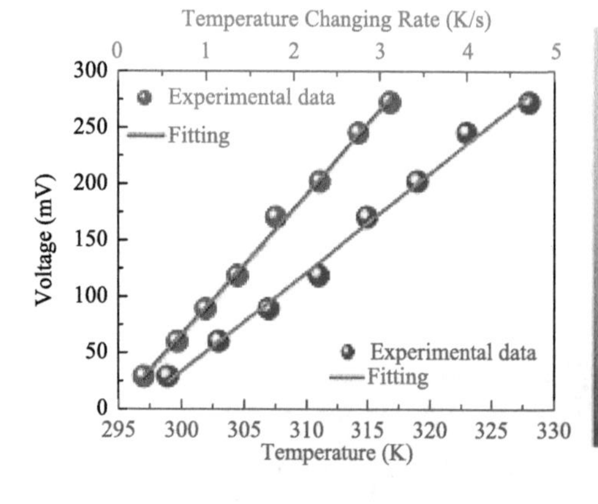

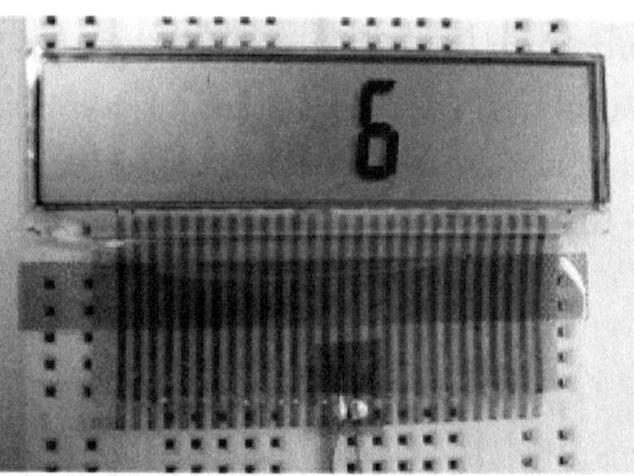

扫码阅全文

http://www.binn.cas.cn/book/202106/W020210623376913565485.pdf

3. Light-Induced Pyroelectric Effect as an Effective Approach for Ultrafast Ultraviolet Nanosensing

Zhaona Wang, Ruomeng Yu, Caofeng Pan, Zhaoling Li, Jin Yang, Fang Yi *and* **Zhong Lin Wang***

Nature Communications, 6 (2015) 8401.

简介：首次报道了利用热势效应来实现高灵敏度 UV 传感的方法。

ABSTRACT Zinc oxide is potentially a useful material for ultraviolet detectors; however, a relatively long response time hinders practical implementation. Here by designing and fabricating a self-powered ZnO/perovskite-heterostructured ultraviolet photodetector, the pyroelectric effect, induced in wurtzite ZnO nanowires on ultraviolet illumination, has been utilized as an effective approach for high-performance photon sensing. The response time is improved from 5.4 s to 53 μs at the rising edge, and 8.9 s to 63 μs at the falling edge, with an enhancement of five orders in magnitudes. The specific detectivity and the responsivity are both enhanced by 322%. This work provides a novel design to achieve ultrafast ultraviolet sensing at room temperature via light-self-induced pyroelectric effect. The newly designed ultrafast self-powered ultraviolet nanosensors may find promising applications in ultrafast optics, nonlinear optics, optothermal detections, computational memories and biocompatible optoelectronic probes.

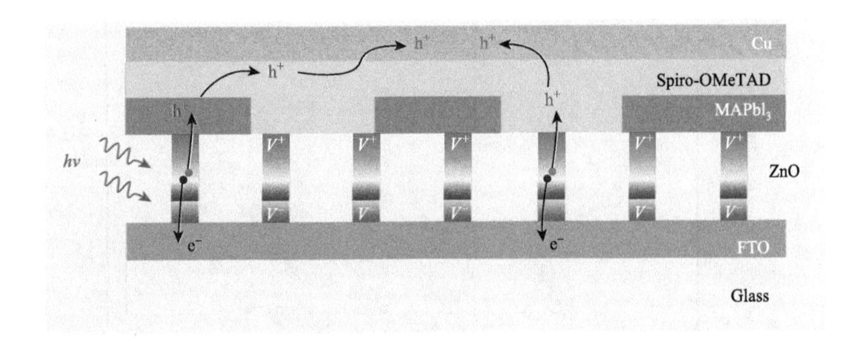

扫码阅全文

http://www.binn.cas.cn/book/202106/W020210623376913748599.pdf

4. Standard and Figure-of-Merit for Quantifying the Performance of Pyroelectric Nanogenerators

Kewei Zhang, Yuanhao Wang*, Zhong Lin Wang*, Ya Yang*

Nano Energy, 55 (2019) 534-540.

简介：首次建立了表征热势纳米发电机的方法和标准。

ABSTRACT Pyroelectric nanogenerators have emerged as an effective approach for the energy of the new-era, the era of internet of things, with a great potential for harvesting waste heat energy. Here, we present the figure-of-merit for quantifying the output of a pyroelectric nanogenertor. The charge-voltage cycle during temperature cycling has been experimentally performed on a PVDF film and PZT ceramic material. Output energy of 2.53 μJ per cycle and 2.59 mJ per cycle is presented at temperature cycling of 23 K (from 300 to 323 K) and 8 K (from 300 to 308 K) for the PVDF and PZT, respectively. The practicability of the derived figure-of-merit as a universal standard for pyroelectric nanogenerator is further experimentally confirmed and compared with previous results. We believe that it could be useful for optimizing the performance of pyroelectric nanogenerators from both material and structural aspects as well as estimating the generated electricity based on maximum temperature cycling.

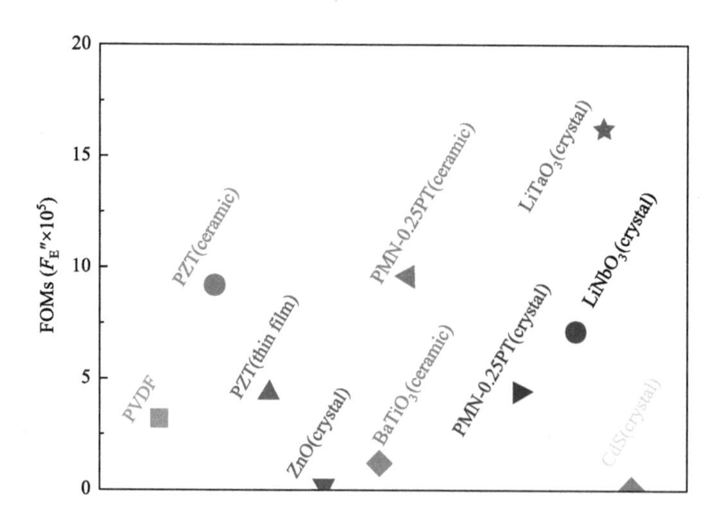

扫码阅全文

http://www.binn.cas.cn/book/202106/W020210623376914086351.pdf

5. Conjuncted Pyro-Piezoelectric Effect for Self-Powered Simultaneous Temperature and Pressure Sensing

Kai Song, Rudai Zhao, Zhong Lin Wang* *and* **Ya Yang***

Advanced Materials, 31 (2019) 1902831.

简介：首次报道了利用热势效应与压阻效应相结合的方法来实现温度与压力传感的方法。

ABSTRACT Ferroelectric materials use both the pyroelectric effect and piezoelectric effect for energy conversion. A ferroelectric $BaTiO_3$-based pyro-piezoelectric sensor system is demonstrated to detect temperature and pressure simultaneously. The voltage signal of the device is found to enhance with increasing temperature difference with a sensitivity of about 0.048 V·℃$^{-1}$ and with applied pressure with a sensitivity of about 0.044 V·kPa^{-1}. Moreover, no interference appears in the output voltage signals when piezoelectricity and pyroelectricity are conjuncted in the device. A novel 4×4 array sensor system is developed to sense real-time temperature and pressure variations induced by a finger. This system has potential applications in machine intelligence and man-machine interaction.

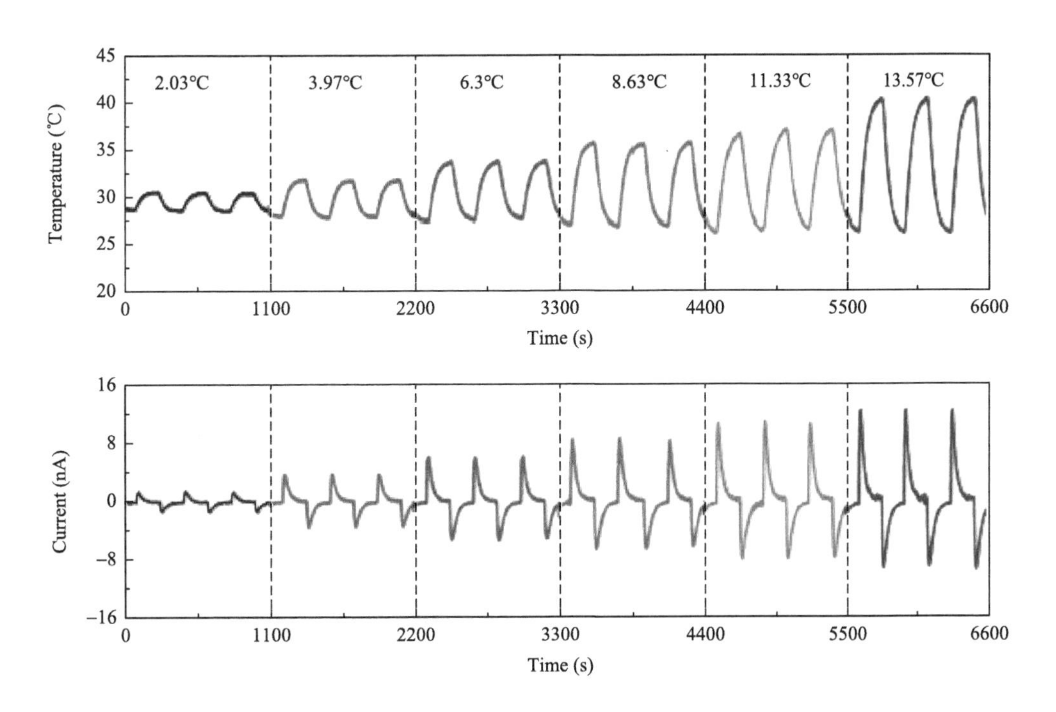

扫码阅全文

http://www.binn.cas.cn/book/202106/W020210623376914587715.pdf

Ⅲ.7 自驱动传感与系统
Self-Powered Sensors and Systems

1. Self-Powered Nanotech

Zhong Lin Wang

Scientific American, 298 (2008) 82-87.

简介：首次提出了自驱动系统的原创大概念，已经得到世界范围的高度认可。

ABSTRACT This paper first proposed the self-powered system based on nanogenerators.

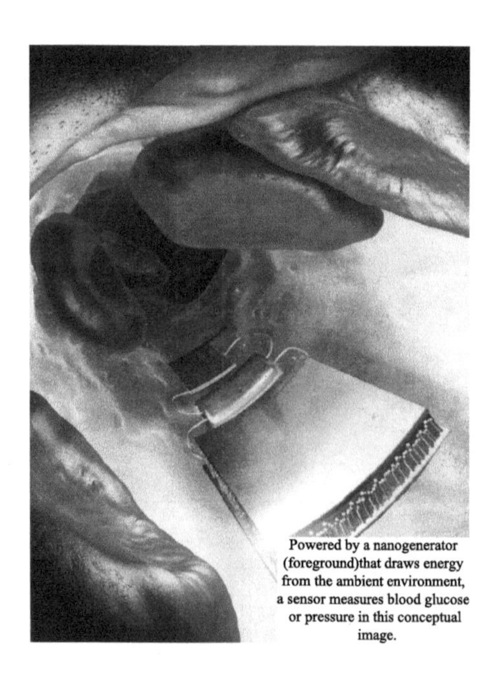
Powered by a nanogenerator (foreground)that draws energy from the ambient environment, a sensor measures blood glucose or pressure in this conceptual image.

扫码阅全文

http://www.binn.cas.cn/book/202106/W020210623484888219539.pdf

2. Self-Powered Nanowire Devices

Sheng Xu, Yong Qin, Chen Xu, Yaguang Wei, Rusen Yang *and* **Zhong Lin Wang***

Nature Nanotechnology, 5 (2010) 366-373.

简介：首次实现了完全基于纳米线阵列的自驱动传感系统。

ABSTRACT The harvesting of mechanical energy from ambient sources could power electrical devices without the need for batteries. However, although the efficiency and durability of harvesting materials such as piezoelectric nanowires have steadily improved, the voltage and power produced by a single nanowire are insufficient for real devices. The integration of large numbers of nanowire energy harvesters into a single power source is therefore necessary, requiring alignment of the nanowires as well as synchronization of their charging and discharging processes. Here, we demonstrate the vertical and lateral integration of ZnO nanowires into arrays that are capable of producing sufficient power to operate real devices. A lateral integration of 700 rows of ZnO nanowires produces a peak voltage of 1.26 V at a low strain of 0.19%, which is potentially sufficient to recharge an AA battery. In a separate device, a vertical integration of three layers of ZnO nanowire arrays produces a peak power density of 2.7 mW·cm^{-3}. We use the vertically integrated nanogenerator to power a nanowire pH sensor and a nanowire UV sensor, thus demonstrating a self-powered system composed entirely of nanowires.

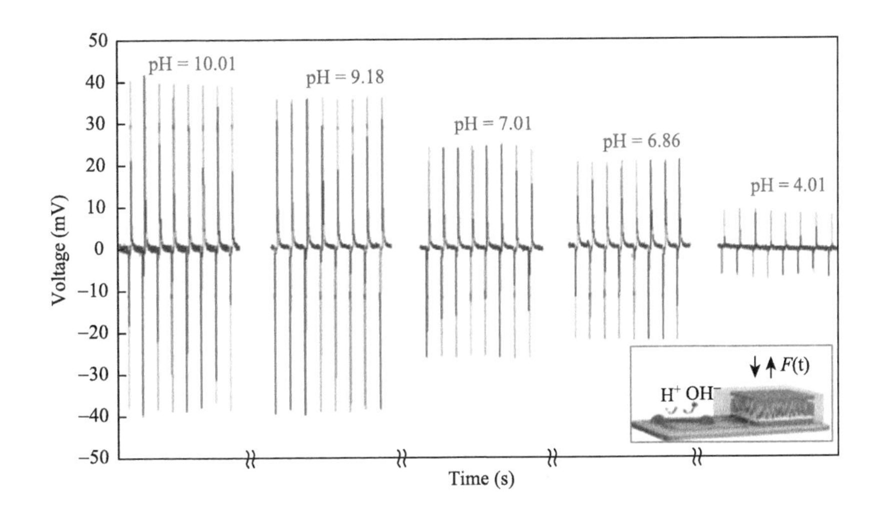

扫码阅全文

http://www.binn.cas.cn/book/202106/W020210623484888922702.pdf

3. Toward Self-Powered Sensor Network

Zhong Lin Wang

Nano Today, 5 (2010) 512-514.

简介：首次定义了自驱动传感系统的概念和组成。

ABSTRACT The future of nanotechnology is likely to focus on the areas of integrating individual nanodevices into a nanosystem that acts like living specie with sensing, communicating, controlling and responding. A nanosystem requires a nano-power source to make the entire package extremely small and high performance. The goal is to make self-powered nanosystem that can operate wirelessly, independently and sustainably. The self-powering approach developed here is a new paradigm in nanotechnology for truly achieving sustainable self-sufficient micro/nano-systems, which are of critical importance for sensing, medical science, infrastructure/environmental monitoring, defense technology and even personal electronics.

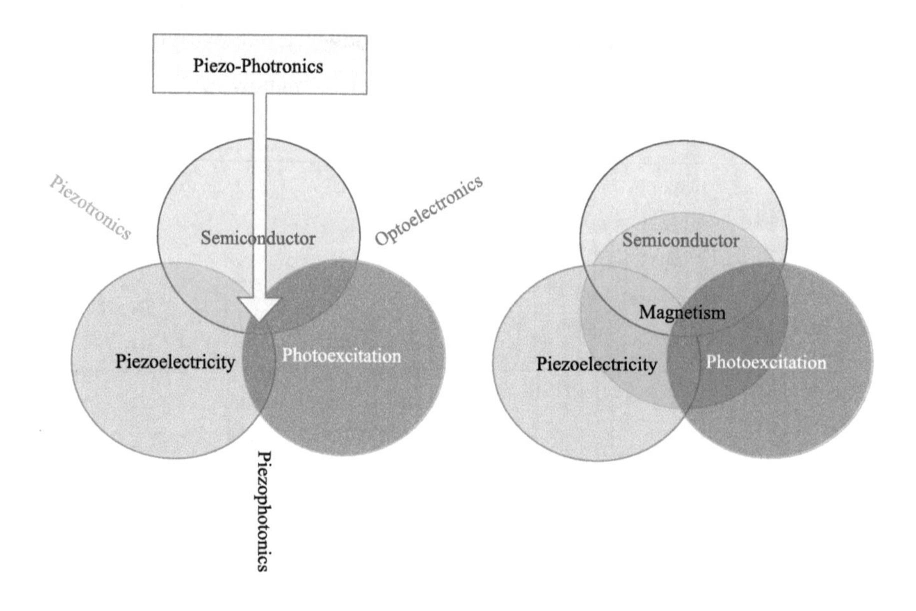

扫码阅全文

http://www.binn.cas.cn/book/202106/W020210623484889610768.pdf

4. Self-Powered System with Wireless Data Transmission

Youfan Hu, Yan Zhang, Chen Xu, Long Lin, Robert L. Snyder *and* Zhong Lin Wang*

Nano Letters, 11 (2011) 2572-2577.

简介：首次利用压电纳米发电机实现了能够无线发射信号的自驱动传感系统。

ABSTRACT We demonstrate the first self-powered system driven by a nanogenerator (NG) that works wirelessly and independently for long-distance data transmission. The NG was made of a free cantilever beam that consisted of a five-layer structure: a flexible polymer substrate, ZnO nanowire textured films on its top and bottom surfaces, and electrodes on the surfaces. When it was strained to 0.12% at a strain rate of 3.56% S^{-1}, the measured output voltage reached 10 V, and the output current exceeded 0.6 μA (corresponding power density 10 mW/cm^3). A system was built up by integrating a NG, rectification circuit, capacitor for energy storage, sensor, and RF data transmitter. Wireless signals sent out by the system were detected by a commercial radio at a distance of 5-10 m. This study proves the feasibility of using ZnO nanowire NGs for building self-powered systems, and its potential application in wireless biosensing, environmental/infrastructure monitoring, sensor networks, personal electronics, and even national security.

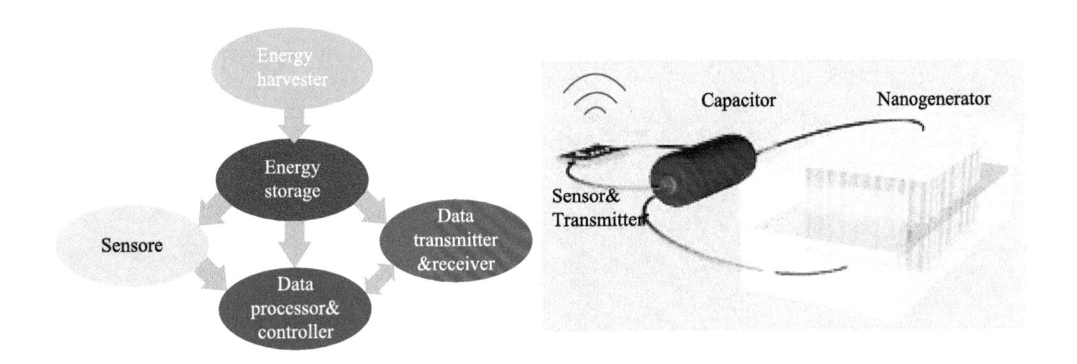

扫码阅全文

http://www.binn.cas.cn/book/202106/W020210623484889740489.pdf

5. Self-Powered Environmental Sensor System Driven by Nanogenerators

Minbaek Lee, Joonho Bae, Joohyung Lee, Churl-Seung Lee, Seunghun Hong *and* **Zhong Lin Wang**

Energy & Environmental Science, 4, (2011) 3359-3363.

简介：首次利用压电纳米发电机驱动环境传感系统。

ABSTRACT We demonstrate a fully stand-alone, self-powered environmental sensor driven by nanogenerators with harvesting vibration energy. Such a system is made of a ZnO nanowire-based nanogenerator, a rectification circuit, a capacitor for charge storage, a signal transmission LED light and a carbon nanotube-based Hg^{2+} ion sensor. The circuit lights up the LED indicator when it detects mercury ions in water solution. It is the first demonstration of a nanomaterial-based, self-powered sensor system for detecting a toxic pollutant.

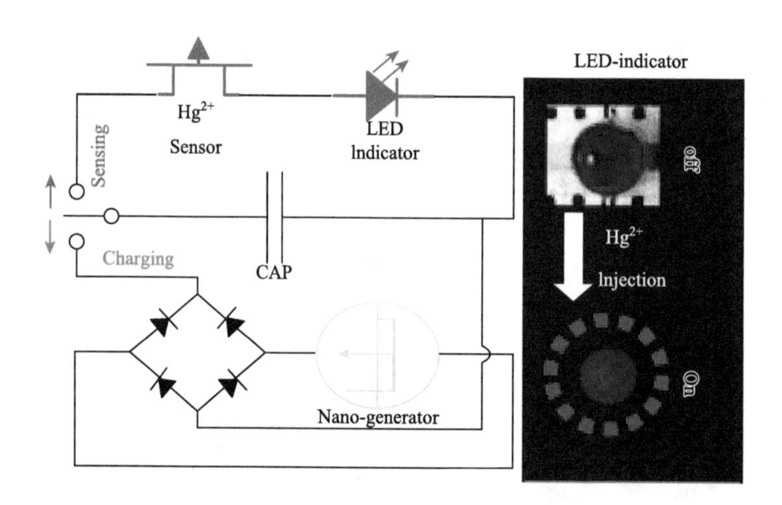

扫码阅全文

http://www.binn.cas.cn/book/202106/W020210623484889993743.pdf

6. Self-Powered Nanosensors and Nanosystems

Zhong Lin Wang

Advanced Materials, 24 (2011) 280-285.

简介：首次定义了自驱动传感系统的概念和组成。

ABSTRACT Sensor networks are a key technological and economic driver for global industries in the near future, with applications in health care, environmental monitoring, infrastructure monitoring, national security, and more. Developing technologies for self-powered nanosensors is vitally important. This paper gives a brief summary about recent progress in the area, describing nanogenerators that are capable of providing sustainable self-sufficient micro/nanopower sources for future sensor networks.

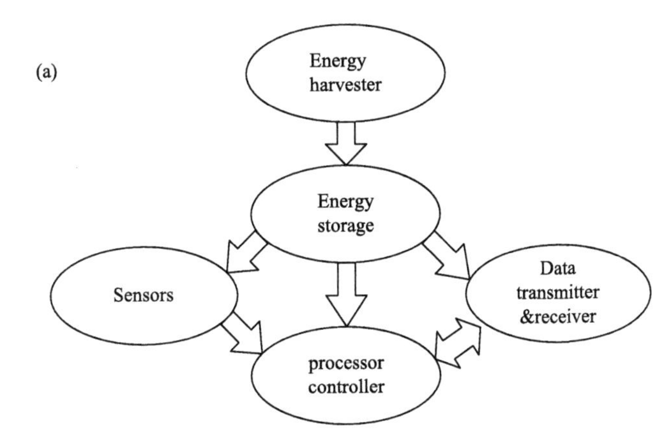

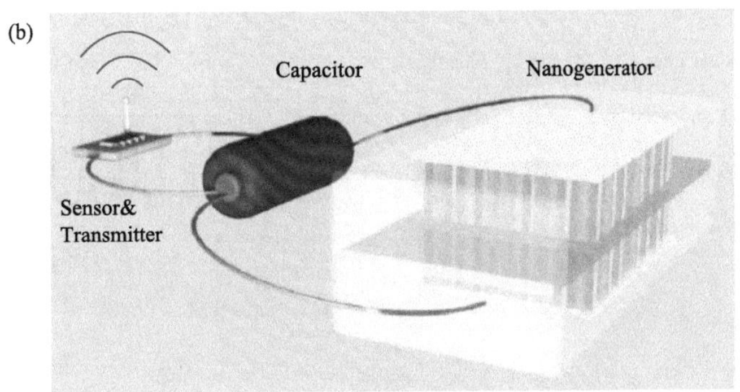

扫码阅全文

http://www.binn.cas.cn/book/202106/W020210623484890126095.pdf

7. Optical Fiber-Based Core-Shell Coaxially Structured Hybrid Cells for Self-Powered Nanosystems

Caofeng Pan, Wenxi Guo, Lin Dong, Guang Zhu *and* Zhong Lin Wang*

Advanced Materials, 24 (2012) 3356-3361.

简介：首次报道了基于光纤的太阳能-机械能复合传感系统。

ABSTRACT An optical fiber-based 3D hybrid cell consisting of a coaxially structured dye-sensitized solar cell (DSSC) and a nanogenerator (NG) for simultaneously or independently harvesting solar and mechanical energy is demonstrated. The current output of the hybrid cell is dominated by the DSSC, and the voltage output is dominated by the NG; these can be utilized complementarily for different applications. The output of the hybrid cell is about 7.65 μA current and 3.3 V voltage, which is strong enough to power nanodevices and even commercial electronic components.

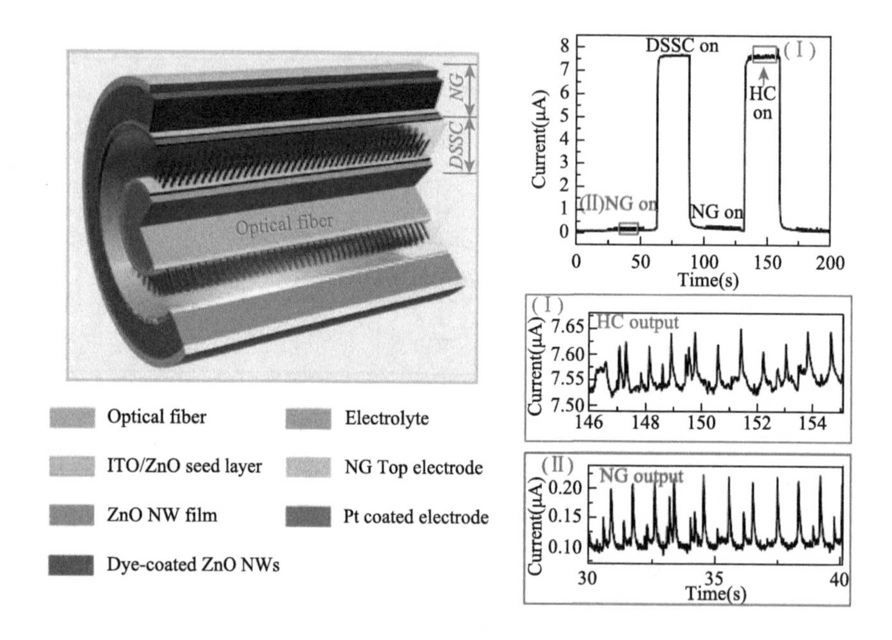

扫码阅全文

http://www.binn.cas.cn/book/202106/W020210623484890327051.pdf

8. Paper-Based Supercapacitors for Self-Powered Nanosystems

Longyan Yuan, Xu Xiao, Tianpeng Ding, Junwen Zhong, Xianghui Zhang, Yue Shen, Bin Hu, Yunhui Huang, Jun Zhou* *and* **Zhong Lin Wang***

Angewandte Chemie, 51 (2012) 4934-4938.

简介：首次报道了应用于自驱动系统的纸质超电容。

ABSTRACT Energy storage on paper: Paper-based, all-solid-state, and flexible supercapacitors were fabricated, which can be charged by a piezoelectric generator or solar cells and then discharged to power a strain sensor or a blue-light-emitting diode, demonstrating its efficient energy management in self-powered nanosystems (see picture).

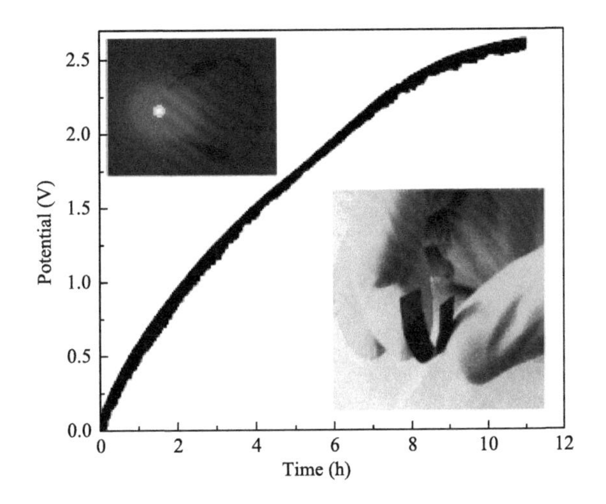

扫码阅全文

http://www.binn.cas.cn/book/202106/W020210623484890852196.pdf

9. Self-Powered Ultrasensitive Nanowire Photodetector Driven by a Hybridized Microbial Fuel Cell

Qing Yang, Ying Liu, Zetang Li, Zongyin Yang, Xue Wang *and* **Zhong Lin Wang***

Angewandte Chemie, 51 (2012) 6443-6446.

简介：首次报道了微生物燃料电池驱动的超灵敏光探测器。

ABSTRACT An integrated system consisting of a carbon fiber-ZnO hybrid nanowire (NW) multicolor photodetector is driven by a microbial fuel cell (see picture; PMMA = poly(methyl methacrylate), *E* = electrode). The self-powered photodetector can detect at light levels of as little as *n* W·cm^{-2} intensity with a responsivity of more than 300 A·W^{-1}.

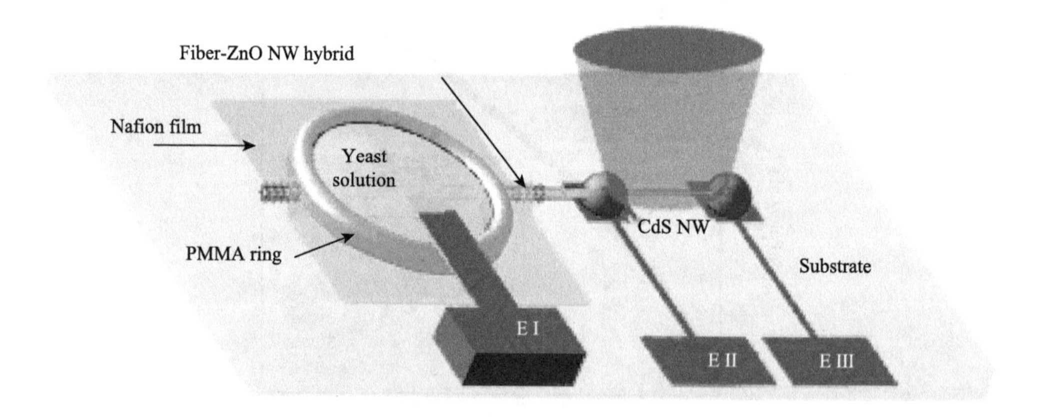

扫码阅全文

http://www.binn.cas.cn/book/202106/W020210623484891138786.pdf

10. A Paper-Based Nanogenerator as a Power Source and Active Sensor

Qize Zhong, Junwen Zhong, Bin Hu, Qiyi Hu, Jun Zhou* *and* Zhong Lin Wang

Energy & Environmental Science, 6 (2013) 1779-1784.

简介：首次报道了基于纸质的纳米发电机。

ABSTRACT Paper-based functional electronic devices endow a new era of applications in radio-frequency identification (RFID), sensors, transistors and microelectromechanical systems (MEMS). As an important component for building an all paper-based system that can work independently and sustainably, a paper-based power source is indispensable. In this study, we demonstrated a paper-based nanogenerator (pNG) that can convert tiny-scale mechanical energy into electricity. The pNG relies on an electrostatic effect, and the electrostatic charges on the paper were generated by the corona method. The instantaneous output power density of a single-layered pNG reached ~90.6 $\mu W \cdot cm^{-2}$ at a voltage of 110 V, and this instantaneously illuminated 70 LEDs. In addition, by sticking the pNG to a movable object, such as the page of a book, the power harvested from the mechanical action of turning the page can drive an LED, which presents its outstanding potential in building paper-based, self-powered systems and as active sensors.

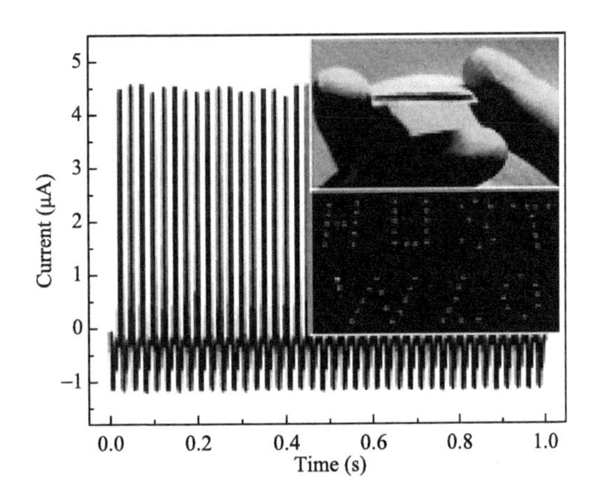

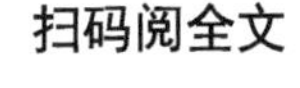

http://www.binn.cas.cn/book/202106/W020210623484891437526.pdf

11. Keystroke Dynamics Enabled Authentication and Identification Using Triboelectric Nanogenerator Array

Changsheng Wu, Wenbo Dingb, Ruiyuan Liu, Jiyu Wang, Aurelia C. Wang, Jie Wang, Shengming Li, Yunlong Zi, Zhong Lin Wang*

Materials Today, 21 (2018) 216-222.

简介：首次报道了利用纳米发电机阵列做的智能键盘来收集个人打字输入特征分析。

ABSTRACT Cyber security has become a serious concern as the internet penetrates every corner of our life over the last two decades. The rapidly developing human-machine interfacing calls for an effective and continuous authentication solution. Herein, we developed a two-factor, pressure-enhanced keystroke-dynamics-based security system that is capable of authenticating and even identifying users through their unique typing behavior. The system consists of a rationally designed triboelectric keystroke device that converts typing motions into analog electrical signals, and a support vector machine (SVM) algorithm-based software platform for user classification. This unconventional keystroke device is self-powered, stretchable and water/dust proof, which makes it highly mobile and applicable to versatile working environments. The promising application of this novel system in the financial and computing industry can push cyber security to the next level, where leaked passwords would possibly be of no concern.

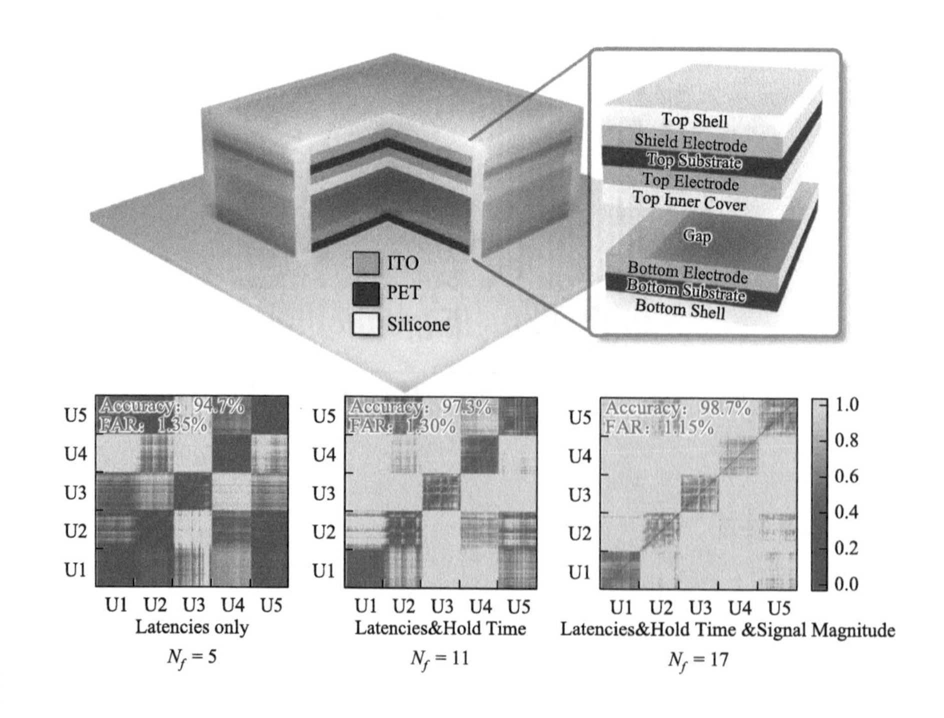

扫码阅全文

http://www.binn.cas.cn/book/202106/W020210623484891723362.pdf

12. A Highly Sensitive, Self-Powered Triboelectric Auditory Sensor for Social Robotics and Hearing Aids

Hengyu Guo, Xianjie Pu, Jie Chen, Yan Meng, Min-Hsin Yeh, Guanlin Liu, Qian Tang, Baodong Chen, Di Liu, Song Qi, Changsheng Wu, Chenguo Hu*, Jie Wang*, Zhong Lin Wang*

Science Robotics, 3 (2018) eaat2516.

简介：首次报道了基于纸质摩擦纳米发电机的麦克风与助听器。

ABSTRACT The auditory system is the most efficient and straightforward communication strategy for connecting human beings and robots. Here, we designed a self-powered triboelectric auditory sensor (TAS) for constructing an electronic auditory system and an architecture for an external hearing aid in intelligent robotic applications. Based on newly developed triboelectric nanogenerator (TENG) technology, the TAS showed ultrahigh sensitivity (110 millivolts/decibel). A TAS with the broadband response from 100 to 5000 hertz was achieved by designing the annular or sectorial inner boundary architecture with systematic optimization. When incorporated with intelligent robotic devices, TAS demonstrated high-quality music recording and accurate voice recognition for realizing intelligent human-robot interaction. Furthermore, the tunable resonant frequency of TAS was achieved by adjusting the geometric design of inner boundary architecture, which could be used to amplify a specific sound wave naturally. On the basis of this unique property, we propose a hearing aid with the TENG technique, which can simplify the signal processing circuit and reduce the power consuming. This work expresses notable advantages of using TENG technology to build a new generation of auditory systems for meeting the challenges in social robotics.

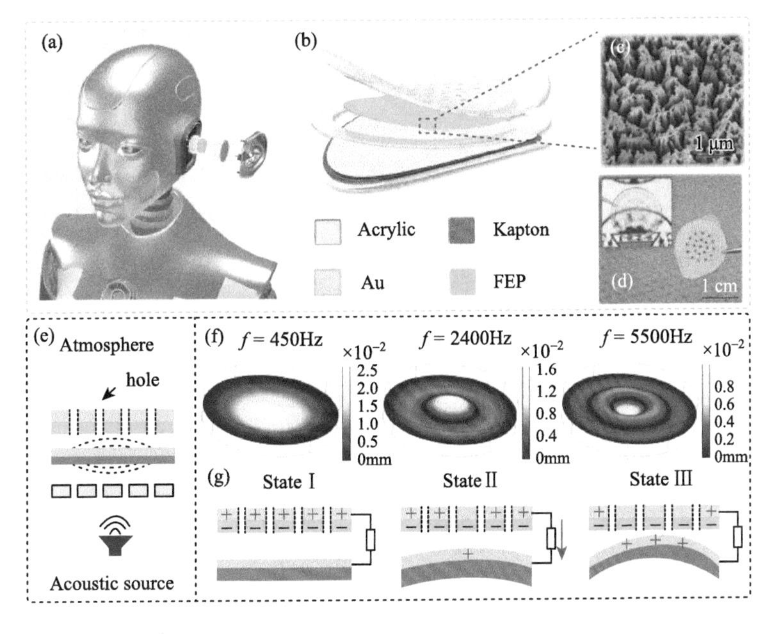

扫码阅全文

http://www.binn.cas.cn/book/202106/W020210623484892210258.pdf

13. Eye Motion Triggered Self-Powered Mechnosensational Communication System Using Triboelectric Nanogenerator

Xianjie Pu, Hengyu Guo, Jie Chen, Xue Wang, Yi Xi, Chenguo Hu*, Zhong Lin Wang*

Science Advances, 3 (2017) 3: e1700694.

简介：首次报道了基于眼球运动驱动的摩擦纳米发电机来实现人机界面信息交互。

ABSTRACT Mechnosensational human-machine interfaces (HMIs) can greatly extend communication channels between human and external devices in a natural way. The mechnosensational HMIs based on biopotential signals have been developing slowly owing to the low signal-to-noise ratio and poor stability. In eye motions, the corneal-retinal potential caused by hyperpolarization and depolarization is very weak. However, the mechanical micromotion of the skin around the corners of eyes has never been considered as a good trigger signal source. We report a novel triboelectric nanogenerator (TENG)-based micromotion sensor enabled by the coupling of triboelectricity and electrostatic induction. By using an indium tin oxide electrode and two opposite tribomaterials, the proposed flexible and transparent sensor is capable of effectively capturing eye blink motion with a super-high signal level (~750 mV) compared with the traditional electrooculogram approach (~1 mV). The sensor is fixed on a pair of glasses and applied in two real-time mechnosensational HMIs—the smart home control system and the wireless hands-free typing system with advantages of super-high sensitivity, stability, easy operation, and low cost. This TENG-based micromotion sensor is distinct and unique in its fundamental mechanism, which provides a novel design concept for intelligent sensor technique and shows great potential application in mechnosensational HMIs.

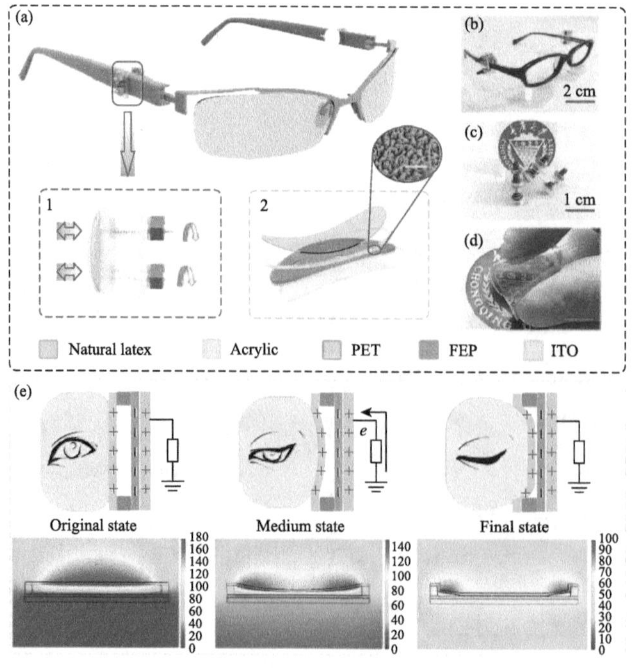

扫码阅全文

http://www.binn.cas.cn/book/202106/W020210623484892649525.pdf

14. Machine-Knitted Washable Sensor Array Textile for Precise Epidermal Physiological Signal Monitoring

Wenjing Fan, Qiang He, Keyu Meng, Xulong Tan, Zhihao Zhou, Gaoqiang Zhang, Jin Yang, Zhong Lin Wang

Science Advances, 6 (2020) eaay2840.

简介：首次报道了基于纺织品的摩擦纳米发电机在体征监测与健康防护方面的应用。

ABSTRACT Wearable textile electronics are highly desirable for realizing personalized health management. However, most reported textile electronics can either periodically target a single physiological signal or miss the explicit details of the signals, leading to a partial health assessment. Furthermore, textiles with excellent property and comfort still remain a challenge. Here, we report a triboelectric all-textile sensor array with high pressure sensitivity and comfort. It exhibits the pressure sensitivity (7.84 mV·Pa^{-1}), fast response time (20 ms), stability (＞100,000 cycles), wide working frequency bandwidth (up to 20 Hz), and machine washability (＞40 washes). The fabricated TATSAs were stitched into different parts of clothes to monitor the arterial pulse waves and respiratory signals simultaneously. We further developed a health monitoring system for long-term and noninvasive assessment of cardiovascular disease and sleep apnea syndrome, which exhibits great advancement for quantitative analysis of some chronic diseases.

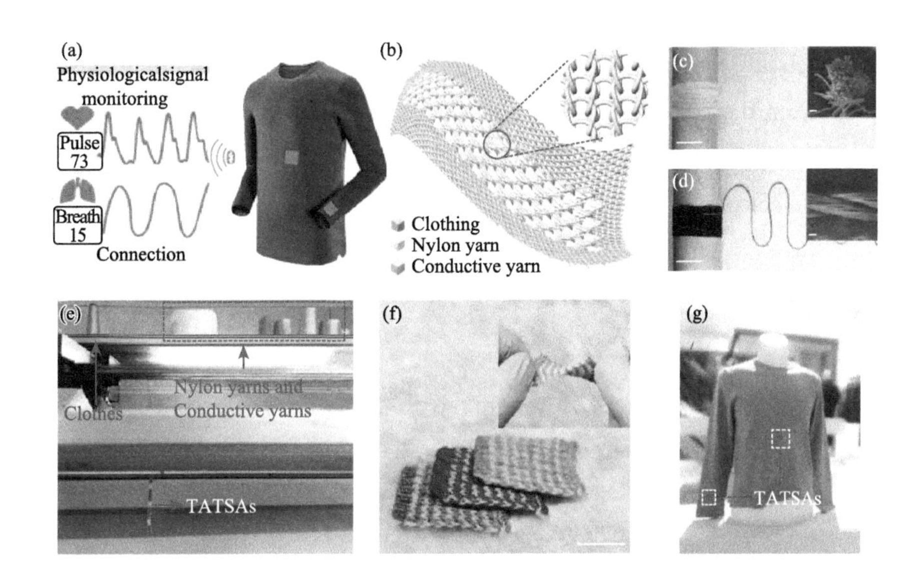

扫码阅全文

http://www.binn.cas.cn/book/202106/W020210623484892941322.pdf

Ⅲ.8 自充电能源包
Self-Charging Power Cell

1. Hybridizing Energy Conversion and Storage in a Mechanical-to-Electrochemical Process for Self-Charging Power Cell

Xinyu Xue, Sihong Wang, Wenxi Guo, Yan Zhang *and* Zhong Lin Wang*

Nano Letters, 12 (2012) 5048-5054.

简介：首次报道了直接把机械能高效转化为电化学能的复合能源转换与存储系统——自充电能源包。这是一个新的研究方向。该研究被评为《物理世界》2012年的十大突破之一，并被美国CNN电台采访报道。

ABSTRACT Energy generation and energy storage are two distinct processes that are usually accomplished using two separated units designed on the basis of different physical principles, such as piezoelectric nanogenerator and Li-ion battery; the former converts mechanical energy into electricity, and the latter stores electric energy as chemical energy. Here, we introduce a fundamental mechanism that directly hybridizes the two processes into one, in which the mechanical energy is directly converted and simultaneously stored as chemical energy without going through the intermediate step of first converting into electricity. By replacing the polyethylene (PE) separator as for conventional Li battery with a piezoelectric poly(vinylidene fluoride) (PVDF) film, the piezoelectric potential from the PVDF film as created by mechanical straining acts as a charge pump to drive Li ions to migrate from the cathode to the anode accompanying charging reactions at electrodes. This new approach can be applied to fabricating a self-charging power cell (SCPC) for sustainable driving micro/nanosystems and personal electronics.

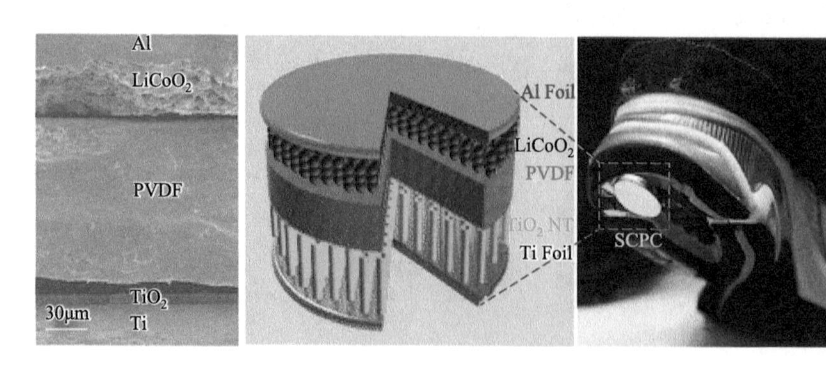

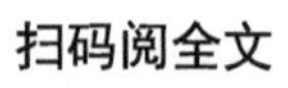
扫码阅全文

http://www.binn.cas.cn/book/202106/W020210623485823237141.pdf

2. A Universal Self-Charging System Driven by Random Biomechanical Energy for Sustainable Operation of Mobile Electronics

Simiao Niu, Xiaofeng Wang, Fang Yi, Yu Sheng Zhou *and* Zhong Lin Wang[*]

Nature Communications, 6 (2015) 8975.

简介：首次报道了由无规则运动所驱动的摩擦纳米发电机与自充电能源包，并展示了在驱动移动与无线小电子产品中的应用。

ABSTRACT Human biomechanical energy is characterized by fluctuating amplitudes and variable low frequency, and an effective utilization of such energy cannot be achieved by classical energy-harvesting technologies. Here we report a high-efficient self-charging power system for sustainable operation of mobile electronics exploiting exclusively human biomechanical energy, which consists of a high-output triboelectric nanogenerator, a power management circuit to convert the random a.c. energy to d.c. electricity at 60% efficiency, and an energy storage device. With palm tapping as the only energy source, this power unit provides a continuous d.c. electricity of 1.044 mW (7.34 $W \cdot m^{-3}$) in a regulated and managed manner. This self-charging unit can be universally applied as a standard 'infinite-lifetime' power source for continuously driving numerous conventional electronics, such as thermometers, electrocardiograph system, pedometers, wearable watches, scientific calculators and wireless radio-frequency communication system, which indicates the immediate and broad applications in personal sensor systems and internet of things.

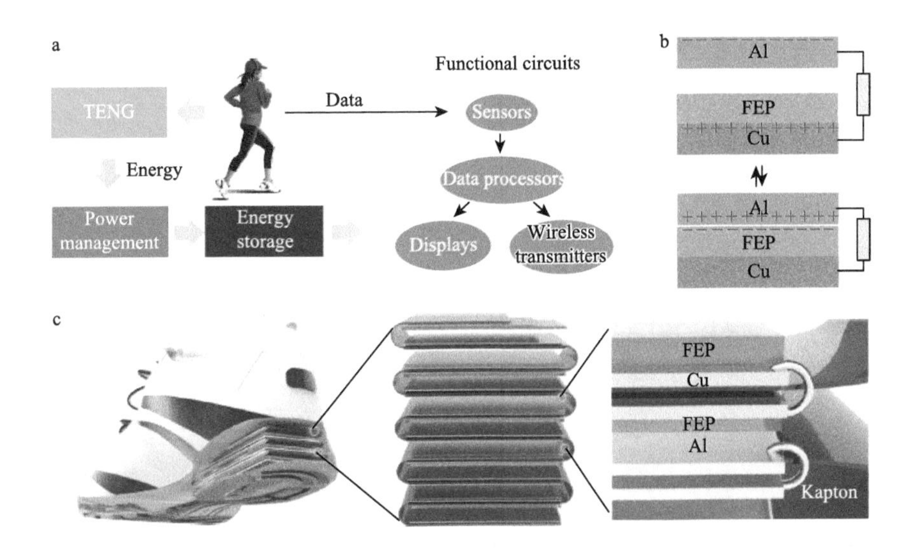

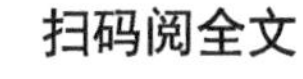

http://www.binn.cas.cn/book/202106/W020210623485824084750.pdf

3. Flexible Self-Charging Power Cell for One-Step Energy Conversion and Storage

Xinyu Xue, Ping Deng, Bin He, Yuxin Nie, Lili Xing, Yan Zhang* *and* **Zhong Lin Wang***

Advanced Energy Materials, 4 (2014) 201301329.

简介：首次报道了一步直接把机械能高效转化为电化学能的自充电能源包。

ABSTRACT Flexible self-charging power cells (SCPCs) are fabricated from flexible graphene electrodes and Kapton shells. The cells can be charged from 500 to 832 mV within 500 s by periodic compressive stress (34 N, 1 Hz). The flexible SCPC can be efficiently charged from a small amount of mechanical energy, such as by finger pressing or a bicycle tire rolling.

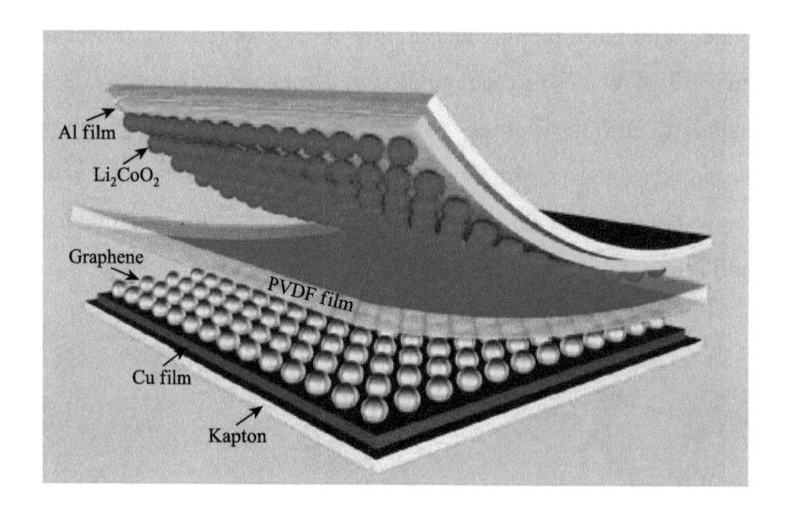

扫码阅全文

http://www.binn.cas.cn/book/202106/W020210623485824475091.pdf

Ⅲ.9 分布式能源的熵理论
Entropy Theory of Distributed Energy

1. Entropy Theory of Distributed Energy for Internet of Things

Zhong Lin Wang

Nano Energy, 58 (2019) 669-672.

简介：我们目前的能源主要是以化石能源为代表的低熵能源，例如煤炭、石油、天然气等，经过燃烧、发电、输运和应用等过程，这些高质量能源最后变为低质量的能量耗散在自然界。化石能源的大量利用必然带来全球气候变暖和分布在环境中无序、低质量的高熵能量的增加。高熵能源指的是散布在自然界中无处不在、无时不有、散乱但总量巨大的动能、势能、热能、电磁辐射能等弥散式能源，它是热力学第二定律（即熵增加原理）的必然表现。该文首次阐述了如何把这些高熵能源有效地转换成分布式电能来满足物联网和传感网络的需求。

ABSTRACT Entropy was first introduced in thermodynamics for describing the statistical behavior of molecules. In this paper, we use the entropy theory to describe the distribution and powering of small electronics in the era of internet of things. We refer the power transmitted from power plants to utility units as "ordered" energy for fixed sites, while refer the energy harvested from environment as "random" energy. Our conclusion is that the "ordered" energy is able to solve part of the power need for distributed electronics for IoTs, but the remaining part has to be supplied by the "random" energy harvested from our living environment using resources such as solar, vibration, motion and even thermal. The entropy idea defines the role played by energy harvesting as it will become increasingly important owing to the ever expanding usage of distributed electronics at the time of intelligence.

扫码阅全文

http://www.binn.cas.cn/book/202106/W020210623486457597746.pdf

第四部分

Part Ⅳ

Ⅳ.1 压电电子学：原理与器件

Piezotronics—Principle and Devices

1. Piezoelectric Field Effect Transistor and Nanoforce Sensor Based on a Single ZnO Nanowire

Xudong Wang, Jun Zhou, Jinhui Song, Jin Liu, Ningsheng Xu *and* **Zhong L. Wang***

Nano Letters, 6 (2006) 2768-2772.

简介：首篇开始认识压电电场在载流子输运中的调制作用的文献。

ABSTRACT Utilizing the coupled piezoelectric and semiconducting dual properties of ZnO, we demonstrate a piezoelectric field effect transistor (PE-FET) that is composed of a ZnO nanowire (NW) (or nanobelt) bridging across two Ohmic contacts, in which the source to drain current is controlled by the bending of the NW. A possible mechanism for the PE-FET is suggested to be associated with the carrier trapping effect and the creation of a charge depletion zone under elastic deformation. This PE-FET has been applied as a force/pressure sensor for measuring forces in the nanonewton range and even smaller with the use of smaller NWs. An almost linear relationship between the bending force and the conductance was found at small bending regions, demonstrating the principle of nanowire-based nanoforce and nanopressure sensors.

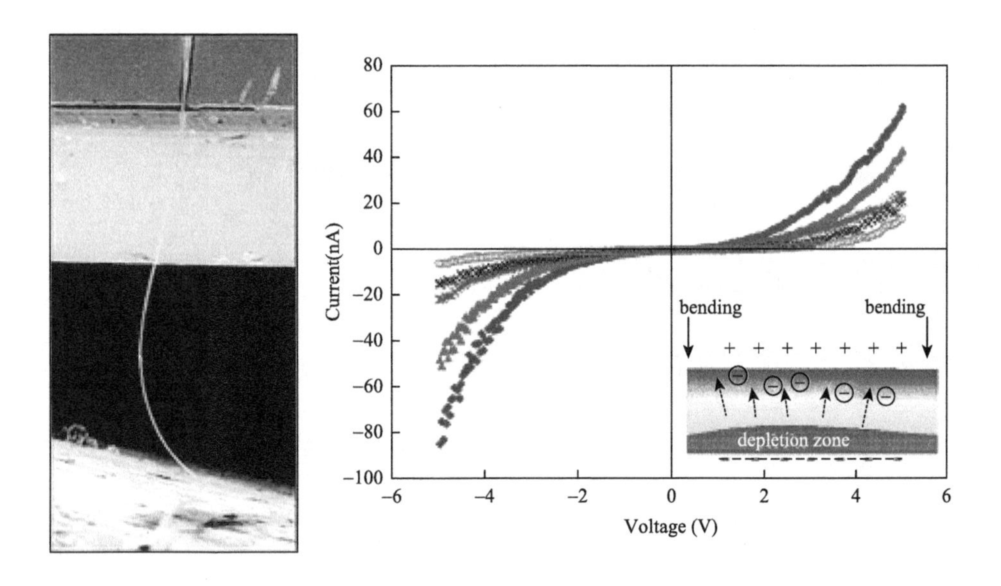

扫码阅全文

http://www.binn.cas.cn/book/202106/W020210623502767583378.pdf

2. Piezoelectric Gated Diode of a Single ZnO Nanowire

Jr H. He, Cheng L. Hsin, Jin Liu, Lih J. Chen* *and* Zhong L. Wang*

Advanced Materials, 19 (2007) 781-784.

简介：第一篇直接报道压电调制的实验观察，从此产生了压电作为门电压进行调控的概念。

ABSTRACT A ZnO nanowire behaves like a rectifier under bending strain, as demonstrated by its current-voltage characteristics (see graph). This is interpreted with the consideration of a piezoelectricity-induced potential energy barrier at the interface of the conductive tip and nanowire (see schematic). Under appropriate bending and voltage control, each NW could correspond to a device element for random-access-memory, diode, and force-sensor applications.

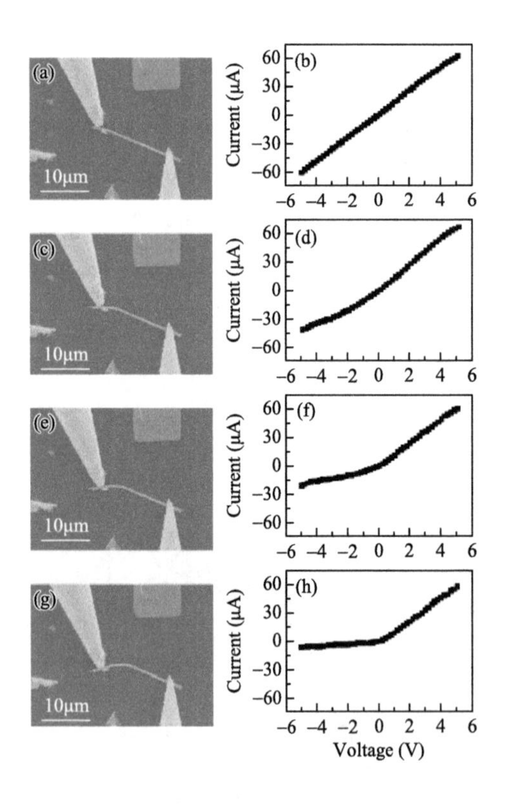

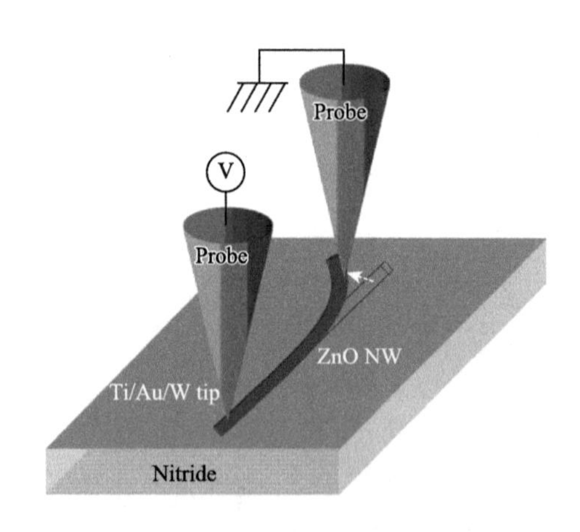

扫码阅全文

http://www.binn.cas.cn/book/202106/W020210623502767814420.pdf

3. Nanopiezotronics

Zhong Lin Wang

Advanced Materials, 19 (2007) 889-892.

简介：本文首次明确提出并定义了压电电子学的概念，即在应力作用下压电产生的界面极化电荷调制半导体界面的载流子输运过程的电子学。第一代半导体（硅和锗基）和第二代半导体（GaAs）的原子结构具有立方对称性，而第三代半导体(GaN, ZnO, SiC)是六方结构，因此由于非中心对称性所产生的压电效应是第三代半导体所不可避免的。所以压电电子学是第三代半导体所特有的效应和领域。

ABSTRACT This article introduces the fundamental principle of nanopiezotronics, which utilizes the coupled piezoelectric and semiconducting properties of nanowires and nanobelts for designing and fabricating electronic devices and components, such as field-effect transistors and diodes. The physics of nanopiezotronics is based on the principle of a nanowire nanogenerator that converts mechanical energy into electric energy. It is anticipated to have a wide range of applications in electromechanical coupled electronics, sensing, harvesting/recycling energy from the environment, and self-powered nanosystems.

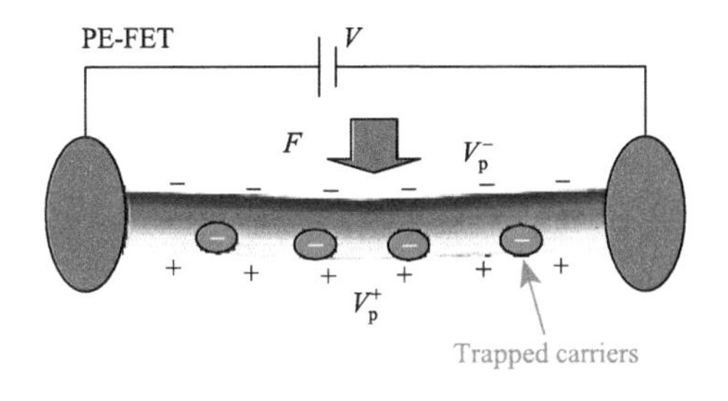

扫码阅全文

http://www.binn.cas.cn/book/202106/W020210623502767952666.pdf

4. Piezoelectric-Potential-Controlled Polarity-Reversible Schottky Diodes and Switches of ZnO Wires

Jun Zhou, Peng Fei, Yudong Gu, Wenjie Mai, Yifan Gao, Rusen Yang, Gang Bao *and* Zhong Lin Wang*

Nano Letters, 8 (2008) 3973-3977.

简介：首次解释了压电电子学效应的物理内涵和基本特征，是压电电子学的核心文献之一。

ABSTRACT Using a two-end bonded ZnO piezoelectric-fine-wire (PFW) (nanowire, microwire) on a flexible polymer substrate, the strain-induced change in I-V transport characteristic from symmetric to diode-type has been observed. This phenomenon is attributed to the asymmetric change in Schottky-barrier heights at both source and drain electrodes as caused by the strain-induced piezoelectric potential-drop along the PFW, which have been quantified using the thermionic emission-diffusion theory. A new piezotronic switch device with an "on" and "off" ratio of ~120 has been demonstrated. This work demonstrates a novel approach for fabricating diodes and switches that rely on a strain governed piezoelectric-semiconductor coupling process.

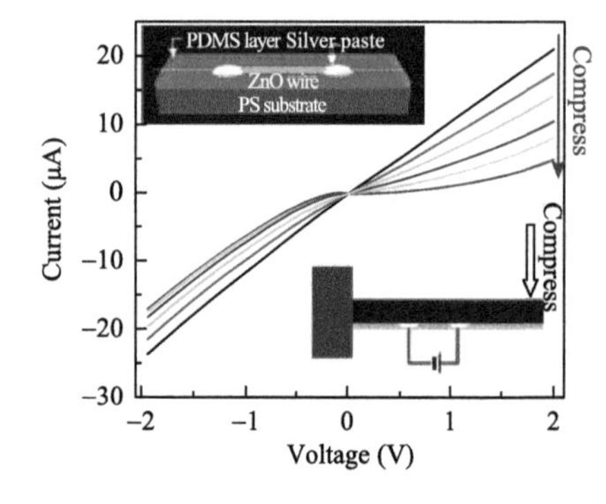

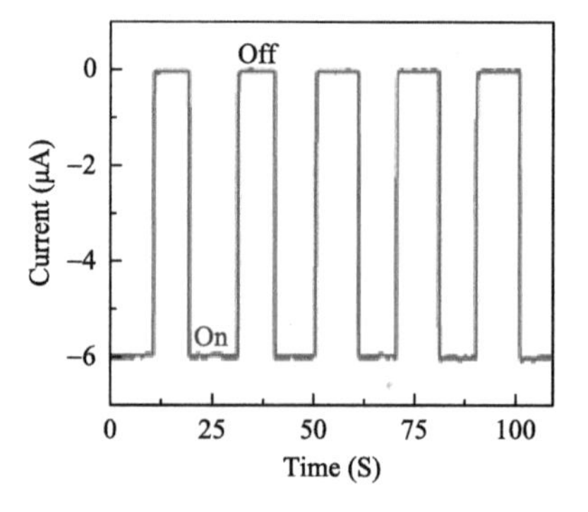

扫码阅全文

http://www.binn.cas.cn/book/202106/W020210623502768065723.pdf

5. Effects of Piezoelectric Potential on the Transport Characteristics of Metal-ZnO Nanowire-Metal Field Effect Transistor

Zhiyuan Gao, Jun Zhou, Yudong Gu, Peng Fei, Yue Hao, Gang Bao *and* **Zhong Lin Wang***

Journal of Applied Physics, 105 (2009) 113707 .

简介：本文进一步深化了压电电子学效应的物理内涵。

ABSTRACT We have investigated the effects of piezoelectric potential in a ZnO nanowire on the transport characteristics of the nanowire based field effect transistor through numerical calculations and experimental observations. Under different straining conditions including stretching, compressing, twisting, and their combination, a piezoelectric potential is created throughout the nanowire to modulate/alternate the transport property of the metal-ZnO nanowire contacts, resulting in a switch between symmetric and asymmetric contacts at the two ends, or even turning an Ohmic contact type into a diode. The commonly observed natural rectifying behavior of the as-fabricated ZnO nanowire can be attributed to the strain that was unpurposely created in the nanowire during device fabrication and material handling. This work provides further evidence on piezopotential governed electronic transport and devices, e.g., piezotronics.

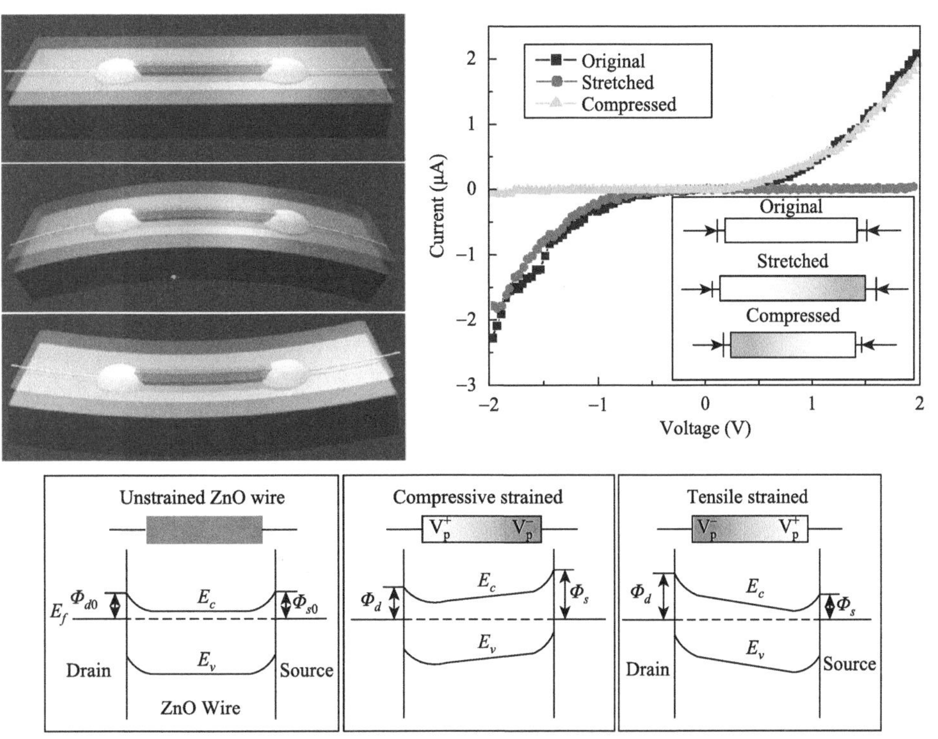

扫码阅全文

http://www.binn.cas.cn/book/202106/W020210623502768356415.pdf

6. Piezotronic and Piezophototronic Effects

Zhong Lin Wang

The Journal of Physical Chemistry Letters, 1 (2010) 1388-1393.

简介：首次明确定义了压电电子学效应与压电光电子学效应。这两个效应都是室温下的效应。

ABSTRACT Owing to the polarization of ions in a crystal that has noncentral symmetry, a piezoelectric potential (piezopotential) is created in the material by applying a stress. The creation of piezopotential together with the presence of Schottky contacts are the fundamental physics responsible for a few important nanotechnologies. The nanogenerator is based on the piezopotential-driven transient flow of electrons in the external load. On the basis of nanomaterials in the wurtzite semiconductors, such as ZnO and GaN, electronics fabricated by using a piezopotential as a gate voltage are called piezotronics, with applications in strain/force/pressure-triggered/ controlled electronic devices, sensors, and logic gates. The piezophototronic effect is a result of three-way coupling among piezoelectricity, photonic excitation, and semiconductor transport, which allows tuning and controlling of electro-optical processes by a strain-induced piezopotential.

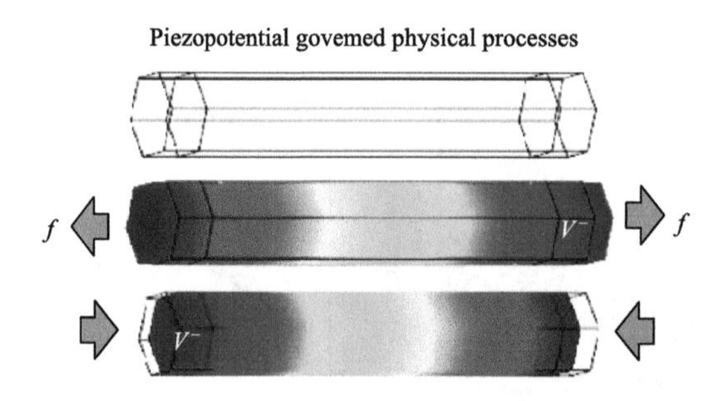

扫码阅全文

http://www.binn.cas.cn/book/202106/W020210623502768513307.pdf

7. Piezopotential Gated Nanowire-Nanotube Hybrid Field-Effect Transistor

Weihua Liu, Minbaek Lee, Lei Ding, Jie Liu *and* Zhong Lin Wang*

Nano Letters, 10 (2010) 3084-3089.

简介：首次证明压电电场可以直接作为门电压来制造场效应三极管。

ABSTRACT We report the first piezoelectric potential gated hybrid field-effect transistors based on nanotubes and nanowires. The device consists of single-walled carbon nanotubes (SWNTs) on the bottom and crossed ZnO piezoelectric fine wire (PFW) on the top with an insulating layer between. Here, SWNTs serve as a carrier transport channel, and a single-crystal ZnO PFW acts as the powerfree, contact-free gate or even an energy-harvesting component later on. The piezopotential created by an external force in the ZnO PFW is demonstrated to control the charge transport in the SWNT channel located underneath. The magnitude of the piezopotential in the PFW at a tensile strain of 0.05% is measured to be 0.4-0.6 V. The device is a unique coupling between the piezoelectric property of the ZnO PFW and the semiconductor performance of the SWNT with a full utilization of its mobility. The newly demonstrated device has potential applications as a strain sensor, force/pressure monitor, security trigger, and analog-signal touch screen.

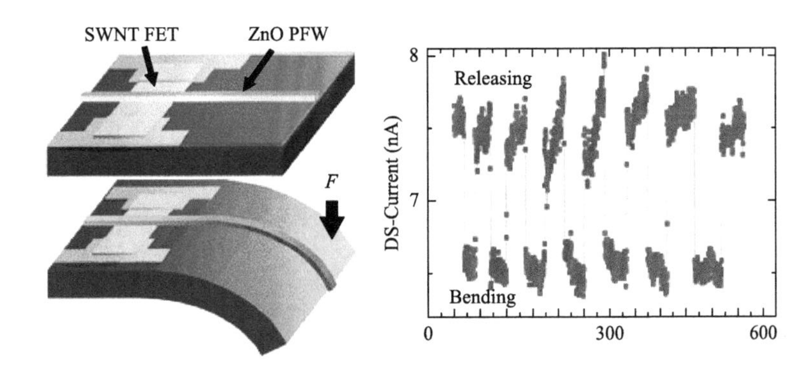

扫码阅全文

http://www.binn.cas.cn/book/202106/W020210623502768815034.pdf

8. Taxel-Addressable Matrix of Vertical-Nanowire Piezotronic Transistors for Active and Adaptive Tactile Imaging

Wenzhuo Wu, Xiaonan Wen, Zhong Lin Wang*

Science, 340 (2013) 952-957.

简介：首次制造出基于纳米线阵列的压电电子学三极管阵列，这是基于两个极而实现三个极功能的电子元件，是压电电子学芯片的开山之作。

ABSTRACT Designing, fabricating, and integrating arrays of nanodevices into a functional system is the key for transferring nanoscale science into applicable nanotechnology. We report large-array three-dimensional (3D) circuitry integration of piezotronic transistors based on vertical zinc oxide nanowires as active taxel-addressable pressure/force-sensor matrix for tactile imaging. Using the piezoelectric polarization charges created at metal-semiconductor interface under strain to gate/modulate transport process of local charge carriers, piezotronic effect has been applied to design independently addressable two-terminal transistor arrays, which convert mechanical stimuli applied on the devices into local electronic controlling signals. The device matrix has been demonstrated for achieving shape-adaptive high-resolution tactile imaging and self-powered, multidimensional active sensing. The 3D piezotronic transistor array may have applications in human-electronics interfacing, smart skin, and micro/nano-electromechanical systems.

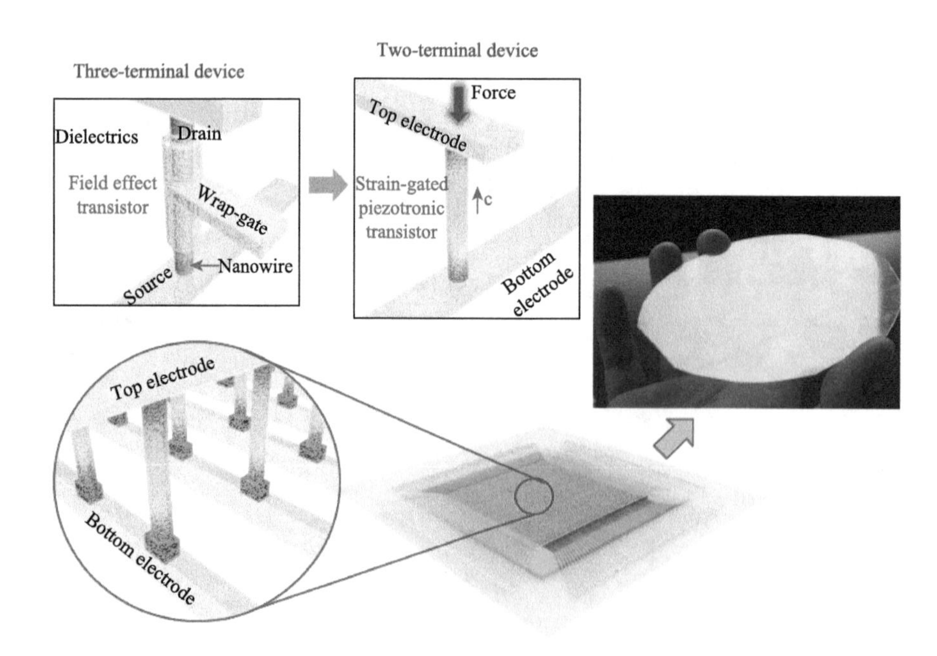

扫码阅全文

http://www.binn.cas.cn/book/202106/W020210623502768964528.pdf

9. Piezotronic Nanowire-Based Resistive Switches as Programmable Electromechanical Memories

Wenzhuo Wu *and* **Zhong Lin Wang***

Nano Letters, 11 (2011) 2779-2885.

简介：首次基于纳米线压电电子学三极管阵列电路来实现应变控制下的逻辑运算，展示了压电电子学三极管在逻辑电路中的基本功能。

ABSTRACT We present the first piezoelectrically modulated resistive switching device based on piezotronic ZnO nanowire (NW), through which the write/read access of the memory cell is programmed via electromechanical modulation. Adjusted by the strain-induced polarization charges created at the semiconductor/metal interface under externally applied deformation by the piezoelectric effect, the resistive switching characteristics of the cell can be modulated in a controlled manner, and the logic levels of the strain stored in the cell can be recorded and read out, which has the potential for integrating with NEMS technology to achieve micro/nanosystems capable for intelligent and self-sufficient multidimensional operations.

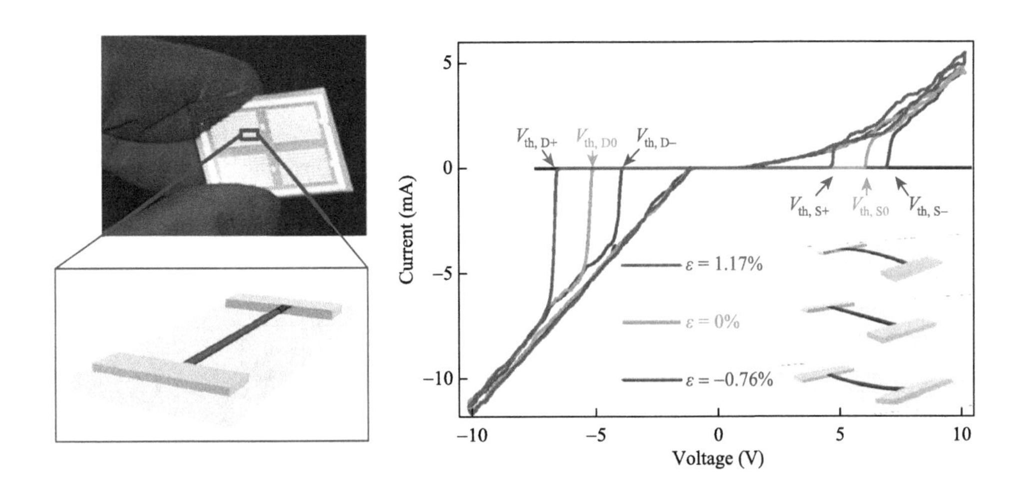

扫码阅全文

http://www.binn.cas.cn/book/202106/W020210623502769428479.pdf

10. Strain-Gated Piezotronic Transistors Based on Vertical Zinc Oxide Nanowires

Weihua Han, Yusheng Zhou, Yan Zhang, Cheng-Ying Chen, Long Lin, Xue Wang, Sihong Wang ***and*** **Zhong Lin Wang***

ACS Nano, 6 (2012) 3760-3766.

简介：首次展示基于单根垂直纳米线的压电电子学三极管的基本特性，为制造压电电子学三极管阵列打下基础。

ABSTRACT Strain-gated piezotronic transistors have been fabricated using vertically aligned ZnO nanowires (NWs), which were grown on GaN/sapphire substrates using a vapor-liquid-solid process. The gate electrode of the transistor is replaced by the internal crystal potential generated by strain, and the control over the transported current is at the interface between the nanowire and the top or bottom electrode. The current-voltage characteristics of the devices were studied using conductive atomic force microscopy, and the results show that the current flowing through the ZnO NWs can be tuned/gated by the mechanical force applied to the NWs. This phenomenon was attributed to the piezoelectric tuning of the Schottky barrier at the Au-ZnO junction, known as the piezotronic effect. Our study demonstrates the possibility of using Au droplet capped ZnO NWs as a transistor array for mapping local strain.More importantly, our design gives the possibility of fabricating an array of transistors using individual vertical nanowires that can be controlled independently by applying mechanical force/pressure over the top. Such a structure is likely to have important applications in highresolution mapping of strain/force/pressure.

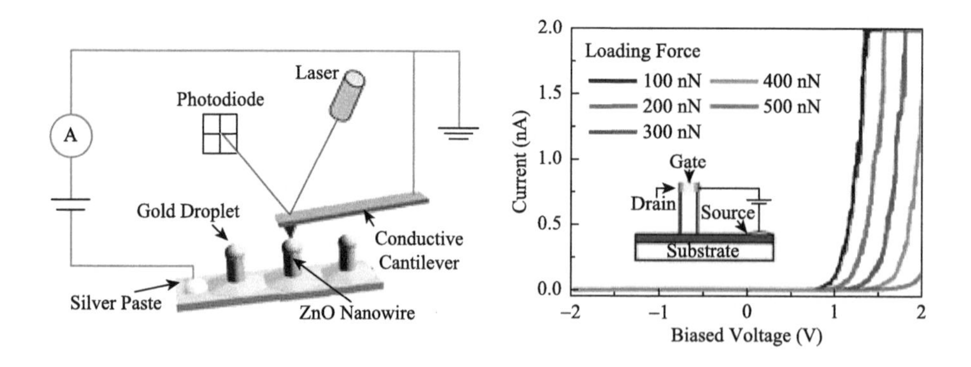

扫码阅全文

http://www.binn.cas.cn/book/202106/W020210623502770009719.pdf

11. Piezotronic Effect on the Transport Properties of GaN Nanobelts for Active Flexible Electronics

Ruomeng Yu, Lin Dong, Caofeng Pan, Simiao Niu, Hongfei Liu, Wei Liu, Soojin Chua, Dongzhi Chi *and* **Zhong Lin Wang***

Advanced Materials, 24 (2012) 3532-3537.

简介：首次研究了在 GaN 纳米带材料中也有压电电子学效应，把可以应用于压电电子学的材料体系从氧化锌拓展到了氮化镓。

ABSTRACT The transport properties of GaN nanobelts (NBs) are tuned using a piezotronic effect when a compressive/tensile strain is applied on the GaN NB. This is mainly due to a change in Schottky barrier height (SBH). A theoretical model is proposed to explain the observed phenomenon.

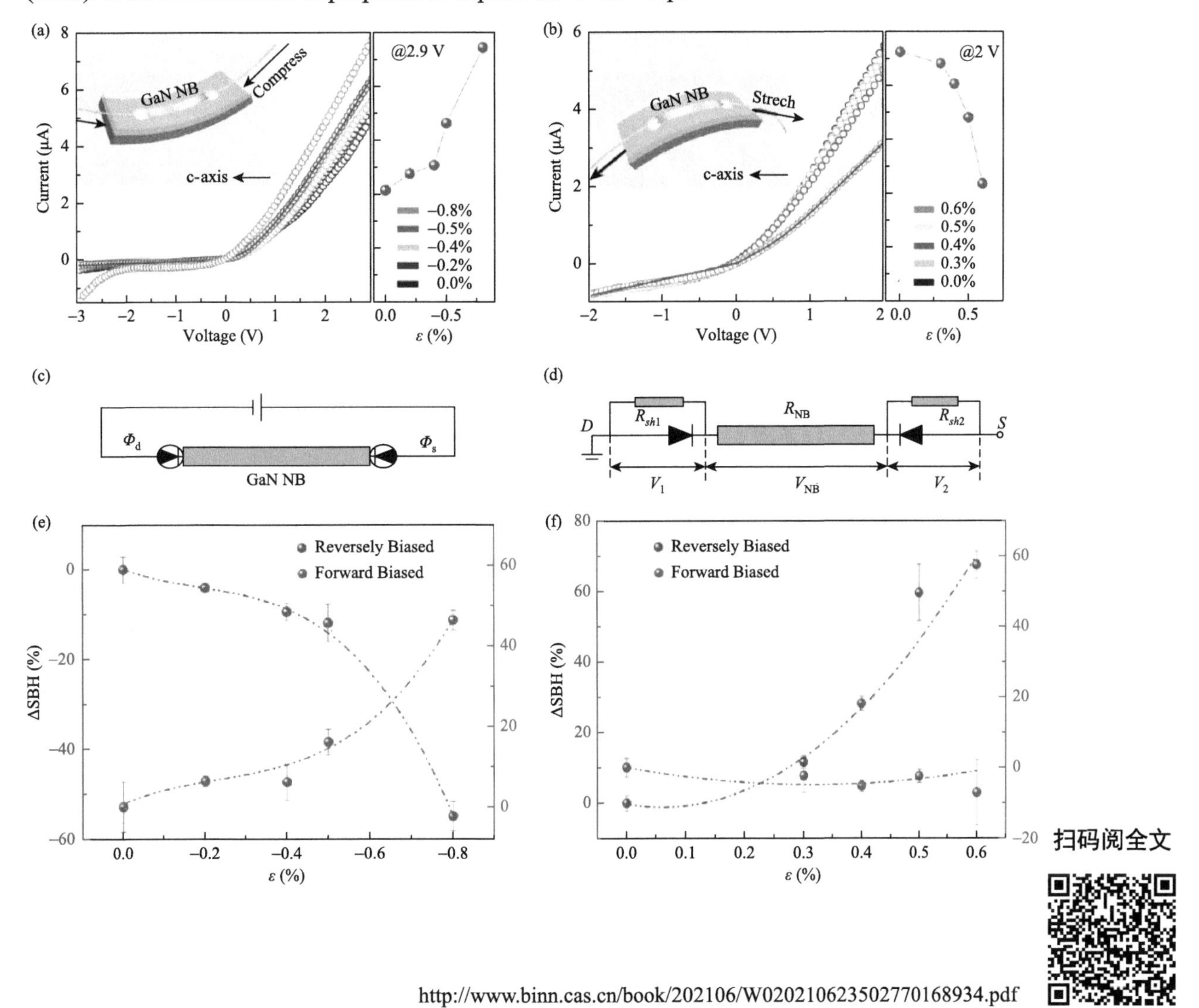

扫码阅全文

http://www.binn.cas.cn/book/202106/W020210623502770168934.pdf

12. Piezo-Phototronic Effect of CdSe Nanowires

Lin Dong, Simiao Niu, Caofeng Pan, Ruomeng Yu, Yan Zhang *and* Zhong Lin Wang*

Advanced Materials, 24 (2012) 5470-5475.

简介：首次研究了在 CdSe 纳米线材料中的压电电子学效应，把可以应用于压电电子学的材料体系扩充到硒化镉。

ABSTRACT The piezo-phototronic effect on transport properties of flexible CdSe NW devices is investigated. An optimum sensitivity of the flexible CdSe NW devices can be achieved by adjusting the applied strain and illumination intensity. The piezo-phototronic effect under compressive strain increases the internal electric field of the Schottky barrier, and assists the separation of the photo-excited electron-hole pairs, resulting in the increase of photocurrent. A trap-mediated mechanism is responsible for the decreased hole separation when the strain is larger than the critical strain.

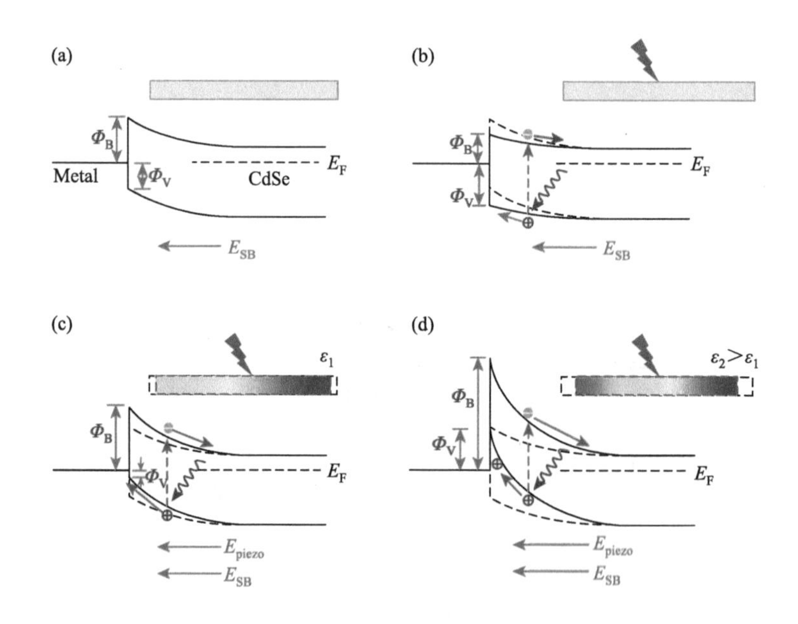

扫码阅全文

http://www.binn.cas.cn/book/202106/W020210623502770451979.pdf

13. Piezotronic Effect in Flexible Thin-film Based Devices

Xiaonan Wen, Wenzhuo Wu, Yong Ding *and* **Zhong Lin Wang***

Advanced Materials, 25 (2013) 3371-3379.

简介：压电电子学效应是一个基本物理效应，不论相关材料是什么形貌的结构都应该具有这样的特性。本文首次证明了薄膜材料中也具有压电电子学效应。

ABSTRACT Flexible piezotronic devices based on RF-sputtered piezoelectric semiconductor thin films have been investigated for the first time. The dominating role of the piezotronic effect over the geometrical and piezoresistive effect in the as-fabricated devices has been confirmed and the modulation effect of the piezopotential on charge carrier transport under different strains is subsequently studied. Moreover, it is also demonstrated that the UV sensing capability of the as-fabricated thin film based piezotronic device can be tuned by the piezopotential, showing significantly enhanced sensitivity and improved reset time under tensile strain.

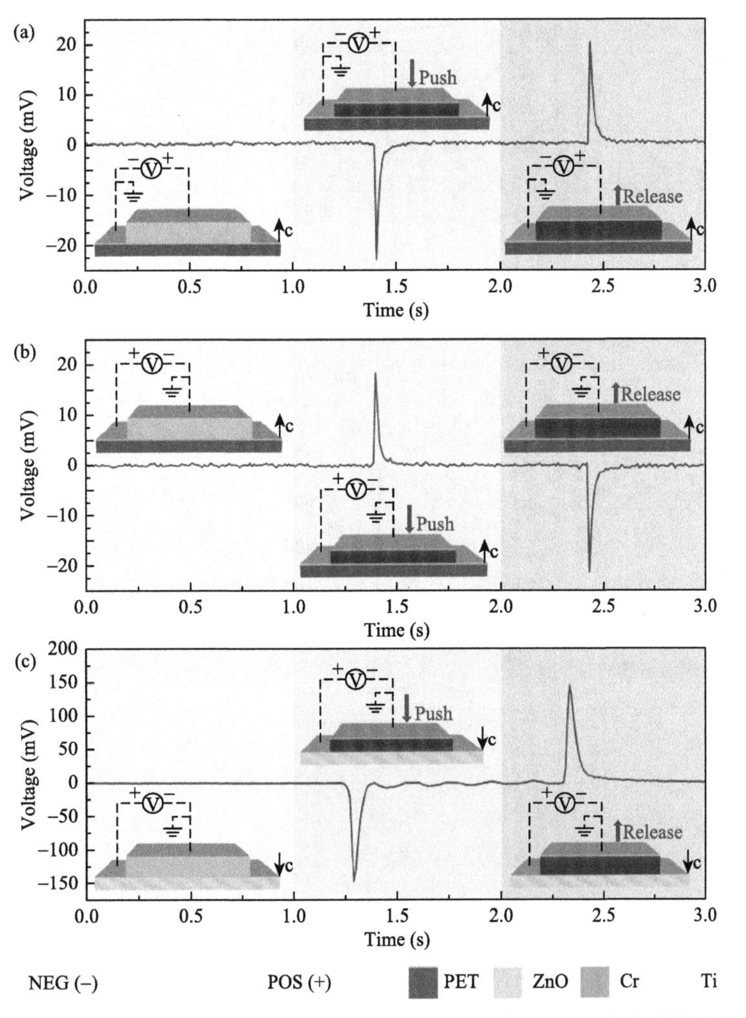

http://www.binn.cas.cn/book/202106/W020210623502771055457.pdf

扫码阅全文

14. Enhanced Performance of Flexible ZnO Nanowire Based Room-Temperature Oxygen Sensors by Piezotronic Effect

Simiao Niu, Youfan Hu, Xiaonan Wen, Yusheng Zhou, Fang Zhang, Long Lin, Sihong Wang *and* **Zhong Lin Wang***

Advanced Materials, 25 (2013), 3701-3706.

简介：利用压电电子学效应，首次报道了在室温下的高灵敏度氧气传感器。

ABSTRACT A flexible oxygen sensor based on individual ZnO nanowires is demonstrated with high sensitivity at room temperature and the influence of the piezotronic effect on the performance of this oxygen sensor is investigated. By applying a tensile strain, the already very high sensitivity due to the Schottky contact and pre-treatment of UV light is even further enhanced.

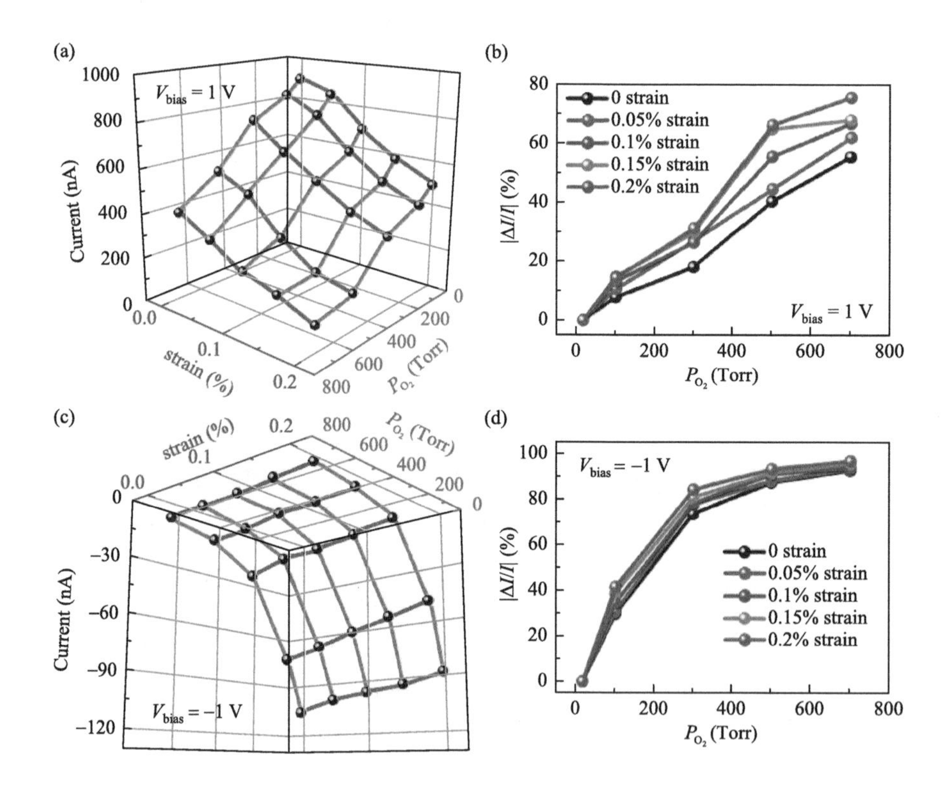

扫码阅全文

http://www.binn.cas.cn/book/202106/W020210623502771591945.pdf

15. Piezotronic Effect in Solution-Grown P-type ZnO Nanowires and Films

Ken C. Pradel[‡], Wenzhuo Wu[‡], Yusheng Zhou, Xiaonan Wen, Yong Ding, Zhong Lin Wang[*]

Nano Letters, 13 (2013) 2647-2654.

简介：不但n型氧化锌中有压电电子学效应，本文首次报道了p型氧化锌纳米线和薄膜材料中也具有压电电子学效应。

ABSTRACT Investigating the piezotronic effect in p-type piezoelectric semiconductor is critical for developing a complete piezotronic theory and designing/fabricating novel piezotronic applications with more complex functionality. Using a low temperature solution method, we were able to produce ultralong (up to 60 μm in length) Sb doped p-type ZnO nanowires on both rigid and flexible substrates. For the p-type nanowire field effect transistor, the on/off ratio, threshold voltage, mobility, and carrier concentration of 0.2% Sb-doped sample are found to be 10^5, 2.1 V, 0.82 $cm^2 \cdot V^{-1} \cdot s^{-1}$, and 2.6×10^{17} cm^{-3}, respectively, and the corresponding values for 1% Sb doped samples are 10^4, 2.0 V, 1.24 $cm^2 \cdot V^{-1} \cdot s^{-1}$, and 3.8×10^{17} cm^{-3}. We further investigated the universality of piezotronic effect in the as-synthesized Sb-doped p-type ZnO NWs and reported for the first time strain-gated piezotronic transistors as well as piezopotential-driven mechanical energy harvesting based on solution-grown p-type ZnO NWs. The results presented here broaden the scope of piezotronics and extend the framework for its potential applications in electronics, optoelectronics, smart MEMS/NEMS, and human-machine interfacing.

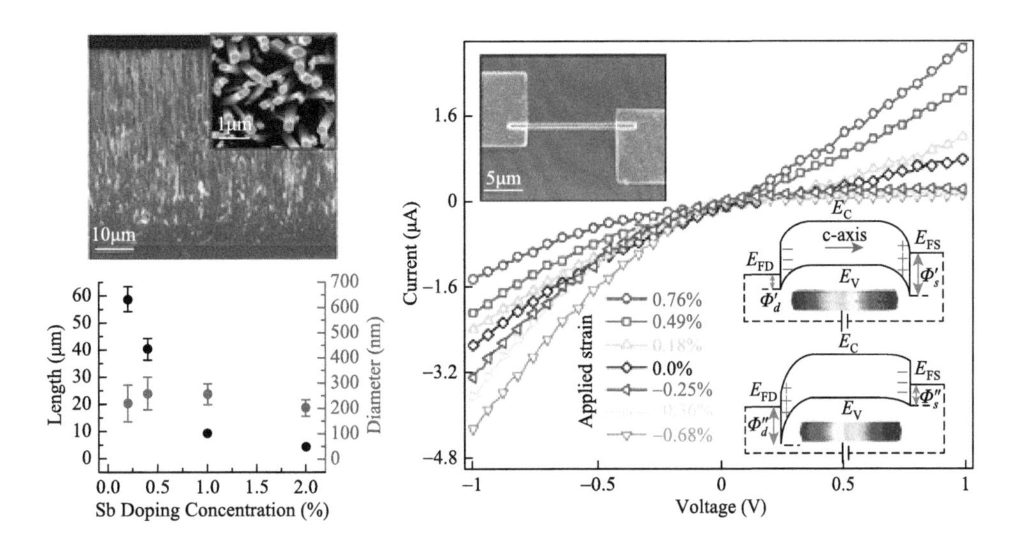

扫码阅全文

http://www.binn.cas.cn/book/202106/W020210623502772067181.pdf

16. Temperature Dependence of the Piezotronic Effect in ZnO Nanowires

Youfan Hu, Benjamin D.B. Klein, Yuanjie Su, Simiao Niu, Ying Liu *and* Zhong Lin Wang*

Nano Letters, 13 (2013) 5026-5032.

简介：本文首次研究了氧化锌纳米线压电电子学效应在 77～300 K 之间对温度的依赖，证明了温度越低效应越大。我们一般的测量都是在室温下进行的。

ABSTRACT A comprehensive investigation was carried out on n-type ZnO nanowires for studying the temperature dependence of the piezotronic effect from 77 to 300 K. In general, lowering the temperature results in a largely enhanced piezotronic effect. The experimental results show that the behaviors can be divided into three groups depending on the carrier doping level or conductivity of the ZnO nanowires. For nanowires with a low carrier density ($<10^{17}/cm^3$ at 77 K), the pieozotronic effect is dominant at low temperature for dictating the transport properties of the nanowires; an opposite change of Schottky barrier heights at the two contacts as a function of temperature at a fixed strain was observed for the first time. At a moderate doping (between $10^{17}/cm^3$ and $10^{18}/cm^3$ at 77 K), the piezotronic effect is only dominant at one contact, because the screening effect of the carriers to the positive piezoelectric polarization charges at the other end (for n-type semiconductors). For nanowires with a high density of carriers ($>10^{18}/cm^3$ at 77 K), the piezotronic effect almost vanishes. This study not only proves the proposed fundamental mechanism of piezotronic effect, but also provides guidance for fabricating piezotronic devices.

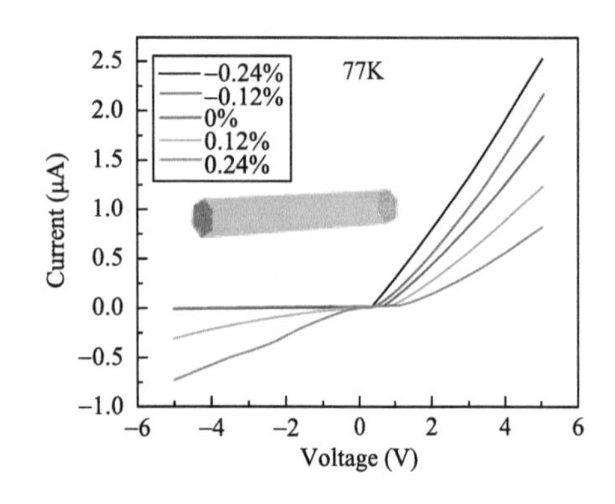

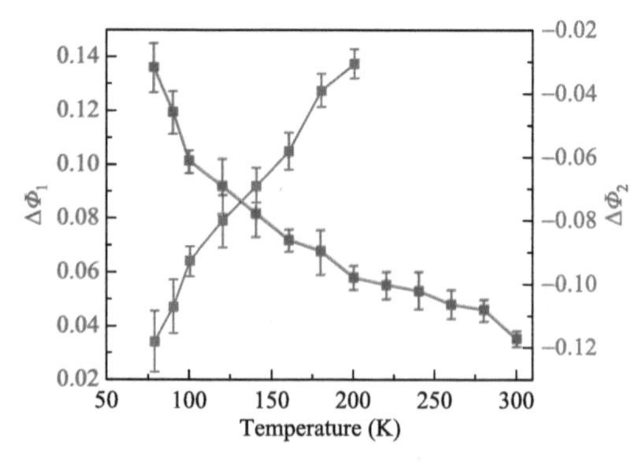

扫码阅全文

http://www.binn.cas.cn/book/202106/W020210623502772211762.pdf

17. Ultrathin Piezotronic Transistors with 2 nm Channel Lengths

Longfei Wang[†], Shuhai Liu[†], Guoyun Gao, Yaokun Pang, Xin Yin, Xiaolong Feng, Laipan Zhu, Yu Bai, Libo Chen, Tianxiao Xiao, Xudong Wang, Yong Qin[*] *and* Zhong Lin Wang[*]

ACS Nano, 12 (2018) 4903-4908.

简介：本文首次证明对于 2 纳米厚的超薄片，压电电子学效应仍然可以有效地调制载流子输运过程。因此，压电电子学三极管有可能打破摩尔定律的限制。

ABSTRACT Because silicon transistors are rapidly approaching their scaling limit due to short-channel effects, alternative technologies are urgently needed for next-generation electronics. Here, we demonstrate ultrathin ZnO piezotronic transistors with a～2 nm channel length using inner-crystal self-generated out-of-plane piezopotential as the gate voltage to control the carrier transport. This design removes the need for external gate electrodes that are challenging at nanometer scale. These ultrathin devices exhibit a strong piezotronic effect and excellent pressure-switching characteristics. By directly converting mechanical drives into electrical control signals, ultrathin piezotronic devices could be used as active nanodevices to construct the next generation of electromechanical devices for human machine interfacing, energy harvesting, and self-powered nanosystems.

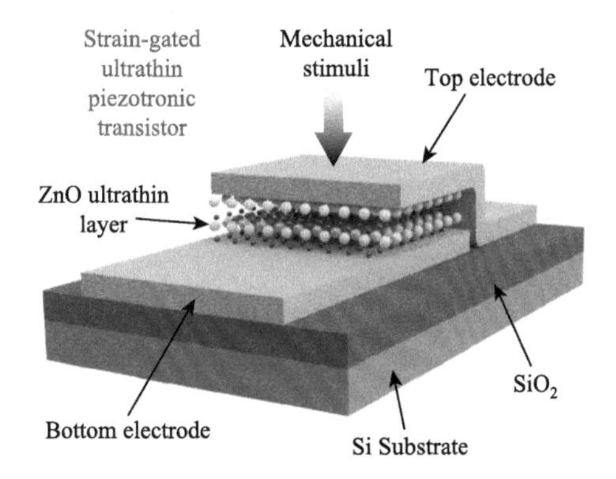

扫码阅全文

http://www.binn.cas.cn/book/202106/W020210623502772890917.pdf

18. Temperature Dependence of the Piezotronic and Piezophototronic Effects in *a*-axis GaN Nanobelts

Xingfu Wang,[+] Ruomeng Yu,[+] Wenbo Peng,[+] Wenzhuo Wu, Shuti Li *and* Zhong Lin Wang[*]

Advanced Materials, 27 (2015) 8067-8074.

简介：本文首次研究了氮化镓纳米带压电电子学效应在 77～300 K 对温度的依赖。

ABSTRACT The temperature dependence of the piezotronic and piezophototronic effects in *a*-axis GaN nanobelts from 77 to 300 K is investigated. The piezotronic effect is enhanced by over 440% under lower temperatures. Two independent processes are discovered to form a competing mechanism through the investigation of the temperature dependence of the piezophototronic effect in *a*-axis GaN nanobelts.

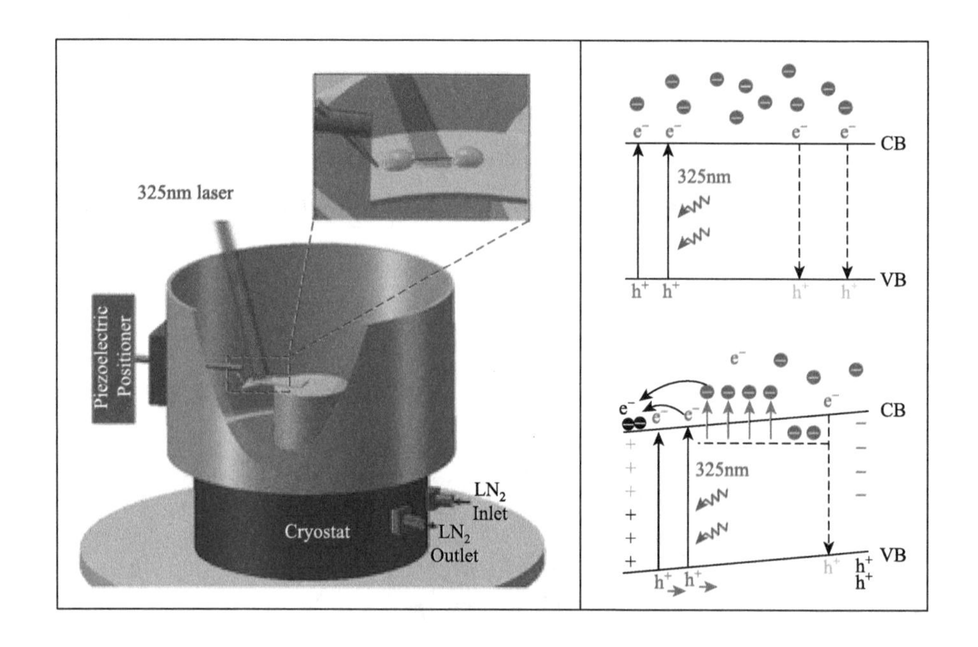

扫码阅全文

http://www.binn.cas.cn/book/202106/W020210623502773435766.pdf

19. Piezo-Phototronic Effect on Selective Electron or Hole Transport through Depletion Region of Vis-NIR Broadband Photodiode

Haiyang Zou, Xiaogan Li, Wenbo Peng, Wenzhuo Wu, Ruomeng Yu, Changsheng Wu, Wenbo Ding, Fei Hu, Ruiyuan Liu, Yunlong Zi *and* **Zhong Lin Wang***

Advanced Materials, 29 (2017) 1701412.

简介：对于 p-Si/n-ZnO 组成的界面，利用压电电子学不但可以提高光探测的效率，还可以改变照射光的波长来分别研究电子和空穴在界面处的输运过程。

ABSTRACT Silicon underpins nearly all microelectronics today and will continue to do so for some decades to come. However, for silicon photonics, the indirect band gap of silicon and lack of adjustability severely limit its use in applications such as broadband photodiodes. Here, a high-performance p-Si/n-ZnO broadband photodiode working in a wide wavelength range from visible to near-infrared light with high sensitivity, fast response, and good stability is reported. The absorption of near-infrared wavelength light is significantly enhanced due to the nanostructured/textured top surface. The general performance of the broadband photodiodes can be further improved by the piezophototronic effect. The enhancement of responsivity can reach a maximum of 78% to 442 nm illumination, the linearity and saturation limit to 1060 nm light are also significantly increased by applying external strains. The photodiode is illuminated with different wavelength lights to selectively choose the photogenerated charge carriers (either electrons or holes) passing through the depletion region, to investigate the piezo-phototronic effect on electron or hole transport separately for the first time. This is essential for studying the basic principles in order to develop a full understanding about piezotronics and it also enables the development of the better performance of optoelectronics.

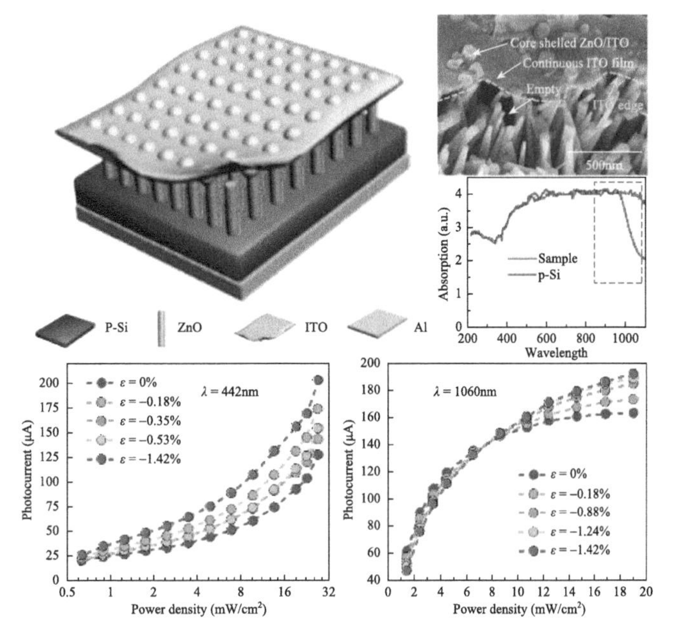

扫码阅全文

http://www.binn.cas.cn/book/202106/W020210623502773933986.pdf

20. Piezotronic Effect on Rashba Spin-Orbit Coupling in a ZnO/P3HT Nanowire Array Structure

Laipan Zhu, Yan Zhang, Pei Lin, Ying Wang, Leijing Yang, Libo Chen, Longfei Wang, Baodong Chen, *and* **Zhong Lin Wang***

ACS Nano, 12 (2018) 1811-1820.

简介：本文首次报道了在室外下压电电子学效应对电子自旋的调控，展示了该效应在自旋电子学中的应用。

ABSTRACT A key concept in the emerging field of spintronics is the voltage-gate control of spin precession via the effective magnetic field generated by the Rashba spin-orbit coupling (SOC). Traditional external gate voltage usually needs a power supply, which can easily bring about background noise or lead to a short circuit in measurement, especially for nanoscale spintronic devices. Here, we present a study on the circular photogalvanic effect (CPGE) in a ZnO/P3HT nanowire array structure with the device excited under oblique incidence. We demonstrate that a strong Rashba SOC is induced by the structure inversion asymmetry of the ZnO/P3HT heterointerface. We show that the Rashba SOC can be effectively tuned by inner-crystal piezo-potential created inside the ZnO nanowires instead of an externally applied voltage. The piezo-potential can not only ensure the stability of future spin-devices under a static pressure or strain but also work without the need of extra energy; hence this room-temperature generation and piezotronic effect control of spin photocurrent demonstrate a potential application in large-scale flexible spintronics in piezoelectric nanowire systems.

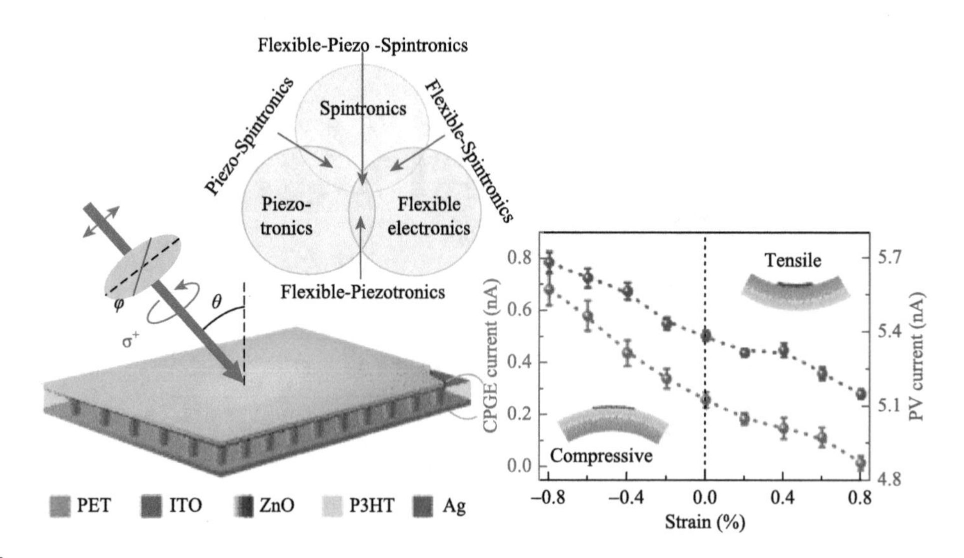

扫码阅全文

http://www.binn.cas.cn/book/202106/W020210623502774540070.pdf

21. Flexoelectronics of Centrosymmetric Semiconductors

Longfei Wang, Shuhai Liu, Xiaolong Feng, Chunli Zhang, Laipan Zhu, Junyi Zhai, Yong Qin* *and* Zhong Lin Wang*

Nature Nanotechnology, 15 (2020) 661-667.

简介：首次报道了对于具有中心对称的半导体，例如硅和锗，仍然具有绕曲电效应，把压电电子学效应应用到了微区局部。

ABSTRACT Interface engineering by local polarization using piezoelectric, pyroelectric and ferroelectric effects has attracted considerable attention as a promising approach for tunable electronics/optoelectronics, human machine interfacing and artificial intelligence. However, this approach has mainly been applied to non-centrosymmetric semiconductors, such as wurtzite-structured ZnO and GaN, limiting its practical applications. Here we demonstrate an electronic regulation mechanism, the flexoelectronics, which is applicable to any semiconductor type, expanding flexoelectricity to conventional semiconductors such as Si, Ge and GaAs. The inner-crystal polarization potential generated by the flexoelectric field serving as a gate can be used to modulate the metal semiconductor interface Schottky barrier and further tune charge-carrier transport. We observe a giant flexoelectronic effect in bulk centrosymmetric semiconductors of Si, TiO_2 and Nb-$SrTiO_3$ with high strain sensitivity (>2,650), largely outperforming state-of-the-art Si-nanowire strain sensors and even piezoresistive, piezoelectric and ferroelectric nanodevices. The effect can be used to mechanically switch the electronics in the nanoscale with fast response (<4 ms) and high resolution (~0.78 nm). This opens up the possibility of realizing strain-modulated electronics in centrosymmetric semiconductors, paving the way for local polarization field-controlled electronics and high-performance electromechanical applications.

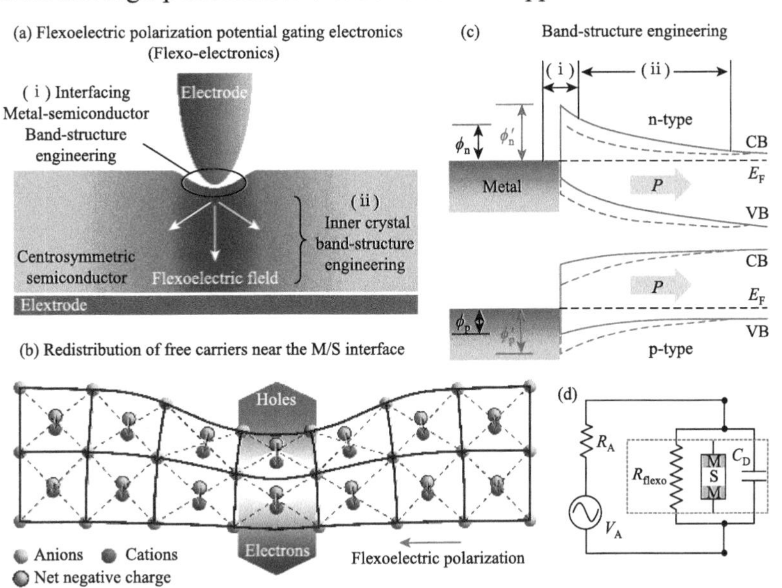

扫码阅全文

http://www.binn.cas.cn/book/202106/W020210623502775185013.pdf

22. Piezotronics and Piezo-Phototronics for Adaptive Electronics and Optoelectronics

Wenzhuo Wu *and* **Zhong Lin Wang***

Nature Reviews Materials, 1 (2016) 16031.

简介：关于压电电子学与压电光电子学的综述。

ABSTRACT Low-dimensional piezoelectric semiconductor nanomaterials, such as ZnO and GaN, have superior mechanical properties and can be integrated into flexible devices that can be subjected to large strain. More importantly, the coupling between piezoelectric polarization and semiconductor properties (for example, electronic transport and photoexcitation) in these materials gives rise to unprecedented device characteristics. This has increased research interest in the emerging fields of piezotronics and piezo-phototronics, which offer new means of manipulating charge-carrier transport, generation, recombination or separation in the controlled operation of flexible devices through the application of external mechanical stimuli. We review the recent progress in advancing our fundamental understanding and in realizing practical applications of piezotronics and piezo-phototronics, and provide an in depth discussion of future research directions.

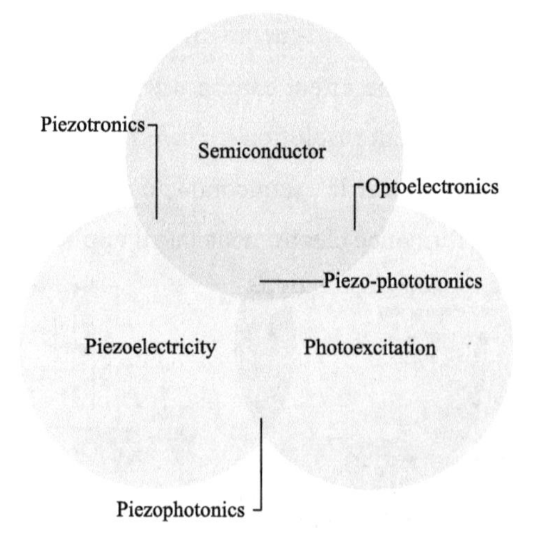

扫码阅全文

http://www.binn.cas.cn/book/202106/W020210623502775944632.pdf

23. Piezotronics and Piezo-Phototronics—Fundamentals and Applications

Zhong Lin Wang* *and* **Wenzhuo Wu**

National Science Review, 1 (2014) 62-90.

简介：关于压电电子学与压电光电子学的综述。

ABSTRACT Technology advancement that can provide new solutions and enable augmented capabilities to complementary metal oxide semiconductor (CMOS)-based technology, such as active and adaptive interaction between machine and human/ambient, is highly desired. Piezotronic nanodevices and integrated systems exhibit potential in achieving these application goals. Utilizing the gating effect of piezopotential over carrier behaviors in piezoelectric semiconductor materials under externally applied deformation, the piezoelectric and semiconducting properties together with optoelectronic excitation processes can be coupled in these materials for the investigation of novel fundamental physics and the implementation of unprecedented applications. Piezopotential is created by the strain-induced ionic polarization in the piezoelectric semiconducting crystal. Piezotronics deal with the devices fabricated using the piezopotential as a gate voltage to tune/control charge-carrier transport across the metal semiconductor contact or the p-n junction. Piezo-phototronics is to use the piezopotential for controlling the carrier generation, transport, separation and/or recombination for improving the performance of optoelectronic devices. This review intends to provide an overview of the rapid progress in the emerging fields of piezotronics and piezo-phototronics. The concepts and results presented in this review show promises for implementing novel nano-electromechanical devices and integrating with micro/nano-electromechanical system technology to achieve augmented functionalities to the state-of-the-art CMOS technology that may find applications in the human machine interfacing, active flexible/stretchable electronics, sensing, energy harvesting, biomedical diagnosis/therapy, and prosthetics.

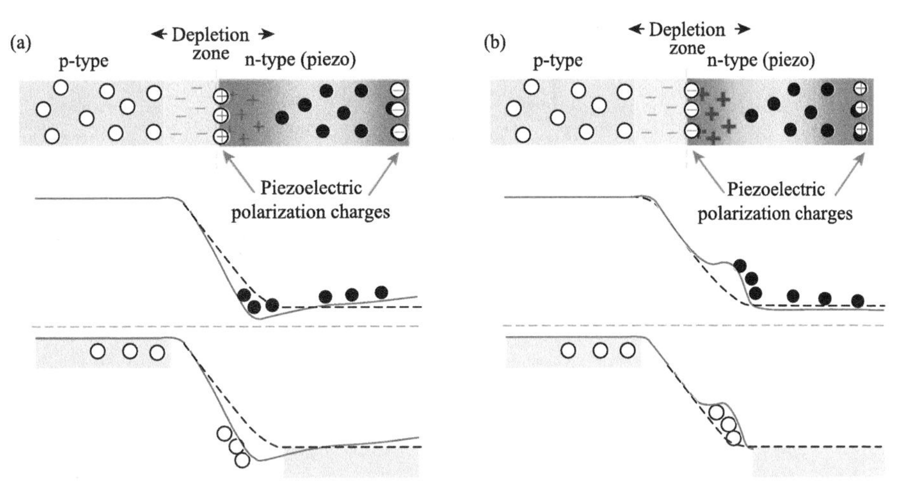

扫码阅全文

http://www.binn.cas.cn/book/202106/W020210623502776404557.pdf

24. Piezotronics and Piezo-Phototronics of Third Generation Semiconductor Nanowires

Caofeng Pan[*], Junyi Zhai[*] *and* Zhong Lin Wang[*]

Chemical Reviews, 119 (2019) 15, 9303-9359.

简介：关于压电电子学与压电光电子学全面的综述。

ABSTRACT With the fast development of nanoscience and nanotechnology in the last 30 years, semiconductor nanowires have been widely investigated in the areas of both electronics and optoelectronics. Among them, representatives of third generation semiconductors, such as ZnO and GaN, have relatively large spontaneous polarization along their longitudinal direction of the nanowires due to the asymmetric structure in their c-axis direction. Two-way or multiway couplings of piezoelectric, photoexcitation, and semiconductor properties have generated new research areas, such as piezotronics and piezo-phototronics. In this review, an in-depth discussion of the mechanisms and applications of nanowire-based piezotronics and piezo-phototronics is presented. Research on piezotronics and piezo-phototronics has drawn much attention since the effective manipulation of carrier transport, photoelectric properties, etc. through the application of simple mechanical stimuli and, conversely, since the design of new strain sensors based on the strain-induced change in semiconductor properties.

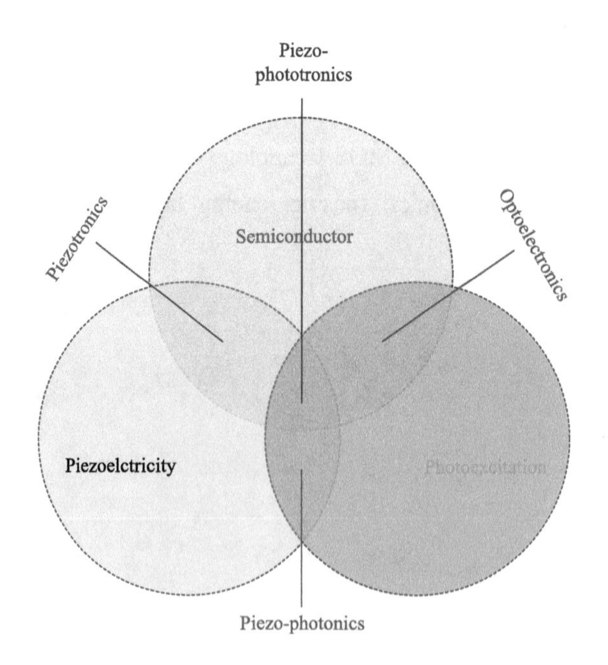

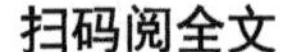
扫码阅全文

http://www.binn.cas.cn/book/202106/W020210623502778500228.pdf

25. Advances in Piezotronic Transistors and Piezotronics

Longfei Wang, Zhong Lin Wang*

Nano Today, 37 (2021) 101108.

简介：关于压电电子学与压电光电子学最新进展的综述。

ABSTRACT Innovative technologies beyond or can augment the conventional metal-oxide-semiconductor (CMOS)-based technology to achieve adaptive and seamless interactions between electronics/machine and human/ambient are urgently needed for smart systems. Piezotronic transistors are a new type of devices designed using completely different principles from CMOS-based technology, which utilize inner-crystal piezoelectric potential as the gate controlling signal to control the electrical transport process, which have been widely demonstrated for the third-generation semiconductors (such as ZnO, GaN). Such devices are innovative in a way that the traditional external channel-width gating is replaced by an inner interface gating or inner channel-width gating. These nanodevices and integrated nanosystems show great potential to achieve complementary functionalities to the state-of-the-art CMOS technology. Coupling the piezo electricity and the electrical transport process in piezoelectric semiconductors results in the emerging field of piezotronics, which show promises for advanced nano-electromechanical devices and micro/nano- electromechanical systems. This manuscript reviews the advances in piezotronic transistors, provides an overview of the development and gives in-depth understandings in the new area of piezotronics. A perspective is given about their further development and potential applications in energy harvesting, active electronics and intelligent sensing.

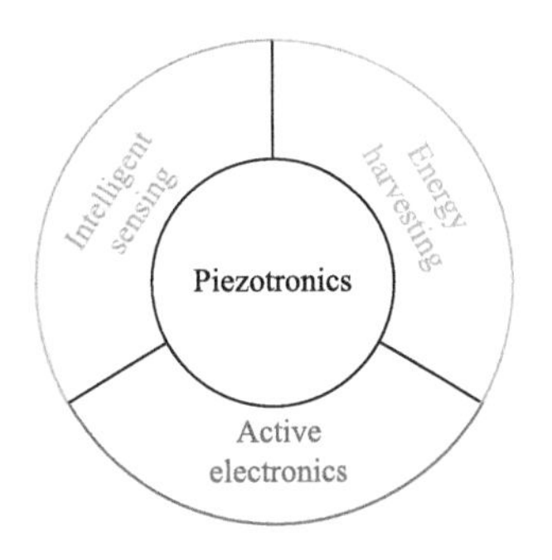

扫码阅全文

http://www.binn.cas.cn/book/202106/W020210623502779055479.pdf

Ⅳ.2 压电电子学：应变传感器

Piezotronics—Strain Sensors

1. Flexible Piezotronic Strain Sensor

Jun Zhou, Yudong Gu, Peng Fei, Wenjie Mai, Yifan Gao, Rusen Yang, Gang Bao *and* **Zhong Lin Wang***

Nano Letters, 8 (2008) 3035-3040.

简介：首篇关于压电电子学效应来调制柔性应变传感器的文章。

ABSTRACT Strain sensors based on individual ZnO piezoelectric fine-wires (PFWs; nanowires, microwires) have been fabricated by a simple, reliable, and cost-effective technique. The electromechanical sensor device consists of a single electrically connected PFW that is placed on the outer surface of a flexible polystyrene (PS) substrate and bonded at its two ends. The entire device is fully packaged by a polydimethylsiloxane (PDMS) thin layer. The PFW has Schottky contacts at its two ends but with distinctly different barrier heights. The Ⅰ-Ⅴ characteristic is highly sensitive to strain mainly due to the change in Schottky barrier height (SBH), which scales linear with strain. The change in SBH is suggested owing to the strain induced band structure change and piezoelectric effect. The experimental data can be well-described by the thermionic emission-diffusion model. A gauge factor of as high as 1,250 has been demonstrated, which is 25% higher than the best gauge factor demonstrated for carbon nanotubes. The strain sensor developed here has applications in strain and stress measurements in cell biology, biomedical sciences, MEMS devices, structure monitoring, and more.

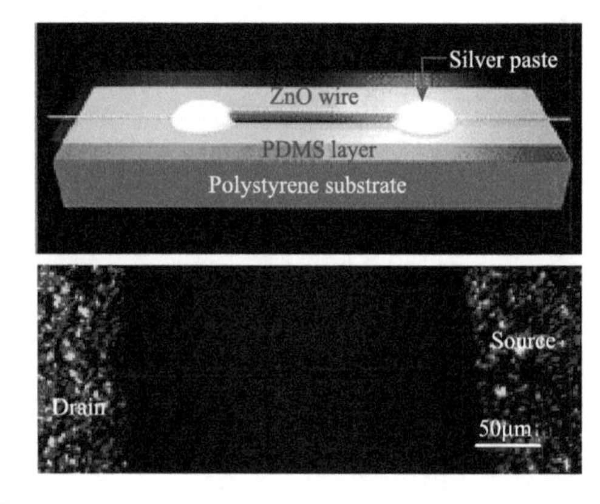

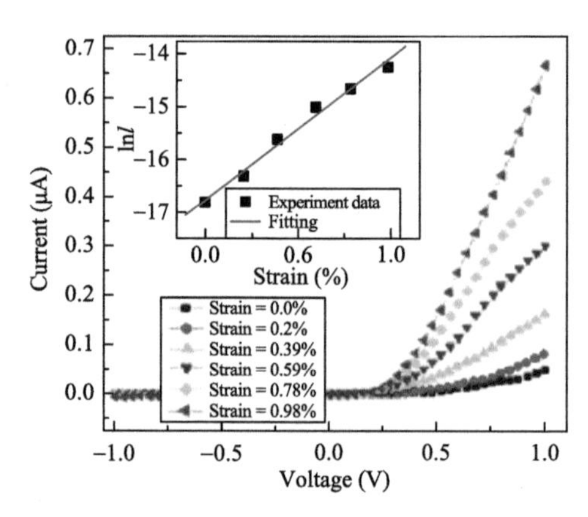

扫码阅全文

http://www.binn.cas.cn/book/202106/W020210623504260973866.pdf

2. High Performance of ZnO Nanowire Protein Sensors Enhanced by the Piezotronic Effect

Ruomeng Yu, Caofeng Pan *and* Zhong Lin Wang*

Energy& Environmental Science, 6 (2013) 494-499.

简介：首篇关于压电电子学效应来调制蛋白质传感器的文章。

ABSTRACT Based on a metal-semiconductor-metal structure, the performance of a ZnO nanowire (NW) based sensor has been studied for detecting Immunoglobulin G (IgG)-targeted protein. By applying a compressive strain, the piezotronic effect on ZnO NW protein sensors can not only increase the resolution of such sensors by tens of times, but also largely improve the detection limit and sensitivity. A theoretical model is proposed to explain the observed behaviors of the sensor. This study demonstrates a prospective approach to raise the resolution, improve the detection limit and enhance the general performance of a biosensor.

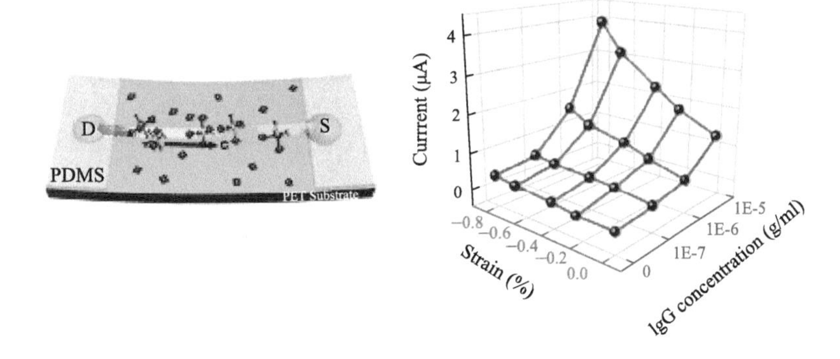

扫码阅全文

http://www.binn.cas.cn/book/202106/W020210623504261436588.pdf

3. Piezotronic Effect on the Sensitivity and Signal Level of Schottky Contacted Proactive Micro/Nanowire Nanosensors

Caofeng Pan[+], Ruomeng Yu[+], Simiao Niu, Guang Zhu *and* Zhong Lin Wang[*]

ACS Nano, 7 (2013) 1803-1810.

简介：首次研究压电电子学效应对基于纳米线具有肖特基接触的传感器信号和选择性的调控。

ABSTRACT We demonstrated the first piezoelectric effect on the performance of a pH sensor using an MSM back-to-back Schottky contacted ZnO micro/nanowire device. When the device is subjected to an external strain, a piezopotential is created in the micro/nanowire, which tunes the effective heights of the Schottky barriers at the local contacts, consequently increasing the sensitivity and signal level of the sensors. Furthermore, the strain-produced piezopotential along the ZnO micro/nanowire will lead to a nonuniform distribution of the target molecules near the micro/nanowire surface owing to electrostatic interaction, which will make the sensor proactive to detect the target molecules even at extremely low overall concentration, which naturally improves the sensitivity and lowers the detection limit. A theoretical model is proposed to explain the observed performance of the sensor using the energy band diagram. This prototype device offers a new concept for designing supersensitive and fast-response micro/nanowire sensors by introducing an external strain and piezotronic effect, which may have great applications in building sensors with fast response and reset time, high selectivity, high sensitivity, and good signal-to-noise ratio for chemical, biochemical, and gas sensing.

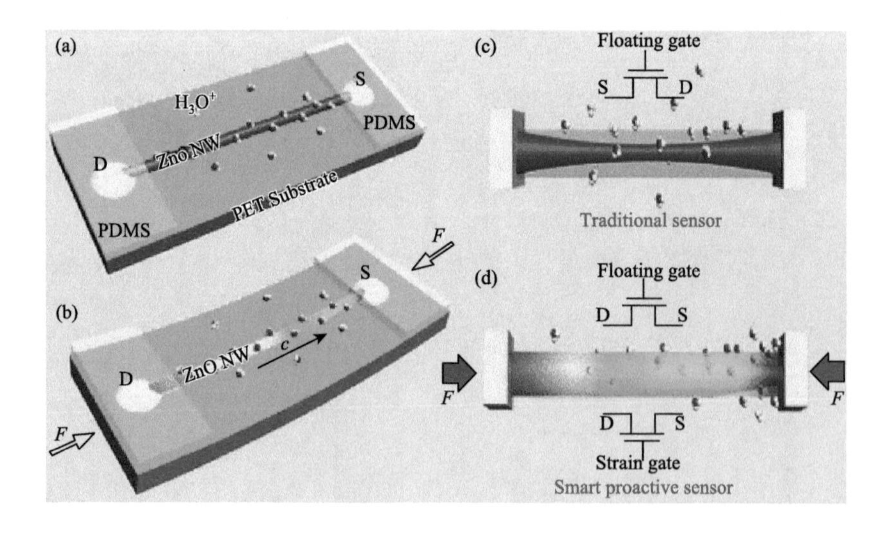

扫码阅全文

http://www.binn.cas.cn/book/202106/W020210623504261647454.pdf

4. Nano-Newton Transverse Force Sensor Using a Vertical GaN Nanowire based on the Piezotronic Effect

Yu Sheng Zhou, Ronan Hinchet, Ya Yang, Gustavo Ardila*, Rudeesun Songmuang, Fang Zhang, Yan Zhang, Weihua Han, Ken Pradel, Laurent Montès, Mireille Mouis *and* Zhong Lin Wang*

Advanced Materials, 25 (2013) 883-888.

简介：首篇利用压电电子学效应来测试纳牛顿级别的力传感器。

ABSTRACT Enhanced transverse force sensitivity of vertically aligned GaN nanowires is demonstrated using the piezotronic effect. The transverse force sensitivity is calculated to be 1.24±0.13 ln(A)/nN, with a resolution better than 16 nN and the response time less than 5 ms. The nano-Newton force resolution shows the potential for piezoelectric semiconductor materials to be used as the main building block for micro-/nanosensor arrays or artificial skin.

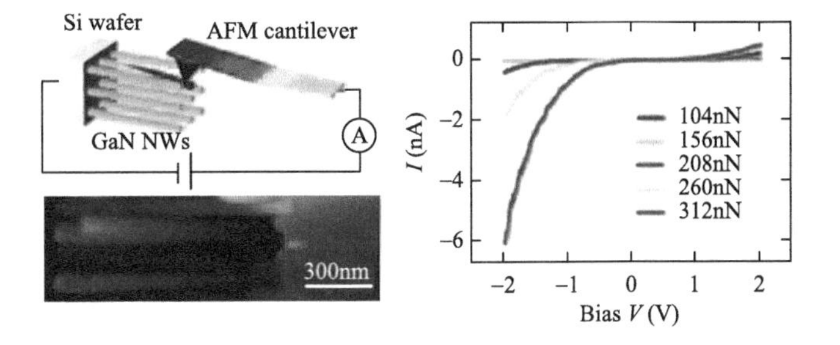

扫码阅全文

http://www.binn.cas.cn/book/202106/W020210623504262537825.pdf

5. GaN Nanobelt-Based Strain-Gated Piezotronic Logic Devices and Computation

Ruomeng Yu, Wenzhuo Wu, Yong Ding *and* Zhong Lin Wang*

ACS Nano, 7 (2013) 6403-6409.

简介：首次基于氮化镓纳米带进行压电电子学逻辑运算。

ABSTRACT Using the piezoelectric polarization charges created at the metal-GaN nanobelt (NB) interface under strain to modulate transport of local charge carriers across the Schottky barrier, the piezotronic effect is utilized to convert mechanical stimuli applied on the wurtzite-structured GaN NB into electronic controlling signals, based on which the GaN NB strain-gated transistors (SGTs) have been fabricated. By further assembling and integrating GaN NB SGTs, universal logic devices such as NOT, AND, OR, NAND, NOR, and XOR gates have been demonstrated for performing mechanical-electrical coupled piezotronic logic operations. Moreover, basic piezotronic computation such as one-bit binary addition over the input mechanical strains with corresponding computation results in an electrical domain by half-adder has been implemented. The strain-gated piezotronic logic devices may find applications in human-machine interfacing, active flexible/stretchable electronics, MEMS, biomedical diagnosis/therapy, and prosthetics.

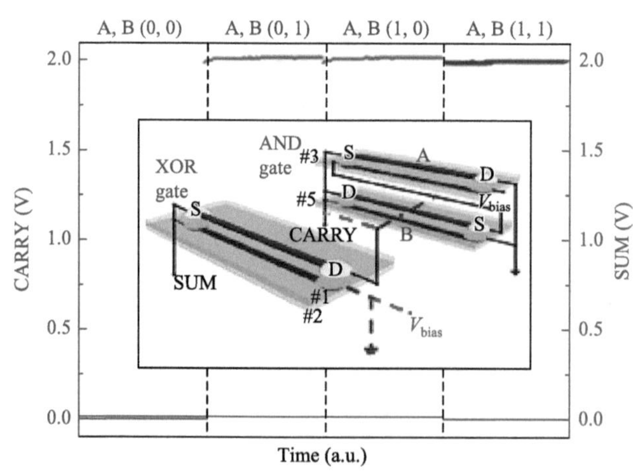

扫码阅全文

http://www.binn.cas.cn/book/202106/W020210623504262900173.pdf

6. Strain-Gated Piezotronic Logic Nanodevices

Wenzhuo Wu, Yaguang Wei *and* **Zhong Lin Wang***

Advanced Materials, 22 (2010) 4711-4715.

简介：首次基于氧化锌纳米线进行压电电子学逻辑运算。

ABSTRACT We present the first piezoelectric triggered mechanical-electronic logic operation using the piezotronic effect, through which the integrated mechanical actuation and electronic logic computation are achieved using only ZnO nanowires (NWs). By utilizing the piezoelectric potential created in a ZnO NW under externally applied deformation, strain-gated transistors (SGTs) have been fabricated. Using the SGTs as building blocks, universal logic components such as inverters, NAND, NOR, and XOR gates have been demonstrated for performing piezotronic logic calculations.

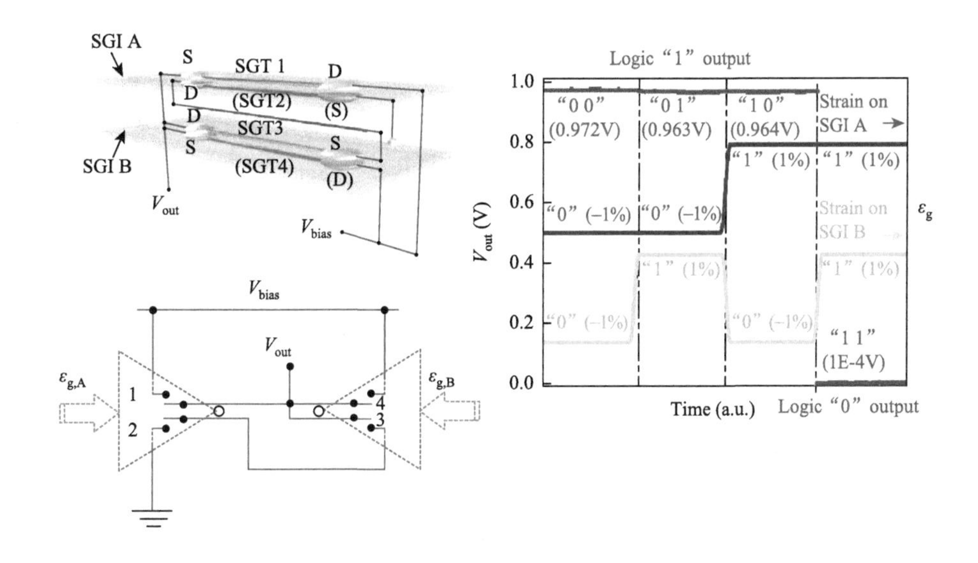

扫码阅全文

http://www.binn.cas.cn/book/202106/W020210623504263539290.pdf

Ⅳ.3 压电电子学：理论
Piezotronics—Theory

1. Equilibrium Potential of Free Charge Carriers in a Bent Piezoelectric Semiconductive Nanowire

Yifan Gao *and* **Zhong Lin Wang***

Nano Letters, 9 (2009) 1103-1110.

简介：首次从理论上计算在压电半导体纳米线中压电电势的分布，是压电电子学的基础。

ABSTRACT We have investigated the behavior of free charge carriers in a bent piezoelectric semiconductive nanowire under thermodynamic equilibrium conditions. For a laterally bent n-type ZnO nanowire, with the stretched side exhibiting positive piezoelectric potential and the compressed side negative piezoelectric potential, the conduction band electrons tend to accumulate at the positive side. The positive side is thus partially screened by free charge carriers while the negative side of the piezoelectric potential preserves as long as the donor concentration is not too high. For a typical ZnO nanowire with diameter 50nm, length 600nm, donor concentration $N_D = 1\times 10^{17}\ cm^{-3}$ under a bending force of 80 nN, the potential in the positive side is <0.05 V and is approximately −0.3 V at the negative side. The theoretical results support the mechanism proposed for a piezoelectric nanogenerator. Degeneracy in the positive side of the nanowire is significant, but the temperature dependence of the potential profile is weak for the temperature range of 100-400 K.

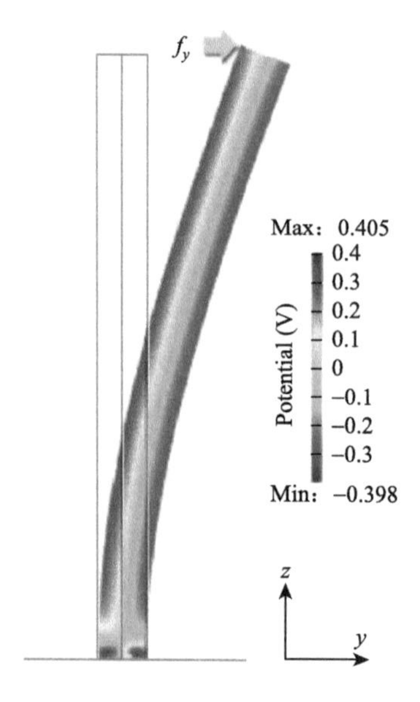

扫码阅全文

http://www.binn.cas.cn/book/202106/W020210623506007999673.pdf

2. Fundamental Theory of Piezotronics

Yan Zhang, Ying Liu *and* **Zhong Lin Wang***

Advanced Materials, 23 (2011) 3004-3013.

简介：首次发展了压电电子学效应的半经典理论，是理解基本器件工作原理的基础。

ABSTRACT Due to polarization of ions in crystals with noncentral symmetry, such as ZnO, GaN, and InN, a piezoelectric potential (piezopotential) is created in the crystal when stress is applied. Electronics fabricated using the inner-crystal piezopotential as a gate voltage to tune or control the charge transport behavior across a metal/semiconductor interface or a p-n junction are called piezotronics. This is different from the basic design of complimentary metal oxide semiconductor (CMOS) field-effect transistors and has applications in force and pressure triggered or controlled electronic devices, sensors, microelectromechanical systems (MEMS), human-computer interfacing, nanorobotics, and touch-pad technologies. Here, the theory of charge transport in piezotronic devices is investigated. In addition to presenting the formal theoretical frame work, analytical solutions are presented for cases including metal semiconductor contact and p-n junctions under simplified conditions. Numerical calculations are given for predicting the current voltage characteristics of a general piezotronic transistor: metal ZnO nanowire metal device. This study provides important insight into the working principles and characteristics of piezotronic devices, as well as providing guidance for device design.

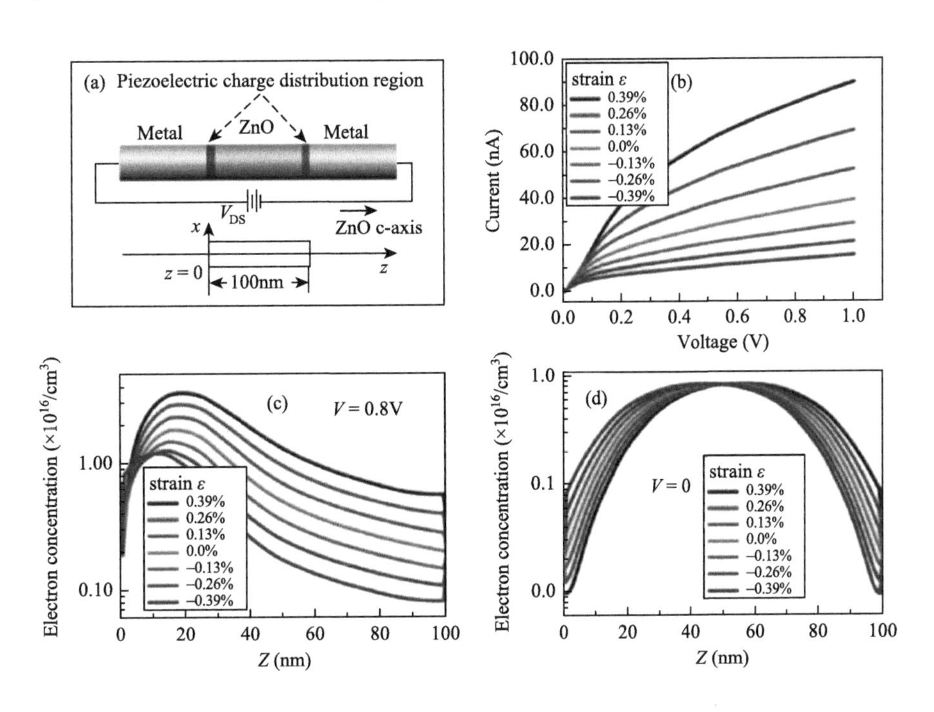

扫码阅全文

http://www.binn.cas.cn/book/202106/W020210623506008391568.pdf

3. Piezo-Phototronics Effect on Nano/Microwire Solar Cells

Yan Zhang, Ya Yang *and* **Zhong Lin Wang***

Energy & Environmental Science, 5 (2012) 6850-6856.

简介：首次发展了基于压电光电子学效应来调制光伏电池效率的半经典理论。

ABSTRACT Wurtzite structures, such as ZnO, GaN, InN and CdS, are piezoelectric semiconductor materials.A piezopotential is formed in the crystal by the piezoelectric charges created by applying a stress. The inner-crystal piezopotential can effectively tune/control the carrier separations and transport processes at the vicinity of a p-n junction or metal semiconductor contact, which is called the piezo-phototronic effect. The presence of piezoelectric charges at the interface/ junction can significantly affect the performances of photovoltaic devices, especially flexible and printed organic/inorganic solar cells fabricated using piezoelectric semiconductor nano/microwires. In this paper, the current voltage characteristics of a solar cell have been studied theoretically and experimentally in the presence of the piezoelectric effect. The analytical results are obtained for a ZnO piezoelectric p-n junction solar cell under simplified conditions, which provide a basic physical picture for understanding the mechanism of the piezoelectric solar cell. Furthermore, the maximum output of the solar cell has been calculated numerically. Finally, the experimental results of organic solar cells support our theoretical model. Using the piezoelectric effect created by external stress, our study not only provides the first basic theoretical understanding about the piezo-phototronic effect on the characteristics of a solar cell but also assists the design for higher performance solar cells.

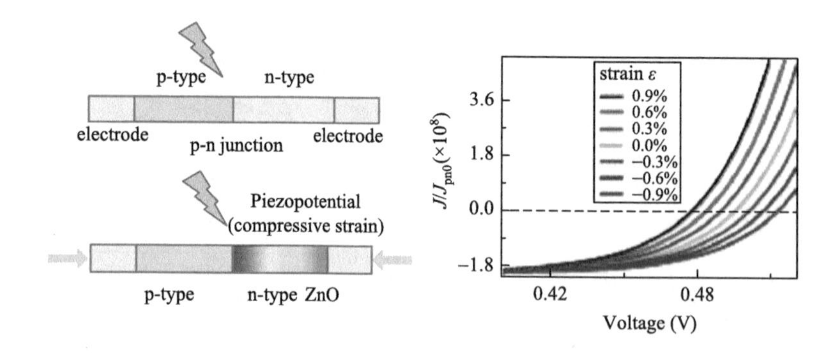

扫码阅全文

http://www.binn.cas.cn/book/202106/W020210623506008748617.pdf

4. First Principle Simulations of Piezotronic Transistors

Wei Liu, Aihua Zhang, Yan Zhang*, Zhong Lin Wang*

Nano Energy, 14 (2015) 355-363.

简介：首次发展了压电电子学效应的第一性原理理论，进一步量化了压电电子学效应在界面的调控作用。

ABSTRACT Piezoelectric semiconductors, such as wurtzite structured ZnO, GaN, and InN, have novel properties owing to piezoelectric polarization tuned/controlled electronic transport characteristics. Under an externally applied strain, piezoelectric charges are created at an interface or junction, which are likely to tune and modulate the local band structure. Taking an Ag-ZnO-Ag two-terminal piezotronic transistor as an example, strain-dependent piezoelectric charge distributions and modulation of Schottky barrier heights (SBHs) at metal/semiconductor interfaces have been investigated by the first principle simulations. The width of piezocharge distribution is calculated by the density function theory and the Poisson equation. The modulations of SBHs at two interfaces show opposite trend under the applied strain. This study not only provides an understanding about the piezotronic effect from quantum theory point of view, but also a new method to calculate the key parameter for optimizing the design of piezotronic devices.

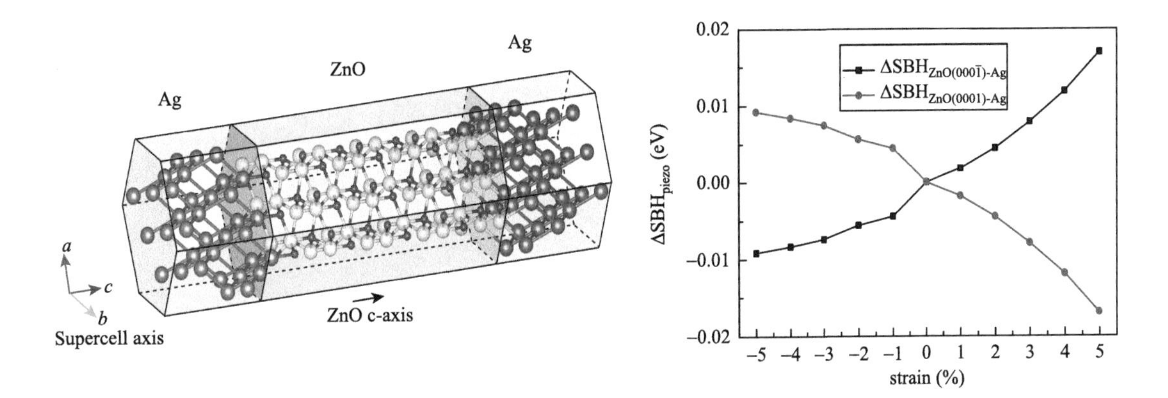

扫码阅全文

http://www.binn.cas.cn/book/202106/W020210623506008960065.pdf

5. Fundamental Theories of Piezotronics and Piezo-Phototronics

Ying Liu, Yan Zhang, Qing Yang, Simiao Niu, Zhong Lin Wang*

Nano Energy, 14 (2015) 257-275.

简介：首次综述了压电电子学效应与压电光电子学效应的基础理论体系。

ABSTRACT Wurtzite structured mateirals such as ZnO, GaN, InN and CdS simultaneously exhibit piezoelectric, semiconducting and photoexcitation properties. The piezotronic effect is to use the inner crystal potential generated by piezoelectric polarization charges for controlling/tuning the charge carrier transport characteristics in these materials. The piezo-phototronic effect is about the use of piezoelectric charges to tube the generation, transport, separation and/or recombination of charge carriers at p-n junction. This article reviews the fundamental theories of piezotronics and piezo-phototronics, forming their basis for electromechancial devices, sensors and energy sciences. Starting from the basic equations for piezoelectricity, semiconductor and photoexcitation, analytical equations for describing the strain-tuned device current were derived. Through analytical calculations and numerical simulations, it was confirmed that the piezoelectric polarization charges can act in the form of inner-crystal charges in the depletion region, resulting in a change in Schottky barrier height, depletion region shift and/or formation of a charge channel, which can be used effectively to enhance the efficiency of LED, solar cell and photon detectors.

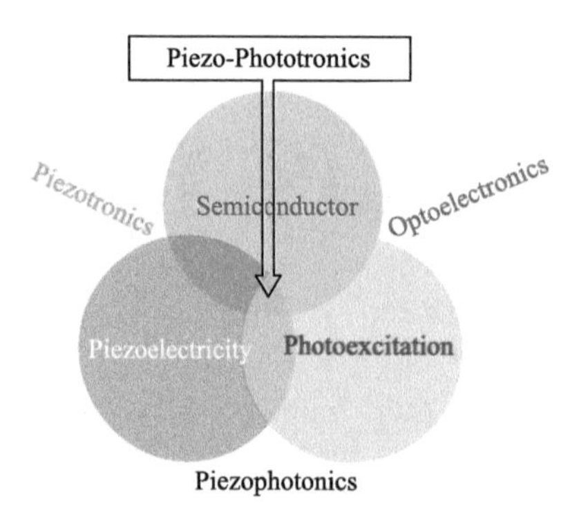

扫码阅全文

http://www.binn.cas.cn/book/202106/W020210623506009361827.pdf

IV.4 压电光电子学：原理与器件

Piezo-Phototronics—Principle and Devices

1. Designing the Electric Transport Characteristics of ZnO Micro/Nanowire Devices by Coupling Piezoelectric and Photoexcitation Effects

Youfan Hu, Yanling Chang, Peng Fei, Robert L. Snyder *and* **Zhong Lin Wang***

ACS Nano, 4 (2010) 1234-1240.

简介：压电光电子学效应是利用压电所产生的界面处极化电荷来调控半导体界面的载流子结合或复合的光电过程的一个效应。本文首次提出了压电光电子学效应的概念。

ABSTRACT The localized coupling between piezoelectric and photoexcitation effects of a ZnO micro/nanowire device has been studied for the first time with the goal of designing and controlling the electrical transport characteristics of the device. The piezoelectric effect tends to raise the height of the local Schottky barrier (SB) at the metal-ZnO contact, while photoexcitation using a light that has energy higher than the band gap of ZnO lowers the SB height. By tuning the relative contributions of the effects from piezoelectricity via strain and photoexcitation via light intensity, the local contact can be tuned step-by-step and/or transformed from Schottky to Ohmic or from Ohmic to Schottky. This study describes a new principle for controlling the coupling among mechanical, photonic, and electrical properties of ZnO nanowires, which could be potentially useful for fabricating piezo-phototronic devices.

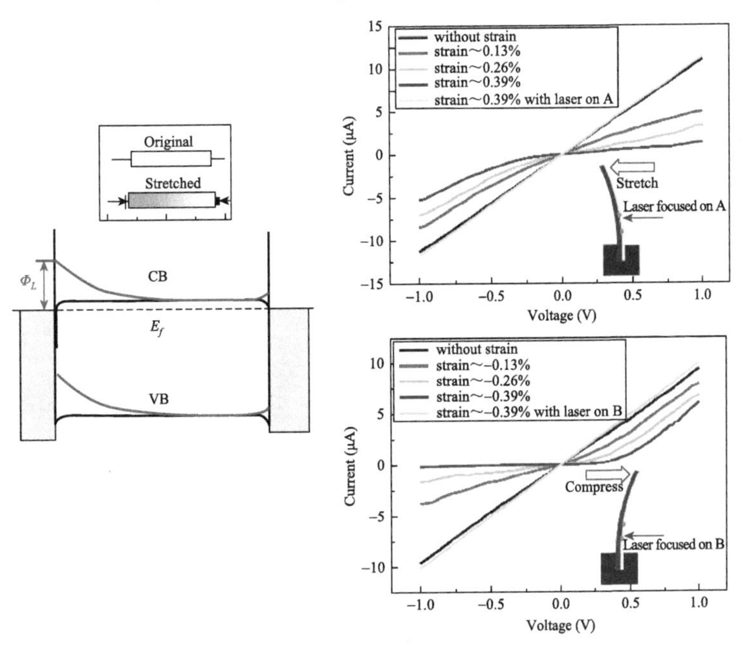

扫码阅全文

http://www.binn.cas.cn/book/202106/W020210623511982639310.pdf

2. Optimizing the Power Output of a ZnO Photocell by Piezopotential

Youfan Hu, Yan Zhang, Yanling Chang, Robert L. Snyder *and* **Zhong Lin Wang***

ACS Nano, 4 (2010) 4220-4224; Corrections: 4 (2010) 4962-4962.

简介：首次提出了压电光电子学效应的基本物理图像和机制。

ABSTRACT Using a metal-semiconductor-metal back-to-back Schottky contacted ZnO microwire device, we have demonstrated the piezoelectric effect on the output of a photocell. An externally applied strain produces a piezopotential in the microwire, which tunes the effective height of the Schottky barrier (SB) at the local contact, consequently changing the transport characteristics of the device. An equivalent circuit model together with the thermionic emission theory has explained the four kinds of relationships observed between the photocurrent and the applied strain. Our study shows the possibility of maximizing the output of a photocell by controlling strain in the device.

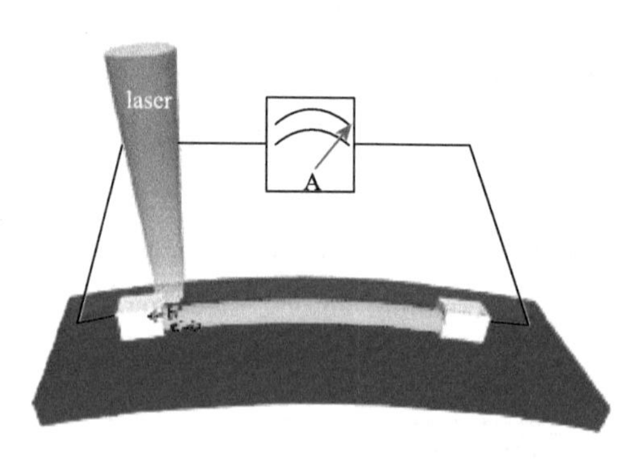

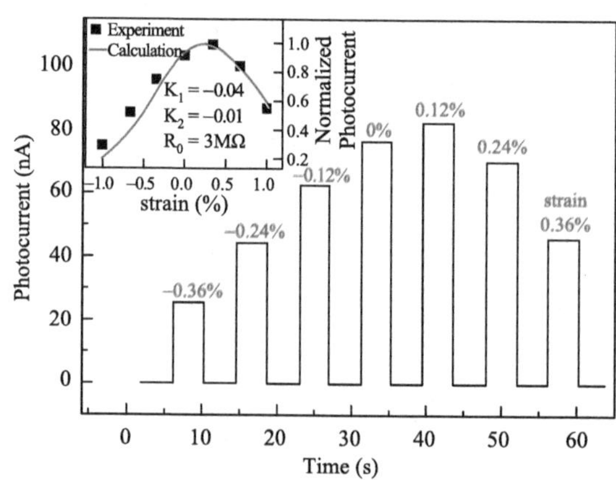

扫码阅全文

http://www.binn.cas.cn/book/202106/W020210623511982835220.pdf

3. Enhancing Sensitivity of a Single ZnO Micro-/Nanowire Photodetector by Piezo-phototronic Effect

Qing Yang, Xin Guo, Wenhui Wang, Yan Zhang, Sheng Xu, Der Hsien Lien *and* **Zhong Lin Wang***

ACS Nano, 4 (2010) 6285-6291.

简介：首次研究了压电光电子学效应如何提高氧化锌纳米线基光探测器的效率。

ABSTRACT We demonstrate the piezoelectric effect on the responsivity of a metal-semiconductor-metal ZnO micro-/nanowire photodetector. The responsivity of the photodetector is respectively enhanced by 530%, 190%, 9%, and 15% upon 4.1 pW, 120.0 pW, 4.1 nW, and 180.4 nW UV light illumination onto the wire by introducing a –0.36% compressive strain in the wire, which effectively tuned the Schottky barrier height at the contact by the produced local piezopotential. After a systematic study on the Schottky barrier height change with tuning of the strain and the excitation light intensity, an in-depth understanding is provided about the physical mechanism of the coupling of piezoelectric, optical, and semiconducting properties. Our results show that the piezo-phototronic effect can enhance the detection sensitivity more than 5-fold for pW levels of light detection.

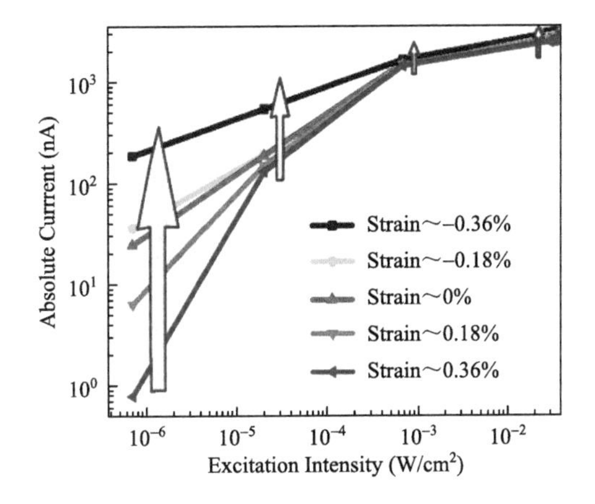

扫码阅全文

http://www.binn.cas.cn/book/202106/W020210623511982960383.pdf

4. Enhanced Cu_2S/CdS Coaxial Nanowire Solar Cells by Piezo-Phototronic Effect

Caofeng Pan, Simiao Niu, Yong Ding, Lin Dong, Ruomeng Yu, Ying Liu, Guang Zhu *and* **Zhong Lin Wang***

Nano Letters, 12 (2012) 3302-3307.

简介：首次研究了压电光电子学效应如何提高 Cu_2S/CdS 光伏电池的效率。

ABSTRACT Nanowire solar cells are promising candidates for powering nanosystems and flexible electronics. The strain in the nanowires, introduced during growth, device fabrication and/or application, is an important issue for piezoelectric semiconductor (like CdS, ZnO, and CdTe) based photovoltaic. In this work, we demonstrate the first largely enhanced performance of n-CdS/p-Cu_2S coaxial nanowire photovoltaic (PV) devices using the piezo-phototronics effect when the PV device is subjected to an external strain. Piezo-phototronics effect could control the electron hole pair generation, transport, separation, and/or recombination, thus enhanced the performance of the PV devices by as high as 70%. This effect offers a new concept for improving solar energy conversation efficiency by designing the orientation of the nanowires and the strain to be purposely introduced in the packaging of the solar cells. This study shed light on the enhanced flexible solar cells for applications in self-powered technology, environmental monitoring, and even defensive technology.

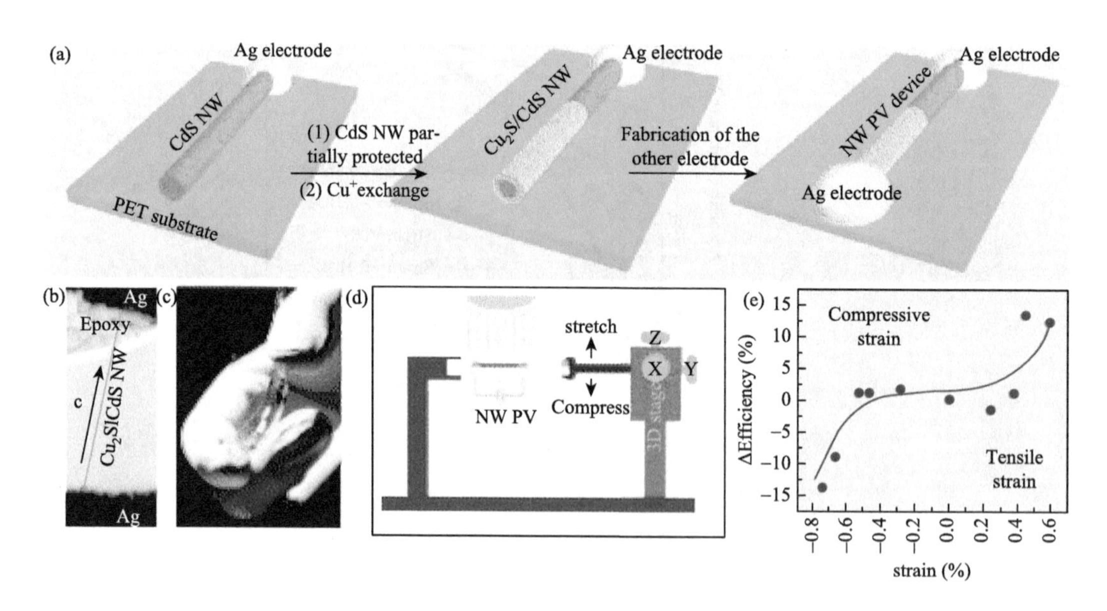

扫码阅全文

http://www.binn.cas.cn/book/202106/W020210623511983110640.pdf

5. Enhancing Light Emission of ZnO Microwire-Based Diodes by Piezo-Phototronic Effect

Qing Yang, Wenhui Wang, Sheng Xu *and* **Zhong Lin Wang***

Nano Letters, 11 (2011) 4012-4017.

简介：首次研究了压电光电子学效应如何提高 n-ZnO/p-GaN LED 发光的效率。这个研究奠定了基本效应的物理图像，是压电光电子学发展历程中一个里程碑性进展。

ABSTRACT Light emission from semiconductors depends not only on the efficiency of carrier injection and recombination but also extraction efficiency. For ultraviolet emission from high band gap materials such as ZnO, nanowires have higher extraction efficiencies than thin films, but conventional approaches for creating a p-n diode result in low efficiency. We exploited the noncentral symmetric nature of n-type ZnO nanowire/p-type GaN substrate to create a piezoelectric potential within the nanowire by applying stress. Because of the polarization of ions in a crystal that has noncentral symmetry, a piezoelectric potential (piezopotential) is created in the crystal under stress. The piezopotential acts as a "gate" voltage to tune the charge transport and enhance carrier injection, which is called the piezo-phototronic effect. We propose that band modification traps free carriers at the interface region in a channel created by the local piezoelectric charges. The emission intensity and injection current at a fixed applied voltage have been enhanced by a factor of 17 and 4, respectively, after applying a 0.093% compressive strain and improved conversion efficiency by a factor of 4.25. This huge enhanced performance is suggested arising from an effective increase in the local "biased voltage" as a result of the band modification caused by piezopotential and the trapping of holes at the interface region in a channel created by the local piezoelectric charges near the interface. Our study can be extended from ultraviolet range to visible range for a variety of optoelectronic devices that are important for today's safe, green, and renewable energy technology.

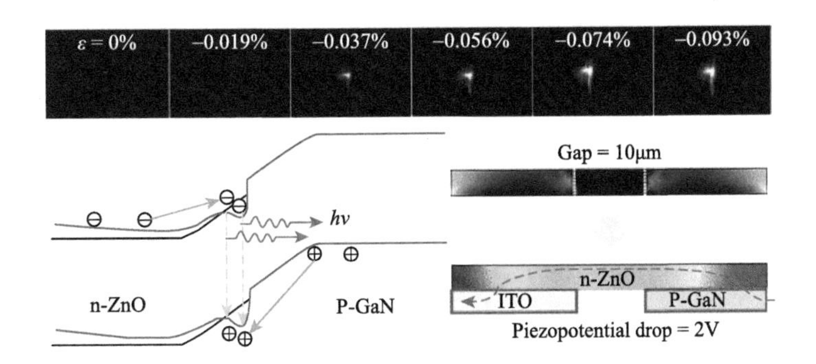

扫码阅全文

http://www.binn.cas.cn/book/202106/W020210623511983827249.pdf

6. Largely Enhanced Efficiency in ZnO Nanowire/P-Polymer Hybridized Inorganic/Organic Ultraviolet Light-Emitting Diode by Piezo-Phototronic Effect

Qing Yang, Ying Liu, Caofeng Pan, Jun Chen, Xiaonan Wen *and* Zhong Lin Wang*

Nano Letters, 13 (2013) 607-613.

简介：首次研究了压电光电子学效应如何提高n-ZnO/p-高分子LED发光的效率，展示了压电光电子学的普适性。

ABSTRACT ZnO nanowire inorganic/organic hybrid ultraviolet (UV) light-emitting diodes (LEDs) have attracted considerable attention as they not only combine the high flexibility of polymers with the structural and chemical stability of inorganic nanostructures but also have a higher light extraction efficiency than thin film structures. However, up to date, the external quantum efficiency of UV LED based on ZnO nanostructures has been limited by a lack of efficient methods to achieve a balance between electron contributed current and hole contributed current that reduces the nonradiative recombination at interface. Here we demonstrate that the piezo-phototronic effect can largely enhance the efficiency of a hybridized inorganic/organic LED made of a ZnO nanowire/p-polymer structure, by trimming the electron current to match the hole current and increasing the localized hole density near the interface through a carrier channel created by piezoelectric polarization charges on the ZnO side. The external efficiency of the hybrid LED was enhanced by at least a factor of 2 after applying a proper strain, reaching 5.92%. This study offers a new concept for increasing organic LED efficiency and has a great potential for a wide variety of high-performance flexible optoelectronic devices.

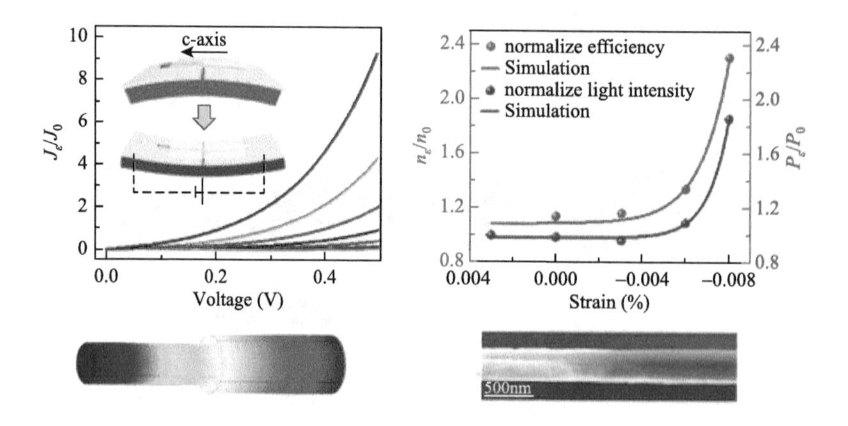

扫码阅全文

http://www.binn.cas.cn/book/202106/W020210623511984371857.pdf

7. High-Resolution Electroluminescent Imaging of Pressure Distribution Using a Piezoelectric Nanowire LED Array

Caofeng Pan, Lin Dong, Guang Zhu, Simiao Niu, Ruomeng Yu, Qing Yang, Ying Liu *and* Zhong Lin Wang*

Nature Photonics, 7 (2013) 752-758.

简介：首次制造了基于压电光电子学效应的 n-ZnO/p-GaN 纳米线 LED 阵列，是发展压电光电子学芯片的一个里程碑性进展。

ABSTRACT Emulation of the sensation of touch through high-resolution electronic means could become important in future generations of robotics and human machine interfaces. Here, we demonstrate that a nanowire light-emitting diode-based pressure sensor array can map two-dimensional distributions of strain with an unprecedented spatial resolution of 2.7 μm, corresponding to a pixel density of 6,350 dpi. Each pixel is composed of a single n-ZnO nanowire/p-GaN light-emitting diode, the emission intensity of which depends on the local strain owing to the piezo-phototronic effect. A pressure map can be created by reading out, in parallel, the electroluminescent signal from all of the pixels with a time resolution of 90 ms. The device may represent a major step towards the digital imaging of mechanical signals by optical means, with potential applications in artificial skin, touchpad technology, personalized signatures, bio-imaging and optical microelectromechanical systems.

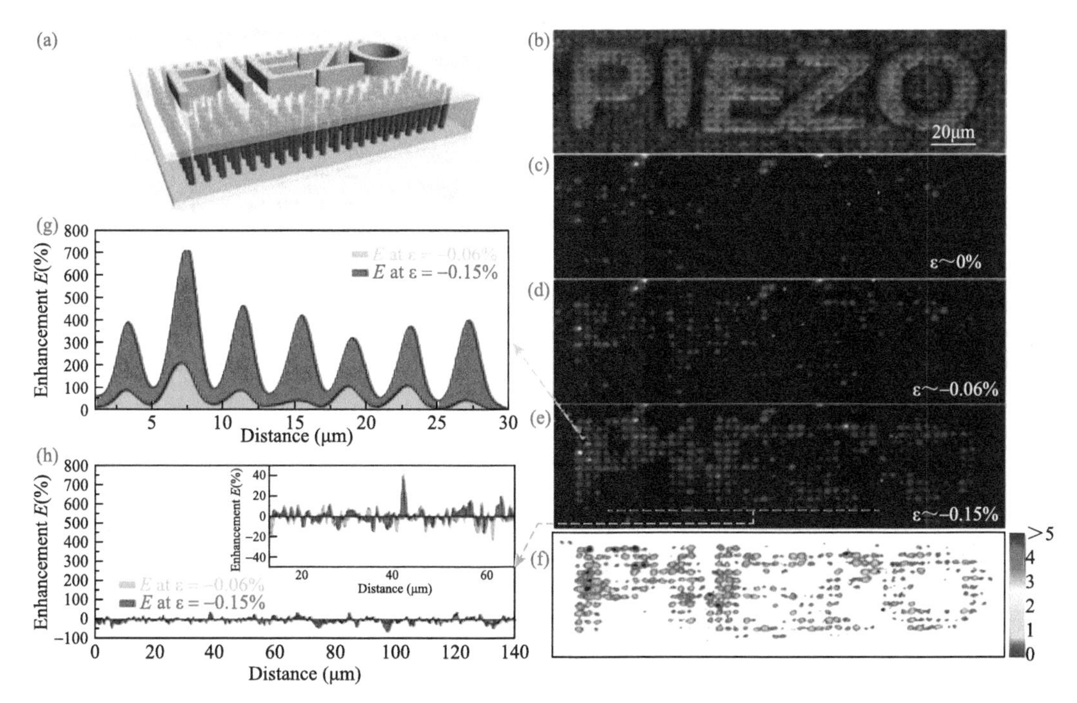

扫码阅全文

http://www.binn.cas.cn/book/202106/W020210623511984518873.pdf

8. Piezo-Phototronic Effect Enhanced Visible/UV Photodetector of a Carbon-Fiber/ZnO-CdS Double-Shell Microwire

Fang Zhang, Simiao Niu, Wenxi Guo, Guang Zhu, Ying Liu, Xiaoling Zhang *and* **Zhong Lin Wang**[*]

ACS Nano, 7 (2013) 4537-4544.

简介：首次报道了压电光电子学效应调制 ZnO-CdS 可见光与紫外光探测器。

ABSTRACT A branched ZnO-CdS double-shell NW array on the surface of a carbon fiber (CF/ZnO-CdS) was successfully synthesized via a facile two-step hydrothermal method. Based on a single CF/ZnO-CdS wire on a polymer substrate, a flexible photodetector was fabricated, which exhibited ultrahigh photon responsivity under illuminations of blue light (1.11×10^5 A/W, 8.99×10^{-8} W/cm^2, 480 nm), green light (3.83×10^4 A/W, 4.48×10^{-8} W/cm^2, 548 nm), and UV light (1.94×10^5 A/W, 1.59×10^{-8} W/cm^2, 372 nm), respectively. The responsivity of this broadband photon sensor was enhanced further by as much as 60% when the device was subjected to a –0.38% compressive strain. This is because the strain induced a piezopotential in ZnO, which tunes the barrier height at the ZnO-CdS heterojunction interface, leading to an optimized optoelectronic performance. This work demonstrates a promising application of piezo-phototronic effect in nanoheterojunction array based photon detectors.

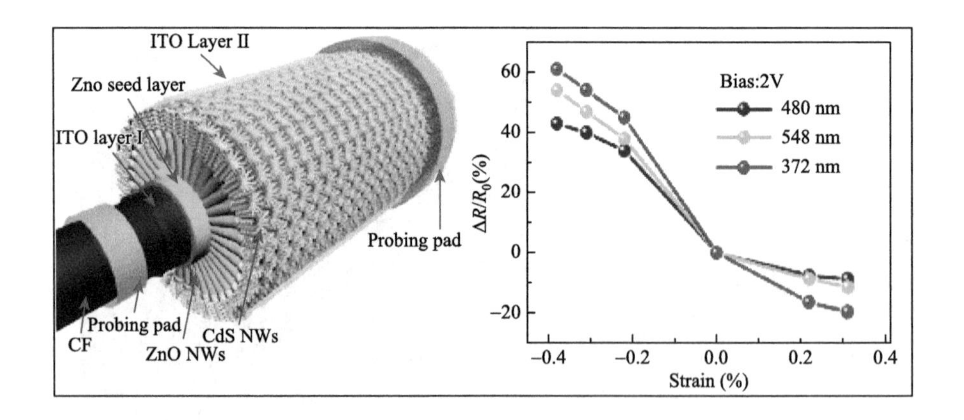

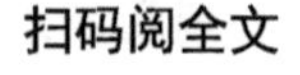

http://www.binn.cas.cn/book/202106/W020210623511985745296.pdf

9. Optimizing Performance of Silicon-Based p-n Junction Photodetectors by the Piezo-Phototronic Effect

Zhaona Wang, Ruomeng Yu, Xiaonan Wen, Ying Liu, Caofeng Pan, Wenzhuo Wu *and* **Zhong Lin Wang***

ACS Nano, 8 (2014) 12866-12873.

简介：首次报道了压电光电子学效应调制硅基光探测器的效率。

ABSTRACT Silicon-based p-n junction photodetectors (PDs) play an essential role in optoelectronic applications for photosensing due to their outstanding compatibility with well-developed integrated circuit technology. The piezo-phototronic effect, a three-way coupling effect among semiconductor properties, piezoelectric polarizations, and photon excitation, has been demonstrated as an effective approach to tune/modulate the generation, separation, and recombination of photogenerated electron-hole pairs during optoelectronic processes in piezoelectric- semiconductor materials. Here, we utilize the strain-induced piezo-polarization charges in a piezoelectric n-ZnO layer to modulate the optoelectronic process initiated in a p-Si layer and thus optimize the performances of p-Si/ZnO NWs hybridized photodetectors for visible sensing via tuning the transport property of charge carriers across the Si/ZnO heterojunction interface. The maximum photoresponsivity R of 7.1 A/W and fastest rising time of 101 ms were obtained from these PDs when applying an external compressive strain of –0.10‰ on the ZnO NWs, corresponding to relative enhancement of 177% in R and shortening to 87% in response time, respectively. These results indicate a promising method to enhance/optimize the performances of non-piezoelectric semiconductor material (e.g., Si) based optoelectronic devices by the piezo-phototronic effect.

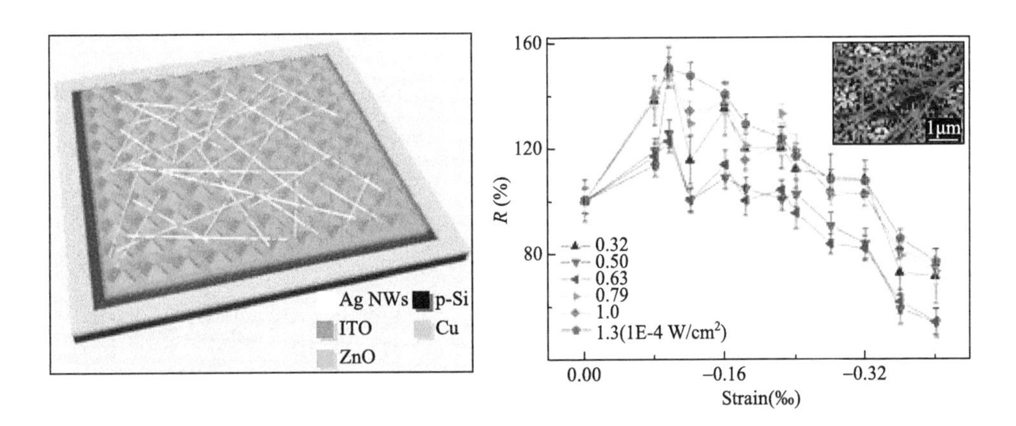

扫码阅全文

http://www.binn.cas.cn/book/202106/W020210623511985969952.pdf

10. Piezo-Phototronic Boolean Logic and Computation Using Photon and Strain Dual-Gated Nanowire Transistors

Ruomeng Yu, Wenzhuo Wu, Caofeng Pan, Zhaona Wang, Yong Ding *and* Zhong Lin Wang*

Advanced Materials, 27 (2015) 940-947.

简介：首次报道了利用压电光电子学器件进行光与应变共同调控的逻辑运算电路。

ABSTRACT Using polarization charges created at the metal-cadmium sulfide interface under strain to gate/modulate electrical transport and optoelectronic processes of charge carriers, the piezo-phototronic effect is applied to process mechanical and optical stimuli into electronic controlling signals. The cascade nanowire networks are demonstrated for achieving logic gates, binary computations, and gated D latches to store information carried by these stimuli.

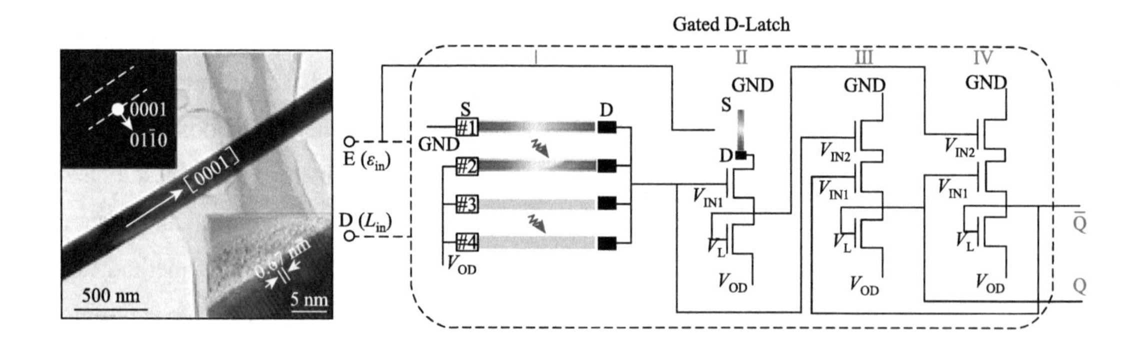

扫码阅全文

http://www.binn.cas.cn/book/202106/W020210623511987006374.pdf

11. Ultrafast Response p-Si/n-ZnO Heterojunction Ultraviolet Detector Based on Pyro-Phototronic Effect

Zhaona Wang, Ruomeng Yu, Xingfu Wang, Wenzhuo Wu *and* Zhong Lin Wang*

Advanced Materials, 28 (2016) 6880.

简介：首次报道了热势压电电子学效应来调制 p-Si/n-ZnO 紫外光探测器的响应时间。

ABSTRACT A light-self-induced pyro-phototronic effect in wurtzite ZnO nanowires is proposed as an effective approach to achieve ultrafast response ultraviolet sensing in p-Si/n-ZnO heterostructures. The relatively long response/recovery time of zinc-oxide-based ultraviolet sensors in air/vacuum has long been an obstacle to developing such detectors for practical applications. The response/recovery time and photoresponsivity are greatly improved by the pyro-phototronic effect.

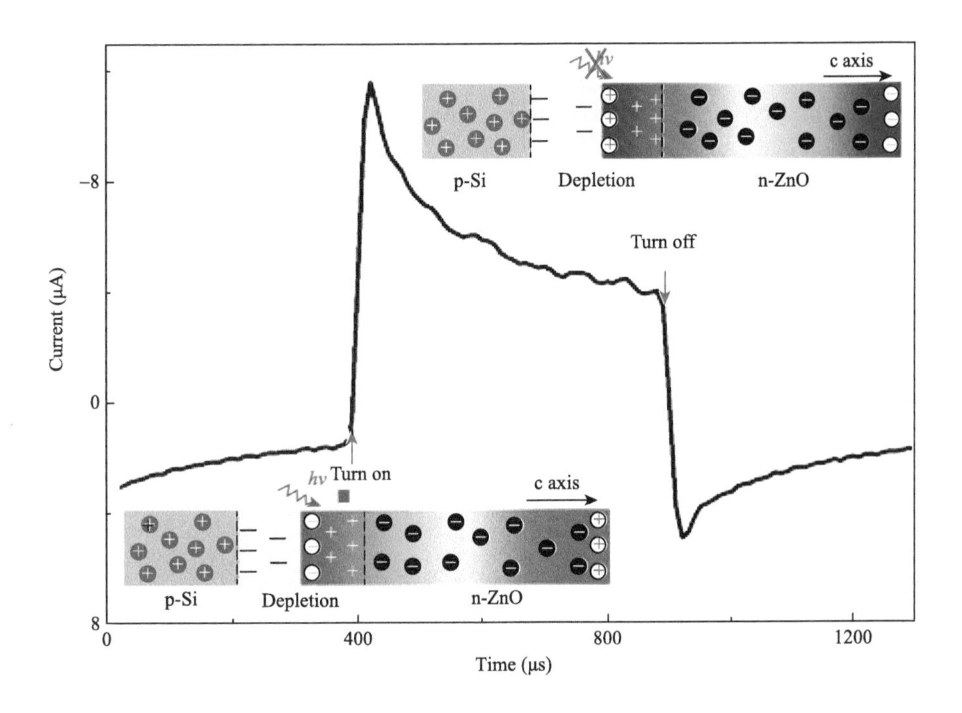

扫码阅全文

http://www.binn.cas.cn/book/202106/W020210623511987567741.pdf

12. Largely Improved Near-Infrared Silicon-Photosensing by the Piezo-Phototronic Effect

Yejing Dai, Xingfu Wang, Wenbo Peng, Haiyang Zou, Ruomeng Yu, Yong Ding, Changsheng Wu *and* **Zhong Lin Wang***

ACS Nano, 11 (2017) 7118-7125.

简介：首次报道了压电光电子学效应调制硅基近红外光探测器的灵敏度。

ABSTRACT Although silicon (Si) devices are the backbone of modern (opto-)electronics, infrared Si-photosensing suffers from low-efficiency due to its limitation in light-absorption. Here, we demonstrate a large improvement in the performance, equivalent to a 366-fold enhancement in photoresponsivity, of a Si-based near-infrared (NIR) photodetector (PD) by introducing the piezo-phototronic effect via a deposited CdS layer. By externally applying a 0.15 compressive strain to the heterojunction, carrier-dynamics modulation at the local junction can be induced by the piezoelectric polarization, and the photoresponsivity and detectivity of the PD exhibit an enhancement of two orders of magnitude, with the peak values up to 19.4 A/W and 1.8×10^{12} cm·Hz$^{1/2}$/W, respectively. The obtained maximum responsivity is considerably larger than those of commercial Si and InGaAs PDs in the NIR waveband. Meanwhile, the rise time and fall time are reduced by 84.6% and 76.1% under the external compressive strain. This work provides a cost-effective approach to achieve high-performance NIR photosensing by the piezo-phototronic effect for high-integration Si-based optoelectronic systems.

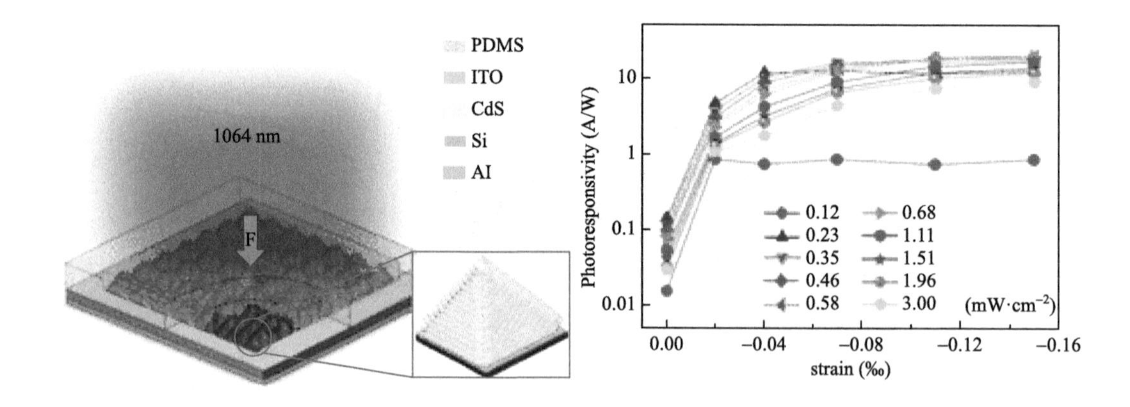

扫码阅全文

http://www.binn.cas.cn/book/202106/W020210623511988047233.pdf

13. Self-Powered Si/CdS Flexible Photodetector with Broadband Response from 325 to 1550 nm Based on Pyro-Phototronic Effect: An Approach for Photosensing below Bandgap Energy

Yejing Dai, Xingfu Wang, Wenbo Peng, Cheng Xu, Changsheng Wu, Kai Dong, Ruiyuan Liu *and* **Zhong Lin Wang***

Advanced Materials, 30 (2018) 1705893.

简介：首次报道了压电光电子学效应调制 Si/CdS 光探测器从紫外到红外大范围的光响应灵敏度。

ABSTRACT Cadmium sulfide (CdS) has received widespread attention as the building block of optoelectronic devices due to its extraordinary optoelectronic properties, low work function, and excellent thermal and chemical stability. Here, a self-powered flexible photodetector (PD) based on p-Si/n-CdS nanowires heterostructure is fabricated. By introducing the pyro-phototronic effect derived from wurtzite structured CdS, the self-powered PD shows a broadband response range, even beyond the bandgap limitation, from UV (325 nm) to near infrared (1550 nm) under zero bias with fast response speed. The light-induced pyroelectric potential is utilized to modulate the optoelectronic processes and thus improve the photoresponse performance. Lasers with different wavelengths have different effects on the self-powered PDs and corresponding working mechanisms are carefully investigated. Upon 325 nm laser illumination, the rise time and fall time of the self-powered PD are 245 and 277 μs, respectively, which are faster than those of most previously reported CdS-based nanostructure PDs. Meanwhile, the photoresponsivity R and specific detectivity D^* regarding to the relative peak-to-peak current are both enhanced by 67.8 times, compared with those only based on the photovoltaic effect-induced photocurrent. The self-powered flexible PD with fast speed, stable, and broadband response is expected to have extensive applications in various environments.

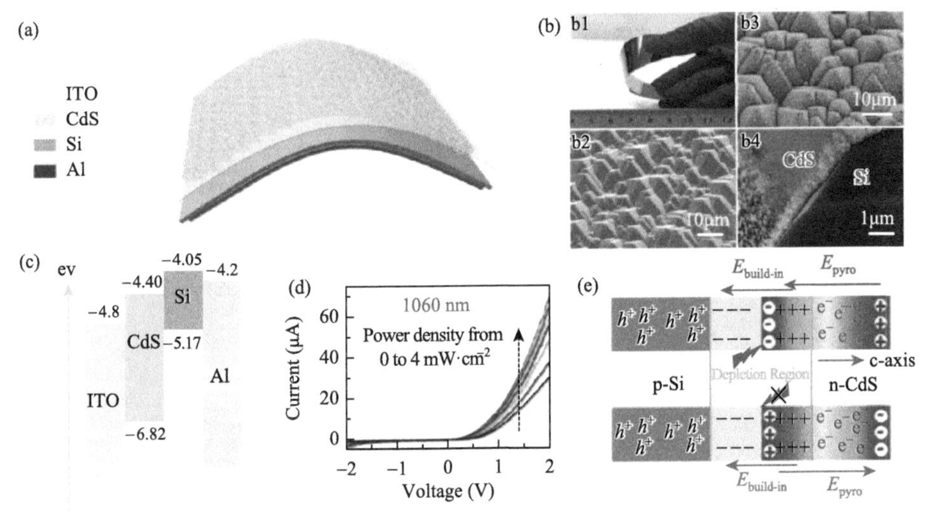

扫码阅全文

http://www.binn.cas.cn/book/202106/W020210623511989021111.pdf

14. Piezo-Phototronic Effect Enhanced Efficient Flexible Perovskite Solar Cells

Junlu Sun, Qilin Hua, Ranran Zhou, Dongmei Li, Wenxi Guo, Xiaoyi Li, Guofeng Hu, Chongxin Shan, Qingbo Meng,*, Lin Dong*, Caofeng Pan* *and* Zhong Lin Wang*

ACS Nano, 13 (2019) 4507-4513.

简介：首次报道了压电光电子学效应调制钙钛矿光伏电池的效率。

ABSTRACT Tremendous work has been made recently to improve the power conversion efficiencies (PCEs) of perovskite solar cells (PSCs); the best reported value is now over 23%. However, further improving the PCEs of PSCs is challenged by material properties, device stability, and packaging technologies. Here, we report a new approach to increase the PCEs of flexible PSCs via introducing the piezophototronic effect in the PSCs by growing an array of ZnO nanowires on flexible plastic substrates, which act as the electron-transport layer for PSCs. From the piezo-phototronic effect, the absolute PCE was improved from 9.3% to 12.8% for flexible perovskite solar cells under a static mechanical strain of 1.88%, with a～40% enhancement but no change in the components of materials and device structure. A corresponding working model was proposed to elucidate the strategy to boost the performance of the PSCs. These findings present a general approach to improve PCEs of flexible PSCs without changing their fundamental materials.

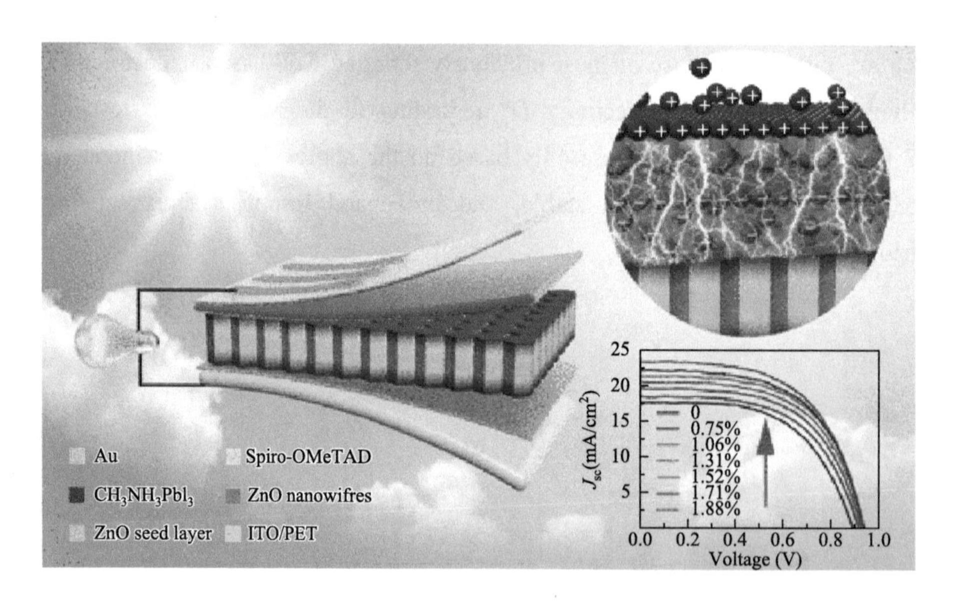

扫码阅全文

http://www.binn.cas.cn/book/202106/W020210623511989432693.pdf

15. Piezo-Phototronic Effect Enhanced Polarization-Sensitive Photodetectors Based on Cation-Mixed Organic-Inorganic Perovskite Nanowires

Laipan Zhu, Qingsong Lai, Wenchao Zhai, Baodong Chen, Zhong Lin Wang*

Materials Today, 37 (2020) 56-63.

简介：首次报道了压电光电子学效应调控基于有机−无机钙钛矿复合光传感器的灵敏度。

ABSTRACT Piezo-phototronic effect has been extensively investigated for the third generation semiconductor nanowires. Here, we present a demonstration that piezo-phototronic effect can even be applied to tune polarization-sensitive photodetectors based on cation-mixed organic inorganic perovskite nanowires. A big anisotropic photoluminescence (PL) with linearly polarized light-excitation was found due to a strong spontaneous piezoelectric polarization besides the anisotropic crystal structure and morphology. The piezo-phototronic effect was utilized to tune the PL intensity, and an improved anisotropic PL ratio from 9.36 to 10.21 for linearly polarized light-excitation was obtained thanks to the modulation by piezo-potential. And a circularly polarization-sensitive PL characterized with circular dichroism ratio was also discovered, which was found to be modulated from 0.085 to 0.555 (with a 5.5-fold improvement) within the range of applied strain. The circular dichroism was resulted from the joint effects of the modulated Rashba spin orbit coupling and the asymmetric carriers separation and recombination for right-and left-handed helicity due to the presence of effective piezo-potential. These findings not only reveal the promising optoelectronic applications of piezo-phototronic effect in perovskite-based polarization-sensitive photodetectors, but also illuminate fundamental understandings of their polarization properties of perovskite nanowires.

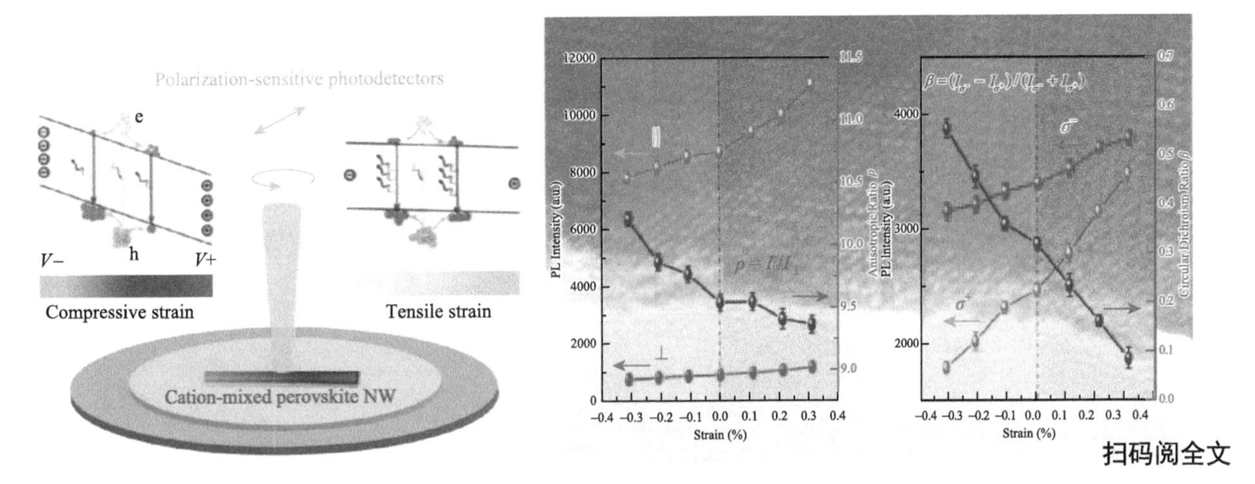

扫码阅全文

http://www.binn.cas.cn/book/202106/W020210623511990595972.pdf

16. Piezo-Phototronic Effect Controlled Dual-Channel Visible Light Communication (PVLC) Using InGaN/GaN Multiquantum Well Nanopillars

Chunhua Du, Chunyan Jiang, Peng Zuo, Xin Huang, Xiong Pu, Zhenfu Zhao, Yongli Zhou, Linxuan Li, Hong Chen,* Weiguo Hu* and Zhong Lin Wang*

Small, 11 (2015) 6071-6077.

简介：首次报道了压电光电子学效应调控基于 InGaN/GaN 的量子阱双通道可见光通信器件。

ABSTRACT Visible light communication (VLC) simultaneously provides illumination and communication via light emitting diodes (LEDs). Keeping a low bit error rate is essential to communication quality, and holding a stable brightness level is pivotal for illumination function. For the first time, a piezo-phototronic effect controlled visible light communication (PVLC) system based on InGaN/GaN multiquantum wells nanopillars is demonstrated, in which the information is coded by mechanical straining. This approach of force coding is also instrumental to avoid LED blinks, which has less impact on illumination and is much safer to eyes than electrical on/off VLC. The two-channel transmission mode of the system here shows great superiority in error self-validation and error self-elimination in comparison to VLC. This two-channel PVLC system provides a suitable way to carry out noncontact, reliable communication under complex circumstances.

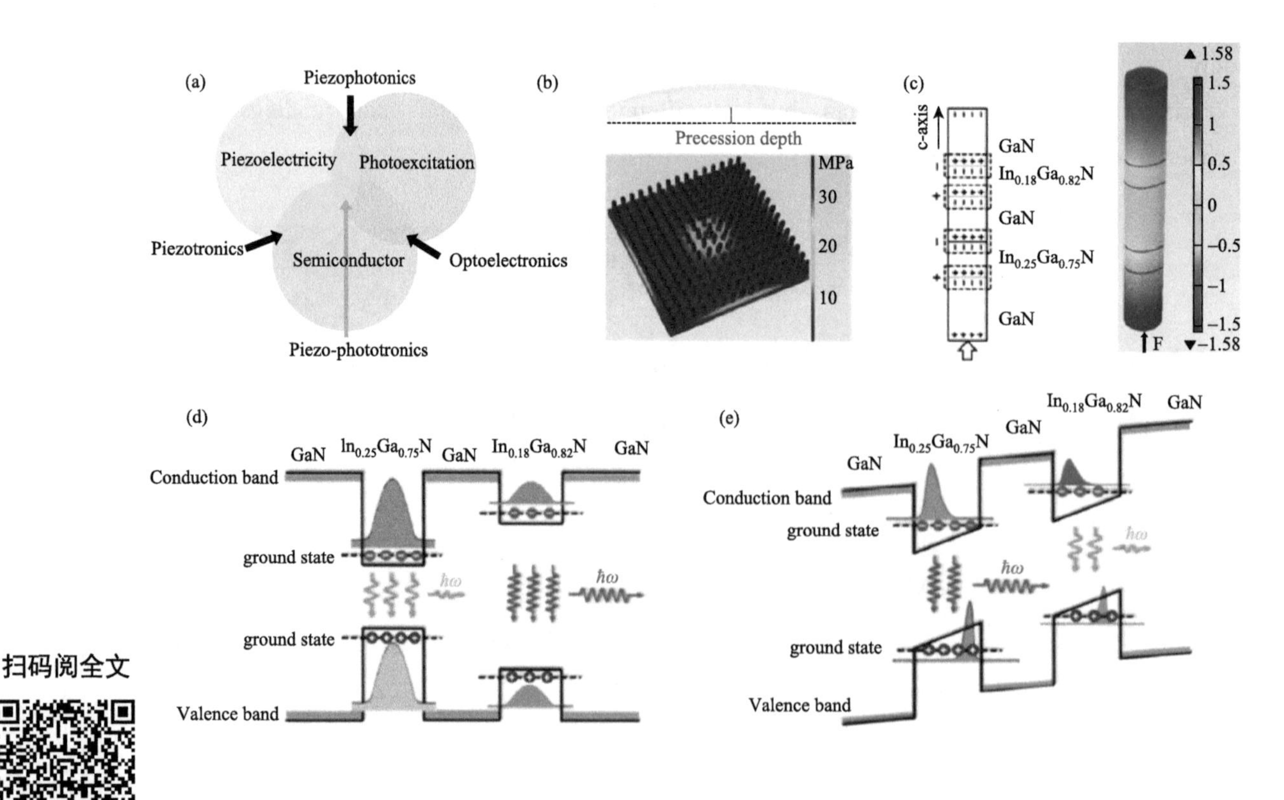

扫码阅全文

http://www.binn.cas.cn/book/202106/W020210623511991013853.pdf

17. Piezophototronic Effect in Single-Atomic-Layer MoS_2 for Strain-Gated Flexible Optoelectronics

Wenzhuo Wu*, Lei Wang*, Ruomeng Yu*, Yuanyue Liu, Su-Huai Wei, James Hone *and* Zhong Lin Wang*

Advanced Materials, 28 (2016) 8463-8468.

简介：首次报道了二维 MoS_2 材料中的压电光电子学效应，给二维材料的研究增加了一个维度。

ABSTRACT Strain-gated flexible optoelectronics are reported based on monolayer MoS_2. Utilizing the piezoelectric polarization created at the metal-MoS_2 interface to modulate the separation/transport of photogenerated carriers, the piezophototronic effect is applied to implement atomic-layer-thick phototransistor. Coupling between piezoelectricity and photogenerated carriers may enable the development of novel optoelectronics.

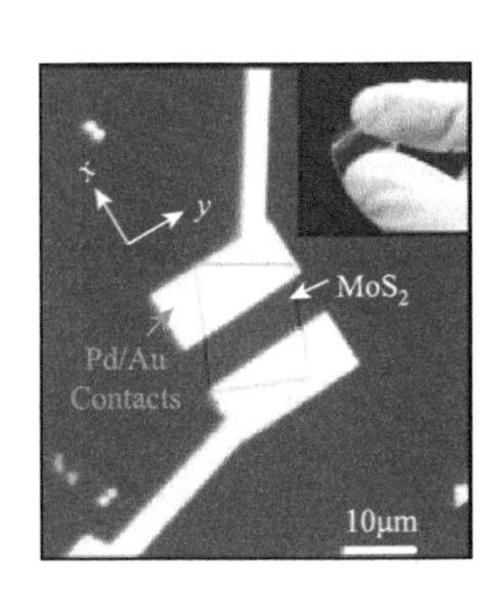

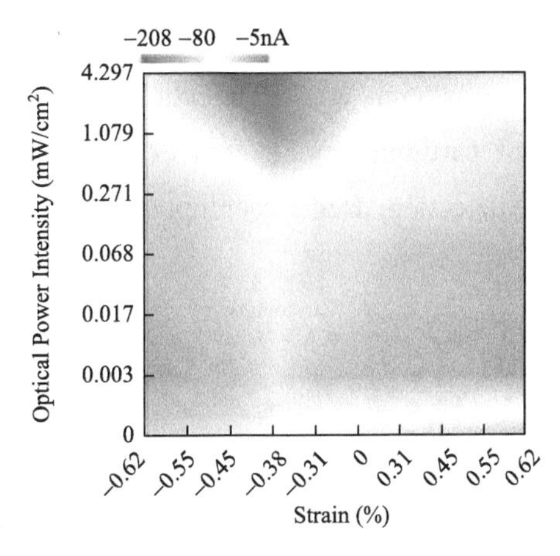

扫码阅全文

http://www.binn.cas.cn/book/202106/W020210623511991392581.pdf

18. Dynamic Real-Time Imaging of Living Cell Traction Force by Piezo-Phototronic Light Nano-Antenna Array

Qiang Zheng, Mingzeng Peng, Zhuo Liu, Shuyu Li, Rongcheng Han, Han Ouyang, Yubo Fan, Caofeng Pan, Weigeo Hu, Junyi Zhai*, Zhou Li*, Zhong Lin Wang*

Science Advances, 7 (2021) eabe7738.

简介：首次利用基于压电光电子效应的纳米线阵列实现实时原位高空间分辨率单细胞牵引力的精确测量。

ABSTRACT Dynamic mapping of the cell-generated force of cardiomyocytes will help provide an intrinsic understanding of the heart. However, a real-time, dynamic, and high-resolution mapping of the force distribution across a single living cell remains a challenge. Here, we established a force mapping method based on a "light nano-antenna" array with the use of piezo-phototronic effect. A spatial resolution of 800 nm and a temporal resolution of 333 ms have been demonstrated for force mapping. The dynamic mapping of cell force of live cardiomyocytes was directly derived by locating the antennas' positions and quantifying the light intensities of the piezo-phototronic light nano-antenna array. This study presents a rapid and ultrahigh-resolution methodology for the fundamental study of cardiomyocyte behavior at the cell or subcellular level. It can provide valuable information about disease detection, drug screening, and tissue engineering for heart-related studies.

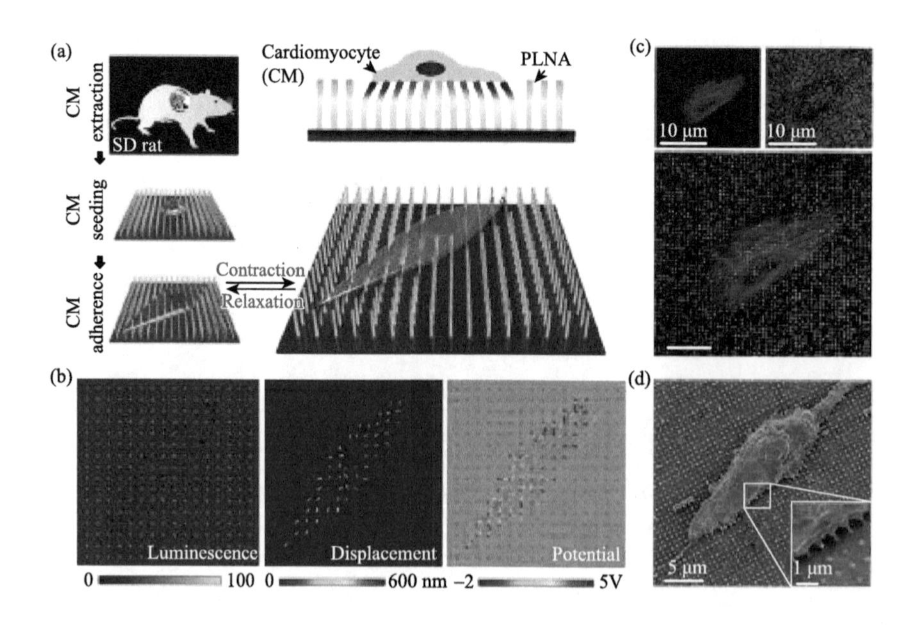

扫码阅全文

http://www.binn.cas.cn/book/202106/W020210701553287761000.pdf

Ⅳ.5 压电光电子学：理论

Piezo-Phototronics—Theory

1. Nanowire Piezo-Phototronic Photodetector: Theory and Experimental Design

Ying Liu, Qing Yang, Yan Zhang, Zongyin Yang *and* **Zhong Lin Wang***

Advanced Materials, 24 (2012) 1410-1417.

简介：首次发展了压电光电子学效应的半经典理论，是理解基本器件工作原理的基础。

ABSTRACT The piezo-phototronic effect is about the use of the inner crystal piezoelectric potential to tune/control charge carrier generation, separation, transport and/or recombination in optoelectronic devices. In this paper, a theoretical model for describing the characteristics of a metal-nanowire-metal structured piezo-phototronic photodetector is constructed. Numerical simulations fit well to the experimental results of a CdS and ZnO nanowire based visible and UV detector, respectively.

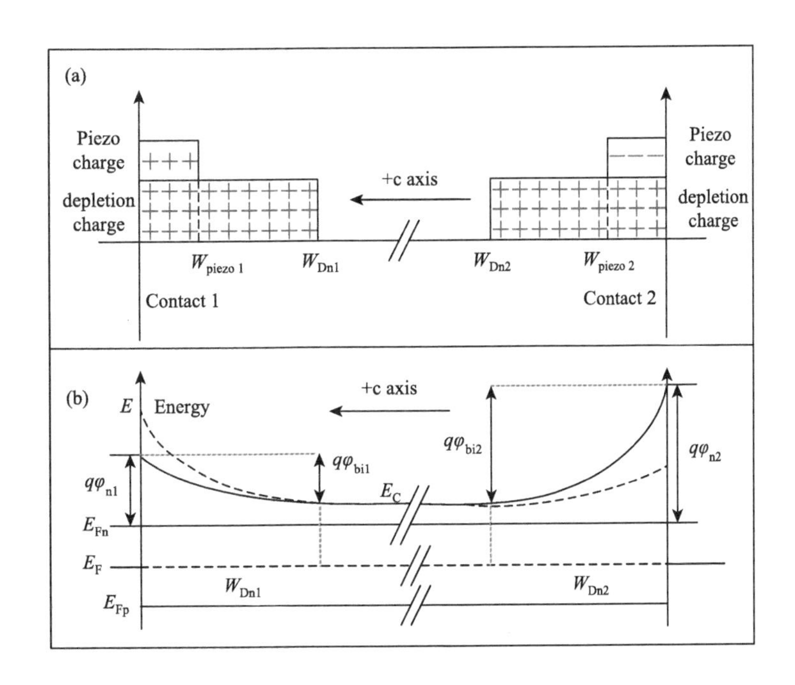

扫码阅全文

http://www.binn.cas.cn/book/202106/W020210623512928100693.pdf

2. Theory of Piezo-Phototronics for Light-Emitting Diodes

Yan Zhang *and* **Zhong Lin Wang***

Advanced Materials, 24 (2012) 4712-4718.

简介：首次发展了基于压电光电子学效应来调制界面载流子输运与结合的半经典理论。

ABSTRACT Devices fabricated by using the inner-crystal piezopotential as a "gate" voltage to tune/control the carrier generation, transport, and recombination processes at the vicinity of a p-n junction are named piezo-phototronics. Here, the theory of the photon emission and carrier transport behavior in piezo-phototronic devices is investigated as a p-n junction light-emitting diode. Numerical calculations are given for predicting the photon emission and current-voltage characteristics of a general piezo-phototronic light-emitting diode.

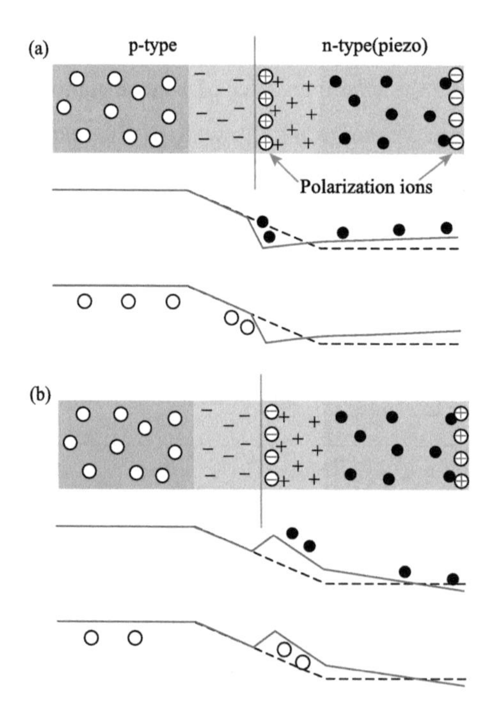

扫码阅全文

http://www.binn.cas.cn/book/202106/W020210623512928369049.pdf

3. Theoretical Study of Piezo-Phototronic Nano-LEDs

Ying Liu, Simiao Niu, Qing Yang, Benjamin D.B. Klein, Yu Sheng Zhou *and* **Zhong Lin Wang***

Advanced Materials, 25 (2014) 7209-7216.

简介：首次发展了基于压电光电子学效应来调制 LED 效率的能带理论。界面处压电电荷的出现不但可以调节载流子复合区的宽度与偏移，还在界面处产生一个牵制电子输运的沟道。

ABSTRACT Two-dimensional finite-element simulation of the piezo-phototronic effect in p-n-junction-based devices is carried out for the first time. A charge channel can be induced at the p-n junction interface when strain is applied, given the n-side is a piezoelectric semiconductor and the p-type side is non-piezoelectric semiconductor. This provides the first simulated evidence supporting the previously suggested mechanism responsible for the experimentally observed gigantic change of light-emission efficiency in piezo-phototronic light-emitting devices.

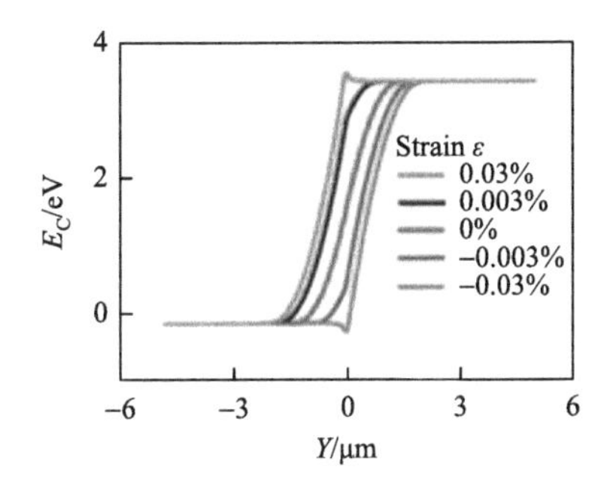

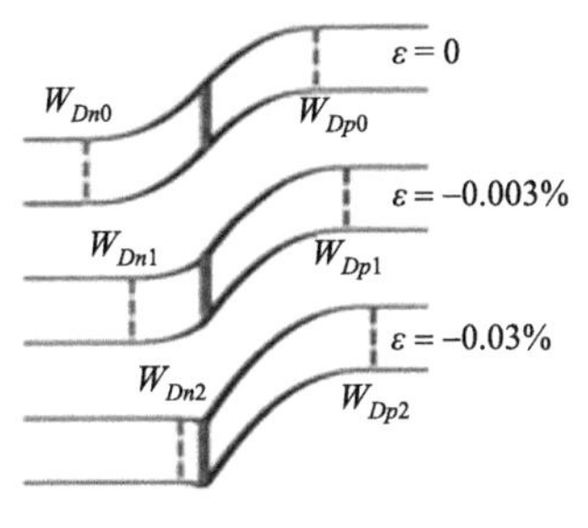

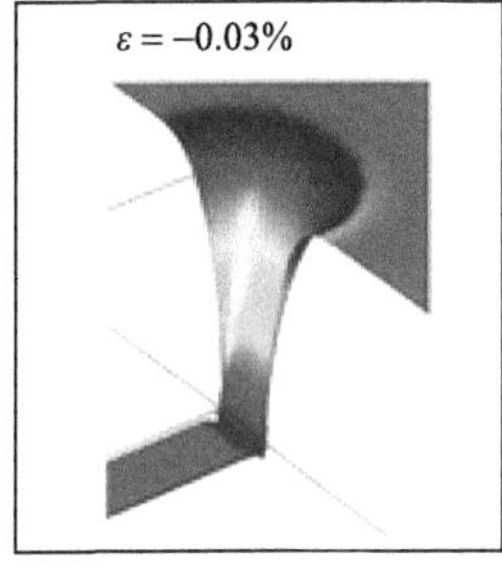

扫码阅全文

http://www.binn.cas.cn/book/202106/W020210623512928753036.pdf

Ⅳ.6 压电光子学

Piezophotonics

1. Dynamic Pressure Mapping of Personalized Handwriting by a Flexible Sensor Matrix Based on the Mechanoluminescence Process

Xiandi Wang, Hanlu Zhang, Ruomeng Yu, Lin Dong*, Dengfeng Peng, Aihua Zhang, Yan Zhang, Hong Liu , Caofeng Pan* *and* Zhong Lin Wang*

Advanced Materials, 27 (2015) 2324-2331.

简介：压电光子学效应是 2008 年从物理模型上首先预言的现象（*Adv. Fun. Mater.*, 18 (2008) 3553）。本文首次报道了压电光子学效应对机械荧光过程的调制及其在应力成像和体征识别方面的应用。

ABSTRACT A self-powered pressure-sensor matrix based on ZnS:Mn particles for more-secure signature collection is presented, by recording both handwritten signatures and the pressure applied by the signees. This large-area, flexible sensor matrix can map 2D pressure distributions *in situ*, either statically or dynamically, and the piezophotonic effect is proposed to initiate the mechanoluminescence process once a dynamic mechanical strain is applied.

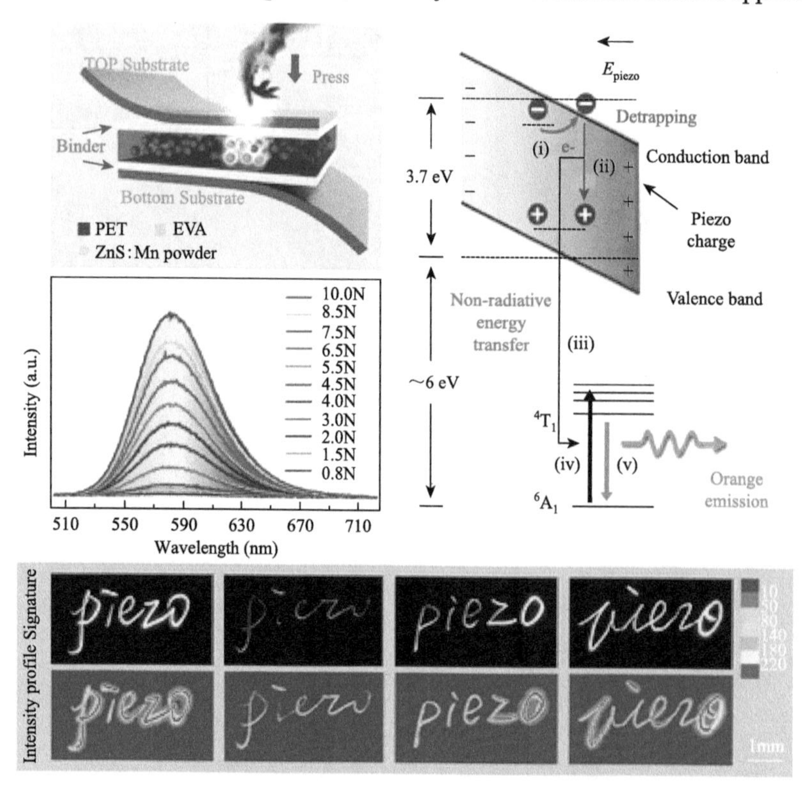

扫码阅全文

http://www.binn.cas.cn/book/202106/W020210623513497482937.pdf

Ⅳ.7 摩擦电子学

Tribotronics

1. Contact Electrification Field-Effect Transistor

Chi Zhang, Wei Tang, Limin Zhang, Changbao Han *and* Zhong Lin Wang*

ACS Nano, 8 (2014) 8702-8709.

简介：利用表面接触起电所产生的局部电场来控制场效应管的载流子输运过程是摩擦电子学（或接触起电电子学）的核心。本文首次介绍了接触起电控制的场效应三极管的概念。

ABSTRACT Utilizing the coupled metal oxide semiconductor field-effect transistor and triboelectric nanogenerator, we demonstrate an external force triggered/controlled contact electrification field-effect transistor (CE-FET), in which an electrostatic potential across the gate and source is created by a vertical contact electrification between the gate material and a foreign object, and the carrier transport between drain and source can be tuned/controlled by the contact-induced electrostatic potential instead of the traditional gate voltage. With the two contacted frictional layers vertically separated by 80μm, the drain current is decreased from 13.4 to 1.9 μA in depletion mode and increased from 2.4 to 12.1 μA in enhancement mode at a drain voltage of 5 V. Compared with the piezotronic devices that are controlled by the strain-induced piezoelectric polarization charged at an interface/junction, the CE-FET has greatly expanded the sensing range and choices of materials in conjunction with semiconductors. The CE-FET is likely to have important applications in sensors, human silicon technology interfacing, MEMS, nanorobotics, and active flexible electronics. Based on the basic principle of the CE-FET, a field of tribotronics is proposed for devices fabricated using the electrostatic potential created by triboelectrification as a gate voltage to tune/control charge carrier transport in conventional semiconductor devices. By the three-way coupling among triboelectricity, semiconductor, and photoexcitation, plenty of potentially important research fields are expected to be explored in the near future.

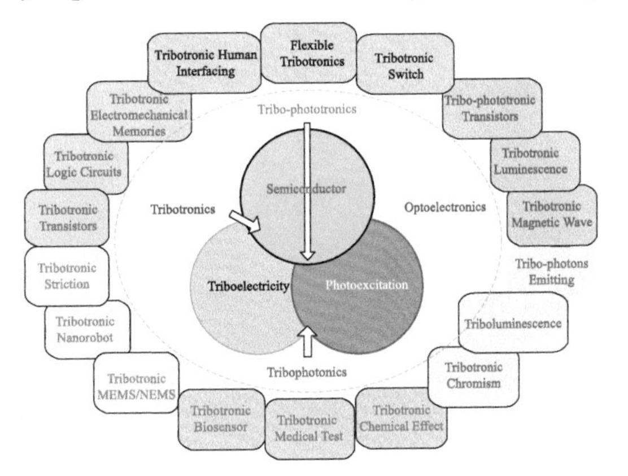

扫码阅全文

http://www.binn.cas.cn/book/202106/W020210623515342519365.pdf

2. Tribotronic Logic Circuits and Basic Operations

Chi Zhang, Li Min Zhang, Wei Tang, Changbao Han *and* Zhong Lin Wang

Advanced Materials, 27 (2015) 3533-3540.

简介：本文首次介绍了利用摩擦电子学原理设计的逻辑电路及其基本运算。

ABSTRACT A tribotronic logic device is fabricated to convert external mechanical stimuli into logic level signals, and tribotronic logic circuits such as NOT, AND, OR, NAND, NOR, XOR, and XNOR gates are demonstrated for performing mechanical-electrical coupled tribotronic logic operations, which realize the direct interaction between the external environment and the current silicon integrated circuits.

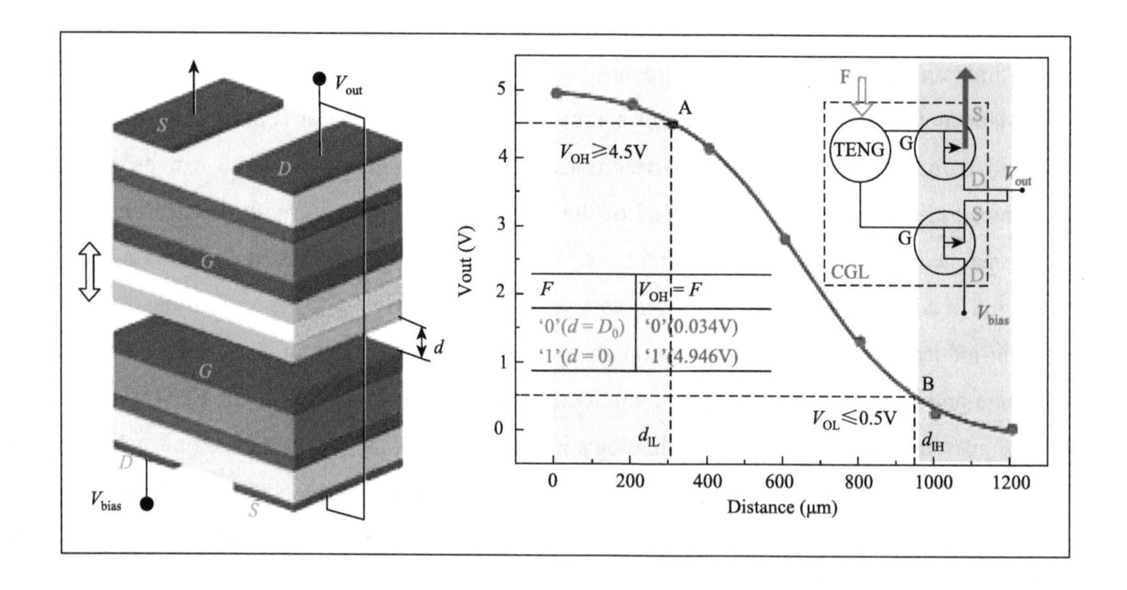

扫码阅全文

http://www.binn.cas.cn/book/202106/W020210623515343512800.pdf

3. Organic Tribotronic Transistor for Contact-Electrification-Gated Light-Emitting Diode

Chi Zhang, Jing Li, Chang Bao Han, Li Min Zhang, Xiang Yu Chen, Li Duo Wang, Gui Fang Dong* *and* **Zhong Lin Wang***

Advanced Functional Materials, 25 (2015) 5625-5632.

简介：本文首次介绍了利用有机材料设计的、接触起电控制的 LED 发光二极管。

ABSTRACT Tribotronics is a new field about the devices fabricated using the electrostatic potential created by contact electrification as a gate voltage to tune/control charge carrier transport in semiconductors. In this paper, an organic tribotronic transistor is proposed by coupling an organic thin film transistor (OTFT) and a triboelectric nanogenerator (TENG) in vertical contact-separation mode. Instead of using the traditional gate voltage for controlling, the charge carrier transportation in the OTFT can be modulated by the contact-induced electrostatic potential of the TENG. By further coupling with an organic light-emitting diode, a contact-electrification-gated light-emitting diode (CG-LED) is fabricated, in which the operating current and light-emission intensity can be tuned/controlled by an external force induced contact electrification. Two different modes of the CG-LED have been demonstrated and the brightness can be decreased and increased by the applied physical contact, respectively. Different from the conventional organic light-emitting transistor controlled by an electrical signal, the CG-LED has realized the direct interaction between the external environment/stimuli and the electroluminescence device. By introducing optoelectronics into tribotronics, the CG-LED has open up a new field of tribophototronics with many potential applications in interactive display, mechanical imaging, micro-opto-electro-mechanical systems, and flexible/touch optoelectronics.

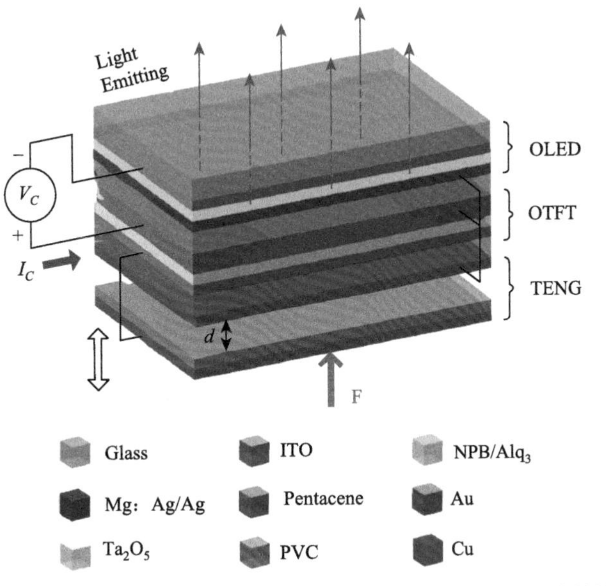

扫码阅全文

http://www.binn.cas.cn/book/202106/W020210623515344388556.pdf

4. Flexible Organic Tribotronic Transistor Memory for a Visible and Wearable Touch Monitoring System

Jing Li, Chi Zhang, Lian Duan, Limin Zhang, Liduo Wang, Gui Fang Dong* *and* **Zhong Lin Wang***

Advanced Materials, 28 (2016) 106-110.

简介：本文首次报道了利用有机材料设计的应用于触摸监控系统中的摩擦电子学记忆单元。

ABSTRACT A new type of flexible organic tribotronic transistor memory is proposed, which can be written and erased by externally applied touch actions as an active memory. By further coupling with an organic light-emitting diode (OLED), a visible and wearable touch monitoring system is achieved, in which touch triggering can be memorized and shown as the emission from the OLED.

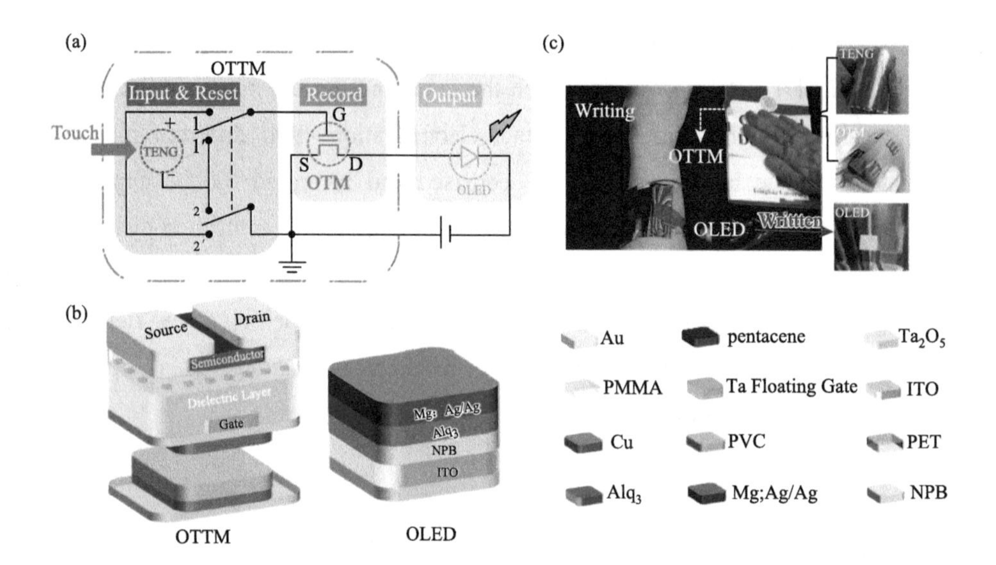

扫码阅全文

http://www.binn.cas.cn/book/202106/W020210623515345141431.pdf

5. Theory of Tribotronics

Ying Liu, Simiao Niu, Zhong Lin Wang*

Advanced Electronic Materials, 1（2015）UNSP 1500124.

简介：本文首次发展了摩擦电子学的基本理论。

ABSTRACT As a potential next generation mechanical-to-electricity power generator, the triboelectric nanogenerator (TENG) has drawn considerable attention in recent years. Its mechanical-to-electrical signal control properties also gave rise to the original idea of “tribotronics”, which utilize triboelectric output to drive/control electronic devices. Using a TENG as input for gate voltage for a field effect transistor, a tribotronic device has potential application in mainly two areas (i) mechanically controlled electronics and (ii) motion/displacement sensing. An experimental study has already been recently reported and therefore a theoretical study is strongly desired as the theoretical basis and optimization strategy for designing such circuits. Here, both analytical calculations and numerical simulations are used to study the tribotronic device, both on the logic operation and on mechanical sensing. The static charge and inherent capacitance of TENG are determined by the structure design of the TENG and have strong coupling with the field effect transistor. Such coupling effect is taken into consideration, thus developing a methodology to effectively optimize the tribotronic device design according to such a coupling effect.

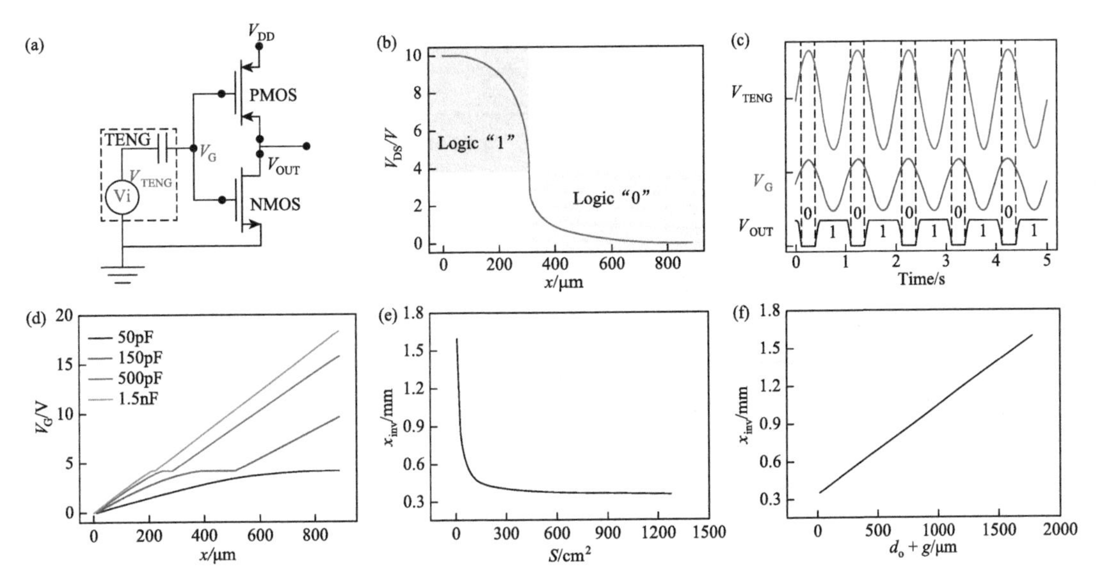

扫码阅全文

http://www.binn.cas.cn/book/202106/W020210623515345462223.pdf

6. Theoretical Study of Triboelectric-Potential Gated/Driven Metal-Oxide-Semiconductor Field-Effect Transistor

Wenbo Pen, Ruomeng Yu, Yongning He *and* Zhong Lin Wang*

ACS Nano, 10 (2016) 4395-4402.

简介：本文首次模拟了接触电势对基于金属-氧化物的场效应管的调制效果。

ABSTRACT Triboelectric nanogenerator has drawn considerable attentions as a potential candidate for harvesting mechanical energies in our daily life. By utilizing the triboelectric potential generated through the coupling of contact electrification and electrostatic induction, the tribotronics has been introduced to tune/control the charge carrier transport behavior of silicon-based metal oxide semiconductor field-effect transistor (MOSFET). Here, we perform a theoretical study of the performances of tribotronic MOSFET gated by triboelectric potential in two working modes through finite element analysis. The drain-source current dependence on contact-electrification generated triboelectric charges, gap separation distance, and externally applied bias are investigated. The in-depth physical mechanism of the tribotronic MOSFET operations is thoroughly illustrated by calculating and analyzing the charge transfer process, voltage relationship to gap separation distance, and electric potential distribution. Moreover, a tribotronic MOSFET working concept is proposed, simulated and studied for performing self-powered FET and logic operations. This work provides a deep understanding of working mechanisms and design guidance of tribotronic MOSFET for potential applications in micro/nanoelectromechanical systems (MEMS/NEMS), human-machine interface, flexible electronics, and self-powered active sensors.

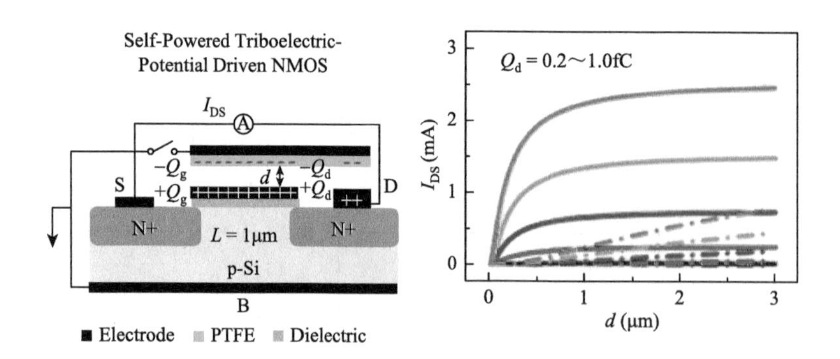

扫码阅全文

http://www.binn.cas.cn/book/202106/W020210623515345929515.pdf

7. Nanoscale Triboelectrification Gated Transistor

Tianzhao Bu, Liang Xu, Zhiwei Yang, Xiang Yang, Guoxu Liu, Yuanzhi Cao, Chi Zhang* *and* **Zhong Lin Wang***

Nature Communications, 11 (2020) 1054.

简介：本文首次把摩擦电子学的器件做到了纳米级尺寸，证明了它的可小型化。

ABSTRACT Tribotronics has attracted great attention owing to the demonstrated triboelectrification-controlled electronics and established direct modulation mechanism by external mechanical stimuli. Here, a nanoscale triboelectrification-gated transistor has been studied with contact-mode atomic force microscopy and scanning Kevin probe microscopy. The detailed working principle was analyzed at first, in which the nanoscale triboelectrification can tune the carrier transport in the transistor. Then with the manipulated nanoscale triboelectrification, the effects of contact force, scan speed, contact cycles, contact region and charge diffusion on the transistor were investigated, respectively. Moreover, the manipulated nanoscale triboelectrification serving as a rewritable floating gate has demonstrated different modulation effects by an applied tip voltage. This work has realized the nanoscale triboelectric modulation on electronics, which could provide a deep understanding for the theoretical mechanism of tribotronics and may have great applications in nanoscale transistor, micro/nano-electronic circuit and nano-electromechanical system.

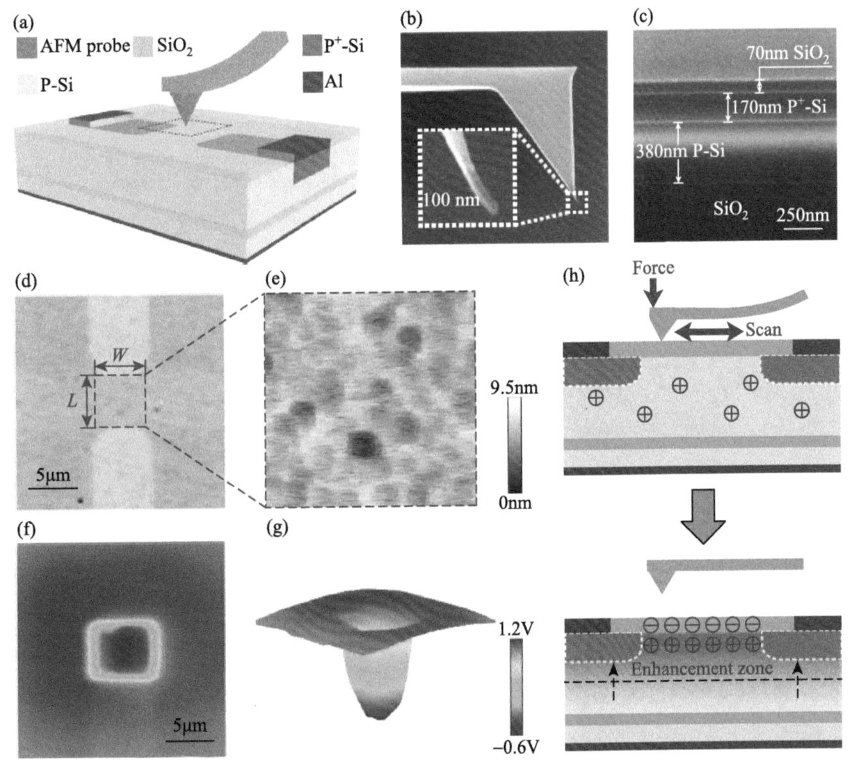

http://www.binn.cas.cn/book/202106/W020210623515346577827.pdf

8. Tribotronics for Active Mechanosensation and Self-Powered Microsystems

Chi Zhang,* Tianzhao Bu, Junqing Zhao, Guoxu Liu, Hang Yang *and* **Zhong Lin Wang***

Advanced Functional Materials, 41 (2019) 1808114.

简介：本文首次综述了摩擦电子学的进展及其在触感电子学与自驱动系统中的应用。

ABSTRACT Tribotronics has attracted great attention as a new research field that encompasses the control and tuning of semiconductor transport by triboelectricity. Here, tribotronics is reviewed in terms of active mechanosensation and human machine interfacing. As a fundamental unit, contact electrification fieldeffect transistors are analyzed, in which the triboelectric potential can be used to control electrical transport in semiconductors. Several tribotronic functional devices have been developed for active control and information sensing, which has demonstrated triboelectricity-controlled electronics and established active mechanosensation for the external environment. In addition, the universal triboelectric power management strategy and the triboelectric nanogenerator-based constant sources are also reviewed, in which triboelectricity can be managed by electronics in the reverse action. With the implantation of triboelectric power management modules, the harvested triboelectricity by various kinds of human kinetic and environmental mechanical energy can be effectively managed as a power supply for self-powered microsystems. In terms of the research prospects for interactions between triboelectricity and semiconductors, tribotronics is expected to demonstrate significant impact and potential applications in microelectro-mechanical systems/nano-electro-mechanical systems (MEMS/NEMS), flexible electronics, robotics, wireless sensor network, and Internet of Things.

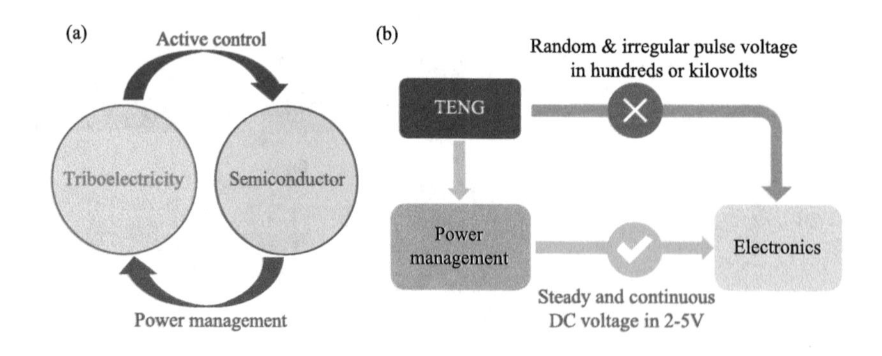

扫码阅全文

http://www.binn.cas.cn/book/202106/W020210623515347003324.pdf

Ⅳ.8　热势光电子效应

Pyro-Phototronic Effect

1. Ultrafast Response p-Si/n-ZnO Heterojunction Ultraviolet Detector Based on Pyro-Phototronic Effect

Zhaona Wang, Ruomeng Yu, Xingfu Wang, Wenzhuo Wu *and* **Zhong Lin Wang**[*]

Advanced Materials, 28 (2016) 6880-6886.

简介：首次提出热势光电子效应：对于脉冲型入射光，由于局部产生的热势效应而导致的电压、电流输出被称为热势光电子效应。该效应具有极高的光探测灵敏度。

ABSTRACT In this paper, the light-self-induced pyro-photoronic effect, a three-way coupling effect among pyroelectric polarization, semiconductor properties, and optical excitation, has been utilized as an effective approach to achieve ultrafast response ultraviolet sensing in p-Si/n-ZnO heterostructures. As a result, the response time is improved from 59 to 19 μs at the rising edge, and 40 to 22 μs at the falling edge. Also, the response time of these PDs decrease as increasing the illumination power intensity, providing a potential opportunity for ultrafast detection of strong light. Besides, the responsivity is enhanced by up to 599% regarding to the relative changes of transient currents. Similar photosensing performances are achieved in both air and vacuum conditions. This newly designed ultrafast ultraviolet nanosensor is compatible to Si-based electronics/optoelectronics and thus may find promising applications in ultrafast photosensing, optothermal detections, health monitoring, and other optoelectronic processes.

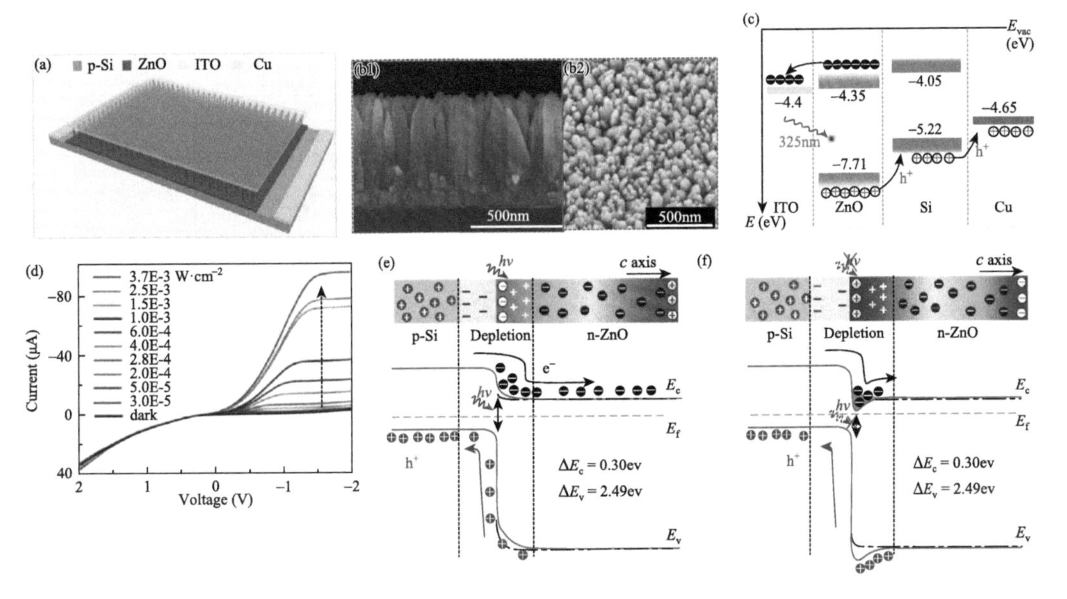

扫码阅全文

http://www.binn.cas.cn/book/202106/W020210629454316950252.pdf

2. Light-Induced Pyroelectric Effect as an Effective Approach for Ultrafast Ultraviolet Nanosensing

Zhaona Wang, Ruomeng Yu, Caofeng Pan, Zhaoling Li, Jin Yang, Fang Yi *and* Zhong Lin Wang*

Nature Communications, 6 (2015) 9401.

简介：首次探讨利用热势效应进行光电子探测。

ABSTRACT Zinc oxide is potentially a useful material for ultraviolet detectors; however, a relatively long response time hinders practical implementation. Here by designing and fabricating a self-powered ZnO/perovskite-heterostructured ultraviolet photodetector, the pyroelectric effect, induced in wurtzite ZnO nanowires on ultraviolet illumination, has been utilized as an effective approach for high-performance photon sensing. The response time is improved from 5.4 s to 53 μs at the rising edge, and 8.9 s to 63 μs at the falling edge, with an enhancement of five orders in magnitudes. The specific detectivity and the responsivity are both enhanced by 322%. This work provides a novel design to achieve ultrafast ultraviolet sensing at room temperature via light-self-induced pyroelectric effect. The newly designed ultrafast self-powered ultraviolet nanosensors may find promising applications in ultrafast optics, nonlinear optics, optothermal detections, computational memories and biocompatible optoelectronic probes.

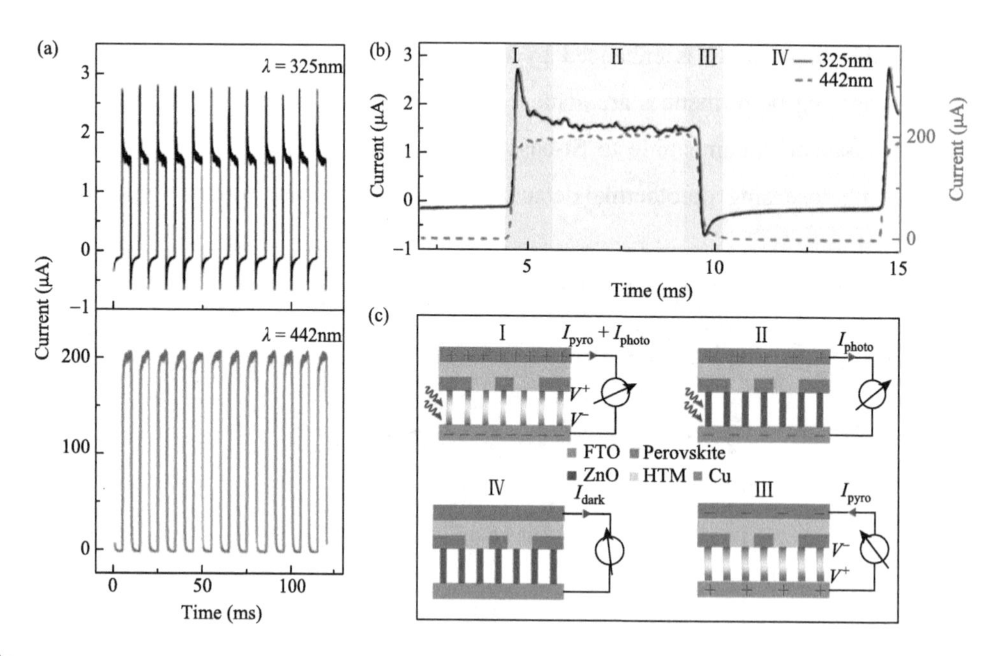

扫码阅全文

http://www.binn.cas.cn/book/202106/W020210629454317580737.pdf

Ⅳ.9 交流光伏效应

AC Photovoltaic Effect

1. Alternating Current Photovoltaic Effect

Haiyang Zou, Guozhang Dai, Aurelia Chi Wang, Xiaogan Li, Steven L. Zhang, Wenbo Ding, Lei Zhang, Ying Zhang *and* Zhong Lin Wang*

Advanced Materials, (2020) 1907249.

简介：基于光伏效应的电池输出一般是直流电，本文首次发现了具有交流特征的光伏电池，其原理可能是动态变化的光照导致的半导体接触面处在非平衡状态下费米面的相对位移。该效应可能在超灵敏光探测传感器方面有独特的应用。

ABSTRACT It is well known that the photovoltaic effect produces a direct current (DC) under solar illumination owing to the directional separation of light-excited charge carriers at the p-n junction, with holes flowing to the p-side and electrons flowing to the n-side. Here, it is found that apart from the DC generated by the conventional p-n photovoltaic effect, there is another new type of photovoltaic effect that generates alternating current (AC) in the nonequilibrium states when the illumination light periodically shines at the junction/interface of materials. The peak current of AC at high switching frequency can be much higher than that from DC. The AC cannot be explained by the established mechanisms for conventional photovoltaics; instead, it is suggested to be a result of the relative shift and realignment between the quasi-Fermi levels of the semiconductors adjacent to the junction/interface under the nonequilibrium conditions, which results in electron flow in the external circuit back and forth to balance the potential difference between two electrodes. By virtue of this effect, the device can work as a high-performance broadband photodetector with extremely high sensitivity under zero bias; it can also work as a remote power source providing extra power output in addition to the conventional photovoltaic effect.

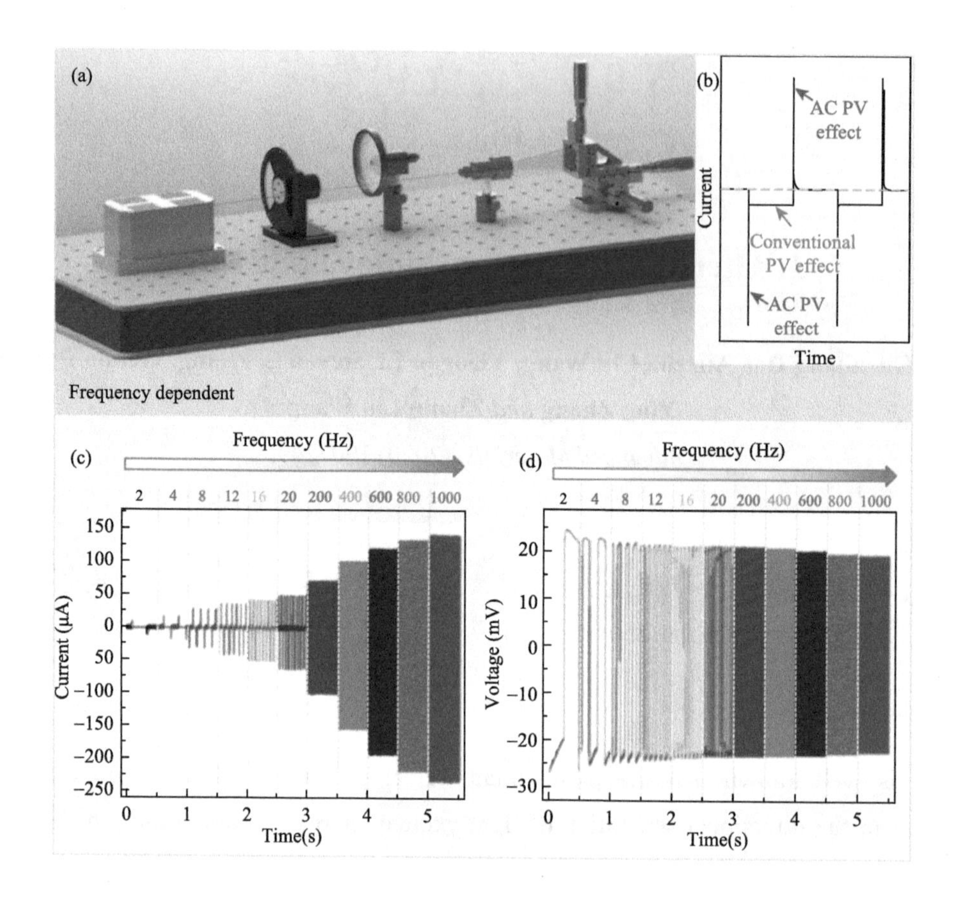

扫码阅全文

http://www.binn.cas.cn/book/202106/W020210626485123164926.pdf